Handbook of
Reinforced Concrete Design

Handbook of Reinforced Concrete Design

S N Sinha
Department of Civil Engineering
Indian Institute of Technology
New Delhi

Tata McGraw-Hill Publishing Company Limited
NEW DELHI

McGraw-Hill Offices

New Delhi New York St Louis San Francisco Auckland Bogotá Guatemala Hamburg Lisbon London Madrid Mexico Milan Montreal Panama Paris San Juan São Paulo Singapore Sydney Tokyo Toronto

Tata McGraw-Hill
*A Division of The **McGraw-Hill** Companies*

> Information contained in this work has been obtained by Tata McGraw-Hill Publishing Company Limited, from sources believed to be reliable. However, neither Tata McGraw-Hill nor its authors guarantee the accuracy or completeness of any information published herein, and neither Tata McGraw-Hill nor its authors shall be responsible for any errors, omissions, or damages arising out of use of this information. This work is published with the understanding that Tata McGraw-Hill and its authors are supplying information but are not attempting to render engineering or other professional services. If such services are required, the assistance of an appropriate professional should be sought.

© 1996, Tata McGraw-Hill Publishing Company Limited

Fourth reprint 2003
RDCYCRDDDLZQL

No part of this publication can be reproduced in any form or by any means without the prior written permission of the publishers

This edition can be exported from India only by the publishers,
Tata McGraw-Hill Publishing Company Limited

ISBN 0-07-462412-1

Published by Tata McGraw-Hill Publishing Company Limited,
7 West Patel Nagar, New Delhi 110 008, typeset in Garamond at
Scribe Consultants, and printed at
Saurabh Print-O-Pack, A-16, Sector IV, NOIDA

To

*My wife Anju
and
sons Vikram and Rohan*

Preface

This book is the culmination of many years of my experience in teaching and designing of reinforced concrete structures. As designing structures is a tedious and time consuming work, design aids are used extensively. A large number of design handbooks are currently available for the design of rectangular beam section and the rectangular and circular column sections. However, for T-beam section and the column sections of other shapes such as L, T and +, design curves are not available. For the rectangular section of column, design curves have been proposed for reinforcement distributed either on two or four faces, which may not necessarily be the optimum reinforcement distribution for a particular geometry of column section. The method of design under axial load and biaxial moments is not only iterative but also involves approximation.

Therefore, there was a definite need to develop a direct method of design which eliminated iteration and provided more exact solutions. Also, the difficulty felt in checking the structural design of L, T and + shapes of the column in Bhikaji Cama Bhawan in New Delhi in 1985 and in many other buildings further spurred the development of design aids for such shapes of the column.

The reinforcement detailing of reinforced concrete structures is an important aspect of structural design which has not been given due emphasis in the literature available on the subject. Therefore, this aspect has been discussed in detail in this book.

This book consists of six chapters. The basic methods of design by working stress and limit state based on IS : 456 – 1978 "Code of Practice for Plain and Reinforced Concrete" have been discussed in Chapter 1. The design for serviceability limit states of deflection, cracks and lateral stability of beams have been dealt with in Chapter 2. The design for flexure of rectangular and T beam sections and slabs have been presented in Chapter 3, while Chapter 4 deals with the design of beam in shear. The design aids for columns of circular, rectangular, L, T and + shapes have been presented in Chapter 5 and the reinforcement detailings have been discussed in Chapter 6.

Although there may be many other aspects of a reinforced concrete design handbook, the scope of this book has been limited to the above topics. While writing this handbook, I was greatly influenced directly or indirectly by various design handbooks and research papers, and this influence is bound to be reflected in my writing. I am very thankful to my students for providing many helpful suggestions and assistance in developing this handbook. I am particularly thankful to A K Sinha, N Kumar and V S Singh who worked with me on design aid for columns for their M Tech Thesis. Their contribution is deeply acknowledged. I am also thankful to S K Sapra, P Saxena, R S Jangid and many other M Tech students who assisted me in this work during their stay at IIT Delhi for their M Tech study. Grateful acknow-

are also due to Raj Veer Agarwal for skilfully preparing the drawings and to V P Gulati for meticulously typing the manuscript. The financial assistance provided by the Curriculum Development Cell of Civil Engineering Department, IIT Delhi for typing the manuscript is also gratefully acknowledged.

Finally, I am specially thankful to my wife Anju and to Vikram and Rohan, my sons, for their help and forbearance, and the book is affectionately dedicated to them.

S N Sinha

Contents

Preface — vii
Notations — xii

1. **Design Concepts** — 1
   1.1 Design Methods 1
   1.2 Permissible Stresses and Factor of Safety 1
   1.3 Limit States 1
   1.4 Characteristic Strength of Materials and Load 3
   1.5 Partial Factor of Safety 3

2. **Serviceability Limit States** — 5
   2.1 Deflection 5
   2.2 Cracks 6
   2.3 Lateral Stability of Beam 12

3. **Flexure** — 13
   3.1 Design Methods 13
   3.2 Working Stress Method 13
      3.2.1 Design of Rectangular Beam Section 13
      3.2.2 Design of Flanged Beam Section 21
      3.2.3 Design of Slab 24
   3.3 Limit State Method 31
      3.3.1 Design of Rectangular Beam Section 44
      3.3.2 Design of Flanged Beam Section 52
      3.3.3 Design of Slab 62

4. **Shear** — 75
   4.1 Design Methods 75
   4.2 Working Stress Method 75
   4.3 Limit State Method 84

5. **Columns** — 93
   5.1 Introduction 93
   5.2 Methods of Design 94
   5.3 Circular Section of Column 108
   5.4 Rectangular Section of Column 117
      5.4.1 Interaction Curves for Axial Force and Uniaxial Moment 145
      5.4.2 Interaction Curves for Axial Load and Biaxial Moments 194

- 5.5 L-Section of Columns 237
- 5.6 T-Section of Columns 294
- 5.7 +-Section of Columns 354

Reinforcement Detailing 404
- 6.1 Introduction 404
- 6.2 General Detailing Requirements 404
  - 6.2.1 Cover to Reinforcement 404
  - 6.2.2 Development of Stress in Reinforcement 405
  - 6.2.3 Reinforcement Splicing 409
  - 6.2.4 Hooks and Bends 412
  - 6.2.5 Change in Direction of Reinforcement 415
- 6.3 Reinforcement Detailing for Foundation 416
  - 6.3.1 Area, Cover, Spacing and Diameter of Reinforcement Bars 416
  - 6.3.2 Reinforcement Detailing for Isolated Footing 417
  - 6.3.3 Reinforcement Detailing for Combined Footing 421
  - 6.3.4 Reinforcement Detailing for Column on Edge of Footing 421
  - 6.3.5 Reinforcement Detailing for Grid and Mat Foundations 426
  - 6.3.6 Reinforcement Detailing of Under-reamed Piles 429
  - 6.3.7 Reinforcement Detailing of Precast-Concrete Pile 430
  - 6.3.8 Pile Group Pattern and Reinforcement Detailing of Pile Cap 431
  - 6.3.9 Reinforcement Detailing of Grade Beams 435
- 6.4 Reinforcement Detailing for Columns 435
  - 6.4.1 Longitudinal Reinforcement Detailing 435
  - 6.4.2 Transverse Reinforcement Detailing 437
- 6.5 Reinforcement Detailing for Beams 446
  - 6.5.1 Flexural Reinforcement Detailing 446
  - 6.5.2 Shear Reinforcement Detailing 454
  - 6.5.3 Torsion Reinforcement Detailing 460
  - 6.5.4 Reinforcement Detailing for Beams of Different Depth at Support 461
  - 6.5.5 Reinforcement Detailing for Intersecting Beams 462
  - 6.5.6 Reinforcement Detailing for Tie Members 463
  - 6.5.7 Reinforcement Detailing at Corners, Junctions of Beams and Columns and in Cranked Beams 463
  - 6.5.8 Reinforcement Detailing for Haunched Beam 464
  - 6.5.9 Reinforcement Detailing at Opening in Web of Beam 467
- 6.6 Reinforcement Detailing for Slabs 469
  - 6.6.1 Reinforcement Detailing for Solid Slabs 469
  - 6.6.2 Reinforcement Detailing for Flat Slabs 481
  - 6.6.3 Reinforcement Detailing for Waffle Slabs 483
- 6.7 Reinforcement Detailing for Staircase 493
  - 6.7.1 Area, Spacing, Cover and Diameter of Reinforcement Bars 493
  - 6.7.2 Reinforcement Detailing for Staircase Supported on Side Beam 493
  - 6.7.3 Reinforcement Detailing for Cantilever Steps from Reinforced Concrete Wall 493
  - 6.7.4 Reinforcement Detailing for Flight Supported on Central Beam 495

6.7.5 Reinforcement Detailing for Staircase Simply Supported at Ends of Landing Slabs 496
6.7.6 Reinforcement Detailing for Staircase Fixed at Ends of Landing Slabs 498
6.7.7 Reinforcement Detailing for Staircase Supported on Landing Slabs which Span Transversely to Flight 498
6.7.8 Reinforcement Detailing for Double Cantilever Staircase with Support at Junction of Landing and Waist Slabs 501
6.7.9 Reinforcement Detailing for Tread-Riser Staircase 501
6.8 Reinforcement Detailing for Ductility Requirements of Earthquake-Resistant Building 504
6.8.1 Reinforcement Detailing for Beams 504
6.8.2 Reinforcement Detailing for Columns 507
6.8.3 Reinforcement Detailing for Beam-Column Joints 508

Appendices 510
A Dead Weight of Materials 510
B Live Loads on Floors and Roofs 512
C Maximum Bending Moments and Reactions in Beams 515
D Area and Parameter of Standard Bars 522
E SI Conversion Factors 525
F Selected References 526

Index 529

Notations

A_c	area of concrete
A_g	gross area of section
A_s	area of steel in column
A_{sc}	area of compression steel in beam
A_{st}	area of tension steel in beam
A_{st1}	area of tension steel in beam for balanced section
A_{st2}	additional area of tension steel in beam for resisting additional moment than that of for balanced section
A_{sv}	area of shear reinforcement
B	dimension of rectangular column along x-axis
b	width of beam
b_f	width of flange of flanged beam
b_w	width of web of flanged beam
D	overall depth of beam or slab or diameter of circular column or dimension of rectangular column along y-axis
d	effective depth of beam or slab
d'	effective cover to reinforcement
d'_b	effective cover to reinforcement along the width B of column section
d'_d	effective cover to reinforcement along the depth D of column section
E_s	Young's modulus of elasticity of steel
E_c	Young's modulus of elasticity of concrete
e_{min}	minimum eccentricity
e_x	eccentricity of load along x-axis
e_y	eccentricity of load along y-axis
f_{ck}	characteristic strength of concrete
f_y	characteristic strength of steel
jd	lever arm
K	coefficient of neutral axis depth
K_{bal}	coefficient of neutral axis depth of balanced section
K_t	end restraint factor at the top of column

K_b	end restraint factor at the bottom of column
K_{ct}	summation of flexural stiffness of columns framing into the top joint
K_{cb}	summation of flexural stiffness of columns framing into the bottom joint
K_{bt}	summation of flexural stiffness of beams framing into the top joint
K_{bb}	summation of flexural stiffness of beams framing into the bottom joint
L_d	development length
l	length of column or span of beam or clear distance between lateral restraint for simply supported and continuous beams and between free end and lateral restraint for cantilever beam
l_{ex}	effective length of column with respect to x-axis
l_{ey}	effective length of column with respect to y-axis
M	allowable moment
M_1	allowable moment of resistance of flanged beam when neutral axis coincides with the bottom of flange
M_{bal}	allowable moment of resistance of balanced section
M_u	ultimate moment
M_{u1}	ultimate moment of resistance of flanged beam when neutral axis coincides with the bottom of flange
M_{u2}	additional moment $M_u - M_{u1}$ in doubly reinforced beam
$M_{u,l}$	ultimate moment of resistance of balanced section
M_{ux}	ultimate moment about x-axis
M_{uy}	ultimate moment about y-axis
M_{uxo}	uniaxial ultimate moment capacity of column section about x-axis
M_{uyo}	uniaxial ultimate moment capacity of column section about y-axis
M_{ax}	additional ultimate moment due to slenderness effect about x-axis
M_{ay}	additional ultimate moment due to slenderness effect about y-axis
m	modular ratio
P	axial load at service state
P_u	ultimate axial load
P_{ub}	ultimate axial load corresponding to condition of maximum compressive strain of 0.0035 in concrete and tensile strain of 0.002 in the outermost layer of tension steel in a compression member
P_{uz}	ultimate concentric load capacity of column
p	percentage of reinforcement
p_c	percentage of compression reinforcement in beam, $100\, A_{sc}/b\, d$
p_t	percentage of tension reinforcement in beam, $100\, A_{st}/b\, d$
p_{t1}	percentage of tension reinforcement in beam for balanced section, $100\, A_{st1}/b\, d$
p_{t2}	percentage of additional tension reinforcement in beam, $100\, A_{st2}/b\, d$

S_v	spacing of reinforcement bars in slab or spacing of stirrups
V	shear force at working load state
V_c	allowable shear strength of concrete
V_s	allowable shear strength of shear reinforcement
V_u	ultimate shear force
V_{uc}	ultimate shear strength of concrete
V_{us}	ultimate shear strength of shear reinforcement
x_o	distance of centre of gravity of the column section along x-axis
x_1	shorter dimension of stirrup
x_u	depth of neutral axis at limit state of collapse
$x_{u,l}$	depth of neutral axis of balanced section at limit state of collapse
y_o	distance of centre of gravity of the column section along y-axis
y_1	longer dimension of stirrup
Z	section modulus
Z_x	section modulus about x-axis
Z_y	section modulus about y-axis
γ_f	partial factor of safety for load
γ_m	partial factor of safety for strength of materials
σ_{cbc}	permissible stress in concrete in compression in bending
σ_{cc}	permissible stress in concrete in dierct compression
σ_s	permissible stress in reinforcement bar
σ_{sc}	permissible stress in reinforcement bar in compression
σ_{st}	permissible stress in reinforcement bar in tension
σ_{sv}	stress in reinforcement bar in shear at working load
τ_{bd}	bond stress
τ_c	shear stress in concrete at working load
τ_{uc}	ultimate shear stress in concrete
$\tau_{c,max}$	maximum allowable shear stress in concrete with shear reinforcement
$\tau_{uc,max}$	maximum ultimate shear strength of concrete with shear reinforcement
τ_v	nominal shear stress at working load
τ_{uv}	nominal ultimate shear stress
τ_{ve}	equivalent shear stress
ϕ	diameter of reinforcement bar

Chapter 1

Design Concepts

1.1 DESIGN METHODS

There are two methods of design of reinforced concrete structures:

1. Working Stress Method
2. Limit State Method

The conventional working stress method of design is based on the behaviour of the structure at working load. It ensures satisfactory behaviour under service load and is assumed to possess adequate safely against collapse. The design is based on linear stress distribution ensuring that the stresses in concrete and steel do not exceed their permissible values at service load.

The limit state method of design ensures adequate safety against the structure being unfit for use when it reaches the acceptable limit states. The two principal limit states are the ultimate limit state of collapse and the serviceability limit state of deflection and cracking. The design of the structure is made for both serviceability limit state of deflection and cracks and the ultimate limit state for collapse. The design for ultimate limit state for collapse is made by considering their behaviour at collapse load.

1.2 PERMISSIBLE STRESSES AND FACTOR OF SAFETY

The permissible stresses in concrete and steel are obtained by dividing their ultimate or yield strength by a suitable factor of safety. They are given in Tables 1.1 and 1.2 for concrete and steel respectively, in accordance with IS: 456–1978.

1.3 LIMIT STATES

The two principal limit states for reinforced concrete structures are:

1. Serviceability limit state

Table 1.1 Permissible stress in concrete

Grade of concrete	Permissible stress in compression (N/mm^2)	
	Bending	Direct
M15	5.0	4.0
M20	7.0	5.0
M25	8.5	6.0
M30	10.0	8.0
M35	11.5	9.0
M40	13.0	10.0

Table 1.2 Permissible stress in steel

Tension/Compression stress	Permissible stress (N/mm^2)				
	Mild steel: Plain/ Deformed	Medium tensile steel: Plain/ Deformed	High strength deformed bars of grades		
			Fe 415	Fe 500	Fe 550
Tension					
diameter ≤ 20 mm	140	0.5 times yield stress or 190 whichever is minimum	230	275	300
diameter > 20 mm	130		230	275	300
Compression					
— in column bars.	130	130	190	190	190
— in bars in beam or slab when the compression resistance of concrete is taken into account.	Compressive stress in the surrounding concrete, multiplied by 1.5 times the modular ratio or permissible compressive stress in column bars whichever is smaller.				
— in bars in beam or slab where the compressive resistance of its concrete is not taken into account.					
diameter ≤ 20 mm	140	0.5 times yield stress or 190 whichever is minimum	190	275	300
diameter > 20 mm	130		190	275	300

2. Limit state for collapse

The serviceability limit states are mainly for deflection and cracking. However, there may be several other serviceability limit states such as durability, fire resistance, excessive vibration, fatigue etc. depending upon the function of the structure. The limit state for collapse is reached when the structure collapses.

1.4 CHARACTERISTIC STRENGTH OF MATERIALS AND LOAD

There is an inevitable variation in the strength of materials and load on the structure. There is, therefore, always the probability that a strength will be obtained which may be less than the specified strength. Therefore, the recent practice is to specify characteristic strength below which, not more than a prescribed percentage of the test results shall fall. IS : 456–1978 specifies the characteristic strength of concrete as that of 28 days cube strength, below which, not more than 5 per cent of the test results may be expected to fall. Similarly the characteristic strength of steel is defined as the value of yield stress in the case of mild steel or 0.2 per cent of proof stress in the case of cold worked deformed bars, below which, strength of not more than 5 per cent of the test results may be expected to fall.

The characteristic load is that value of load which has an accepted probability of not being exceeded during the life span of the structure. IS : 456–1978 recommends the probability of 5 per cent of load not exceeding the characteristic load.

1.5 PARTIAL FACTOR OF SAFETY

The load considered for design for each limit state is known as design load and is obtained from characteristic load with appropriate partial factor of safety. Thus,

$$\text{Design load} = \text{Characteristic load} \times \text{Partial factor of safety} (Y_f)$$

The value of partial factor of safety Y_f depends on the importance of the limit states under consideration and the accuracy in predicting the load. Table 1.3 gives the value of partial factor of safety for ultimate limit state and the serviceability limit state for different load combinations.

Similarly the partial factor of safety is introduced for strength of material to account for the variation in strength of the material. Table 1.4 gives the value of partial factor of safety for ultimate and serviceability limit states for concrete and steel.

Table 1.3 Partial factor of safety for load (Y_f)

Load combinations	Limit states	
	Ultimate	Serviceability
1. Dead	1.5	1.0
Imposed	1.5	1.0
2. Dead	1.5 or 0.9*	1.0
Wind or Earthquake	1.5	1.0
3. Dead	1.2	1.0
Imposed	1.2	0.8
Wind or Earthquake	1.2	0.8

* This value is to be considered when stability against overturning or stress reversal is critical.

Table 1.4 Partial factors of safety for strength of concrete and steel (Y_m)

Limit states	Material	
	Concrete	Steel
Ultimate	1.5	1.15
Serviceability		
Deflection	1.0	1.0
Cracking	1.3	1.0

Chapter 2

Serviceability Limit States

2.1 DEFLECTION

IS 456:1978 recommends the following limiting values of deflection for reinforced concrete structures:

(i) The final deflection of horizontal members below the level of casting should not exceed the effective span/250.
(ii) The deflection taking place after the construction of partitions or application of finishes should not exceed span/350 or 20 mm, whichever is less.

For rectangular beams and slabs, the serviceability limit state for deflection may generally be assumed to be satisfactory provided that the values of effective span to effective depth ratio are not greater than the allowable value of effective span to effective depth ratio obtained by multiplying the basic value of effective span to effective depth ratio with the modification factors for type and percentage of tension and compression steel.

The basic values of effective span to effective depth ratio for rectangular beams and slabs are given in Table 2.1. The modification factors for tension and compression reinforcements are shown in Charts 2.1 and 2.2 respectively.

Table 2.1 Basic values of effective span/effective depth ratio for rectangular beams and slabs

Support Conditions	Effective span/Effective depth ratio	
	span ≤ 10 mm	span > 10 mm
Cantilever	7	Deflection calculation should be made
Simply supported	20	20 × 10/span
Continuous	26	26 × 10/span

The allowable value of effective span to effective depth ratio for T or inverted L beams can be obtained by multiplying the allowable value of effective span to effective depth ratio for rectangular beam with the modification factor for T beam given in Chart 2.3.

Based on above, the allowable values of effective span to effective depth ratio for rectangular beams are obtained which are plotted in Charts 2.4, 2.5 and 2.6 to facilitate design.

2.2 CRACKS

IS : 456–1978 recommends that the width of surface cracks should not exceed 0.3 mm for structures not subjected to aggressive enviornment. If the structure is exposed to aggressive environments, the width of cracks near the main reinforcement should not exceed 0.004 times the nominal cover to the reinforcement. To be within the permissible limit, the width of cracks is controlled by the following provisions for beams and slabs.

Provisions for control of cracks in beams are:

(i) For beams not exposed to aggressive environment, clear spacing between bars or groups of bars, near the tension face of a beam, should not be greater than the value given in Table 2.2. Bars with diameter less than 0.45 times the maximum diameter of a bar in the section, should be ignored.

Table 2.2 Maximum clear spacing of reinforcement bars (mm)

Characteristic strength of steel f_y (N/mm^2)	Percentage redistribution to or from section considered				
	−30	−15	0	+15	+30
250	215	260	300	300	300
415	125	155	180	210	235
500	105	130	150	175	195

(ii) The clear distance from the corner of a beam to the surface of the nearest longitudinal bar should not exceed half the value given in (i).
(iii) When the depth of the web in a beam exceeds 750 mm, side face reinforcement should be provided along the two faces. The total area of such reinforcement should not be less than 0.1 per cent of the web area and should be distributed on the two faces at a spacing not exceeding 300 mm or web thickness, whichever is less.

Provisions for control of cracks in slabs are:

(i) The nominal spacing of main reinforcement bars should not be more than three times the effective depth of a slab or 450 mm, whichever is smaller.
(ii) The nominal spacing of secondary bars provided against shrinkage and temperature effect should not be more than five times the effective depth of a slab or 450 mm, whichever is smaller.

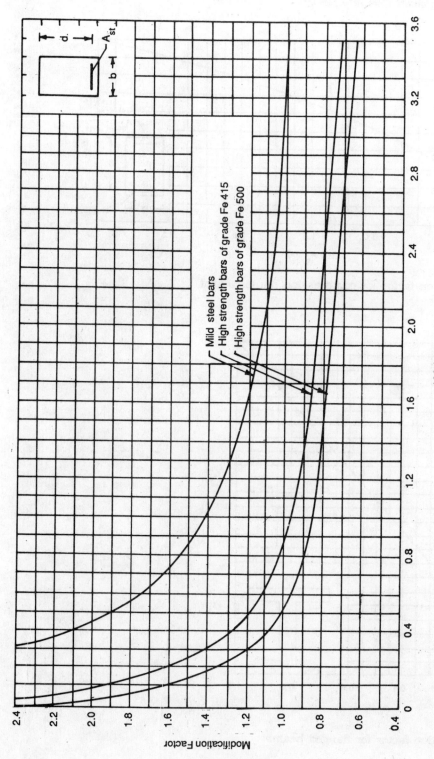

Chart 2.1 Modification factor for tension reinforcement

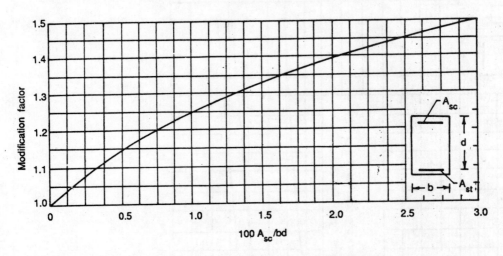

Chart 2.2 Modification factor for compression reinforcement

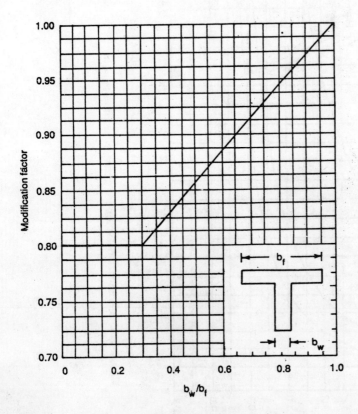

Chart 2.3 Modification factor for flanged beams

Serviceability Limit States

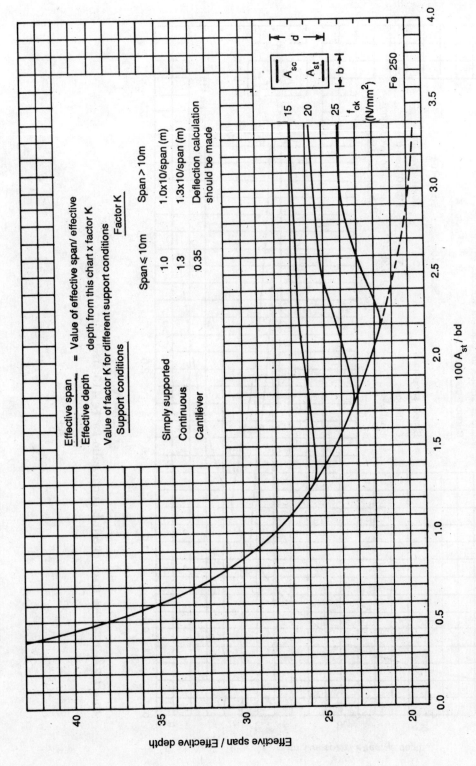

Chart 2.4 Effective span/effective depth ratio for Fe 250 steel

Handbook of Reinforced Concrete Design

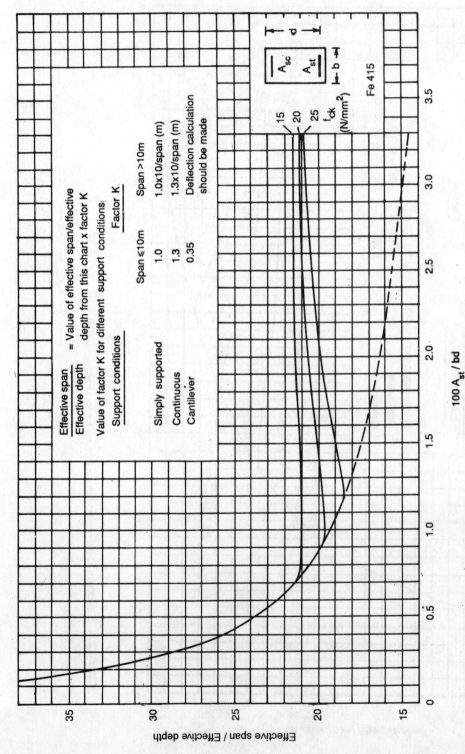

Chart 2.5 Effective span/effective depth ratio for Fe 415 steel

Serviceability Limit States 11

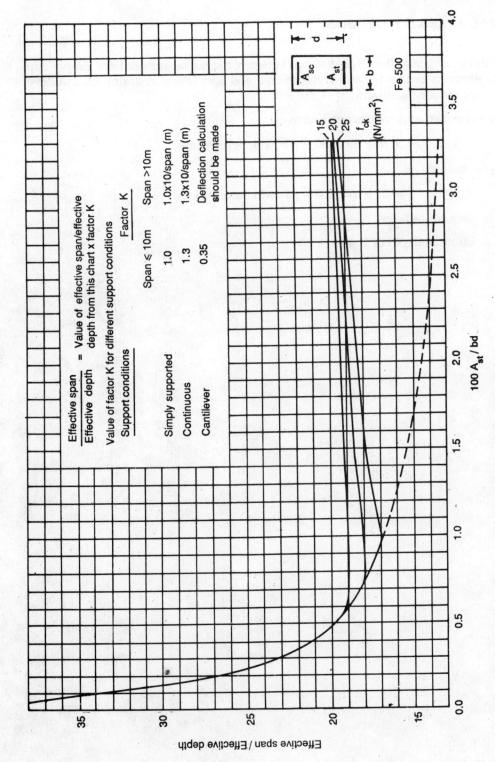

Chart 2.6 Effective span/effective depth ratio for Fe 500 steel

2.3 LATERAL STABILITY OF BEAM

For lateral stability of beam, the clear distance between the lateral restraint for simply supported and continuous beams (l) and between free end and lateral restraint for cantilever beams should not exceed the minimum of the following.

For simply supported and continuous beams,

$$l \ngtr 60\,b \quad \text{or} \quad 250\,b^2/d \quad \text{whichever is less}$$

For cantilever beams,

$$l \ngtr 25\,b \quad \text{or} \quad 100\,b^2/d \quad \text{whichever is less}$$

where b = breadth of beam
   d = effective depth of beam

Chapter 3

Flexure

3.1 DESIGN METHODS

Following are the two methods of design of reinforced concrete structures in flexure:

(i) Working Stress Method
(ii) Limit State Method

Although the limit state method of design is extensively used, some structural elements such as water retaining and bridge structures are designed by the working stress method. Therefore, both the methods of design are discussed below.

3.2 WORKING STRESS METHOD

Following are the basic assumptions of the working stress design method:

(i) At any cross-section, the plane section before bending remains a plane after bending.
(ii) All tensile stresses are taken up by reinforcement and none by concrete except as otherwise specially permitted.
(iii) The stress-strain relationship of steel and concrete under working load is a straight line.
(iv) The modular ratio m has the value $280/3\,\sigma_{cbc}$, where σ_{cbc} is permissible compressive stress of concrete in bending in N/mm^2.

3.2.1 Design of Rectangular Beam Section

The design of rectangular section for the given moment, M and cross-sectional dimensions may result as singly reinforced balanced or under reinforced or doubly reinforced section. This can be ascertained by comparing the given moment, M with the allowable moment of resistance of balanced section, M_{bal} obtained by:

$$M_{bal} = 0.5\,\sigma_{cbc}\,K_{bal}\,(1 - K_{bal}/3)\,bd^2 \tag{3.1}$$

where σ_{cbc} = permissible stress in concrete

K_{bal} = coefficient of neutral axis depth of balanced section

$$= \frac{1}{1 + \sigma_{st}/(m\,\sigma_{cbc})} \qquad (3.2)$$

σ_{st} = permissible stress in steel

m = modular ratio

b = width of section

d = effective depth of section

For balanced section ($M = M_{bal}$), area of steel A_{st1} is obtained by:

$$A_{st1} = \frac{M_{bal}}{\sigma_{st}(1 - K_{bal}/3)\,d} \qquad (3.3)$$

For under reinforced section ($M < M_{bal}$), area of steel A_{st} is obtained by:

$$A_{st} = \frac{M}{\sigma_{st}(1 - K/3)\,d} \qquad (3.4)$$

where K = coefficient of depth of neutral axis

$$= (m^2 p^2 - 2\,m\,p)^{1/2} - m\,p \qquad (3.5)$$

$$p = A_{st}/bd \qquad (3.6)$$

Solution of Eqs. 3.4 and 3.5 gives the area of steel A_{st}. As the solution of above equation is tedious, the area of steel can be determined by iterative procedure as described below:

(i) Assume $K = K_{bal} = \dfrac{1}{1 + \sigma_{st}/(m\,\sigma_{cbc})}$

(ii) Compute $A_{st} = \dfrac{M}{\sigma_{st}(1 - K/3)\,d}$

(iii) For the area of steel A_{st} computed in step (ii), determine

$$K = (m^2 p^2 + 2\,m\,p)^{1/2} - mp$$

(iv) Repeat steps (ii) and (iii) till the values of A_{st} converge.

The doubly reinforced section ($M > M_{bal}$) may be considered equivalent to a singly reinforced balanced section with tension steel A_{st1} and a section with compression steel A_{sc} and additional tension steel A_{st2}. The area of tension steel A_{st} and compression steel A_{sc} are given by:

$$A_{st} = A_{st1} + A_{st2}$$

$$= \frac{M_{bal}}{\sigma_{st}(1 - K_{bal}/3)\,d} + \frac{M - M_{bal}}{\sigma_{st}(d - d')} \qquad (3.7)$$

and
$$A_{sc} = \frac{M - M_{bal}}{(1.5\,m - 1)\,\sigma_{cbc}\,(1 - d'/(K_{bal}\,d))\,(d - d')} \tag{3.8}$$

where d' is the effective cover to compression reinforcement.

Based on above calculations, the following charts have been prepared to facilitate design.

Chart 3.1: To determine the balanced moment of resistance, M_{bal} and the corresponding area of steel, A_{st1}

Chart 3.2: To determine the area of steel, A_{st} of under reinforced section. It also gives the value of $M_{bal}/\sigma_{cbc}\,bd^2$ on the curve for different values of σ_{st}

Chart 3.3: To determine the area of tension steel A_{st2} for moment $M - M_{bal}$

Chart 3.4: To determine the area of compression steel A_{sc} for moment $M - M_{bal}$

Example 3.1: Calculate the area of steel required for a rectangular beam section of 300 mm width and 500 mm effective depth subjected to moments (i) 50 kN.m and (ii) 95 kN.m. Consider concrete of grade M 20 and steel of grade Fe 415.

Solution

Permissible stresses for concrete of grade M 20 and steel of grade Fe 415 are,

$$\sigma_{cbc} = 7\,N/mm^2$$

$$\sigma_{st} = 230\,N/mm^2$$

Case (i) $M = 50$ kN.m

Compute $\quad \dfrac{M}{b\,d^2} = \dfrac{50 \times 10^6}{300 \times 500 \times 500} = 0.6667\,N/mm^2$

To ascertain the section to be singly reinforced balanced or under reinforced or doubly reinforced, consider Chart 3.1 where point P_3 gives,

$$\frac{M_{bal}}{b\,d^2} = 0.91\,N/mm^2$$

Since $\dfrac{M}{bd^2} < \dfrac{M_{bal}}{bd^2}$, it is a singly under reinforced section.

The area of steel is determined with the help of Chart 3.2 as follows.

Compute $\quad \dfrac{M}{\sigma_{cbc}\,b\,d^2} = \dfrac{50 \times 10^6}{7 \times 300 \times 500 \times 500} = 0.09524$

Corresponding to above value of $M/\sigma_{cbc}\,bd^2$, point P_4 on the Chart 3.2 gives,

$$100\,A_{st}/bd = 0.3175$$

$$A_{st} = 0.3175\,bd/100$$

$$A_{st} = 0.3175 \times 300 \times 500 / 100 = 476.25 \text{ mm}^2$$

Case (ii) $M = 95$ kN.m

Compute
$$\frac{M}{bd^2} = \frac{95 \times 10^6}{300 \times 500 \times 500} = 1.267 \text{ N/mm}^2$$

The value of M_{bal}/bd^2 determined in case (i) is,

$$\frac{M_{bal}}{bd^2} = 0.91 \text{ N/mm}^2$$

Since $\frac{M}{bd^2} > \frac{M_{bal}}{bd^2}$, it is a doubly reinforced section.

The area of steel A_{st1} for singly reinforced balanced section is determined by Chart 3.1 where point P_5 gives,

$$100 A_{st1}/bd = 0.44$$

$\therefore$
$$A_{st1} = 0.44 \, bd/100$$
$$= 0.44 \times 300 \times 500 / 100 = 660 \text{ mm}^2$$

Consider effective cover to compressive reinforcement, $d' = 50$ mm.

Compute
$$d'/d = 50/500 = 0.1$$

$$\frac{M - M_{bal}}{bd^2(1 - d'/d)} = \frac{1.267 - 0.91}{0.9} = 0.3967 \text{ N/mm}^2$$

Area of additional tension steel is determined by Chart 3.3 where point P_3 gives,

$$100 A_{st2}/bd = 0.168$$

$\therefore$
$$A_{st2} = 0.168 \, bd/100$$
$$= 0.168 \times 300 \times 500 / 100 = 252 \text{ mm}^2$$

Total area of tension steel is,

$$A_{st} = A_{st1} + A_{st2} = 660.0 + 252.0 = 912.0 \text{ mm}^2$$

The area of compression steel A_{sc} is determined by Chart 3.4 where point P_4 gives,

$$A_{sc}/A_{st2} = 2.525$$

$\therefore$
$$A_{sc} = 2.525 \, A_{st2} = 2.525 \times 252 = 636.3 \text{ mm}^2$$

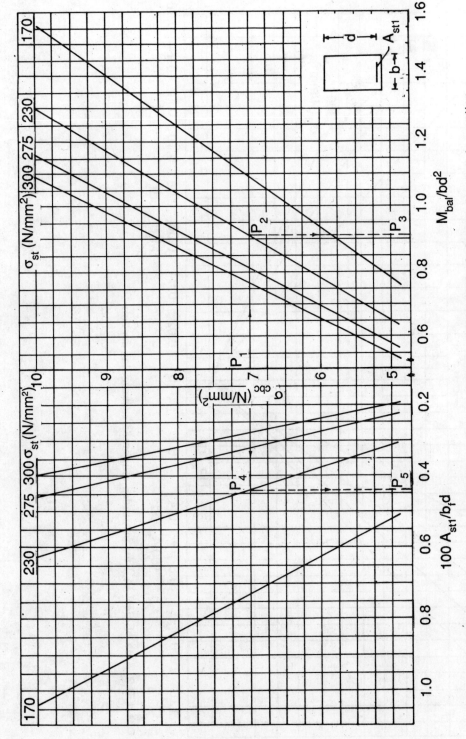

Chart 3.1 Allowable moment of resistance (M_{bal}) and area of steel (A_{st1}) for balanced rectangular section

18 *Handbook of Reinforced Concrete Design*

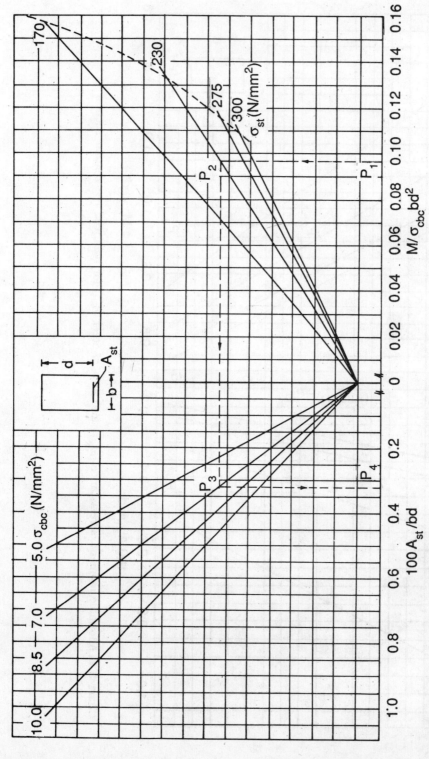

Chart 3.2 Area of steel (A_{st}) for an under reinforced rectangular section

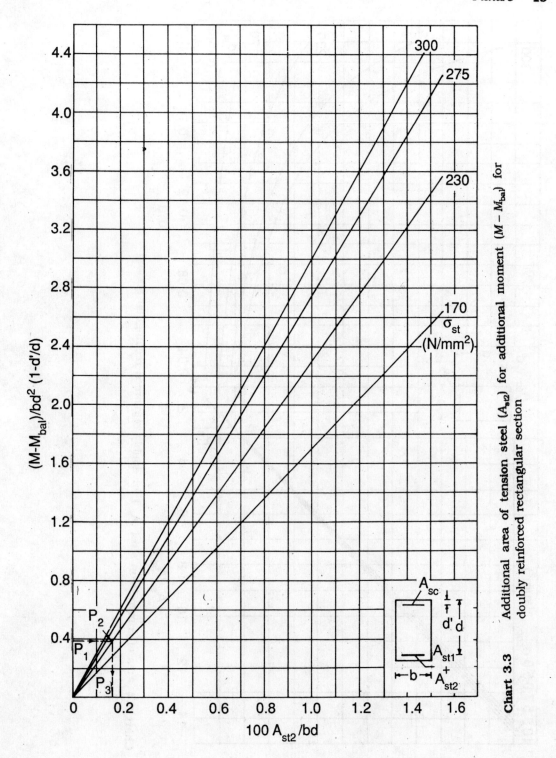

Chart 3.3 Additional area of tension steel (A_{st2}) for additional moment ($M - M_{bal}$) for doubly reinforced rectangular section

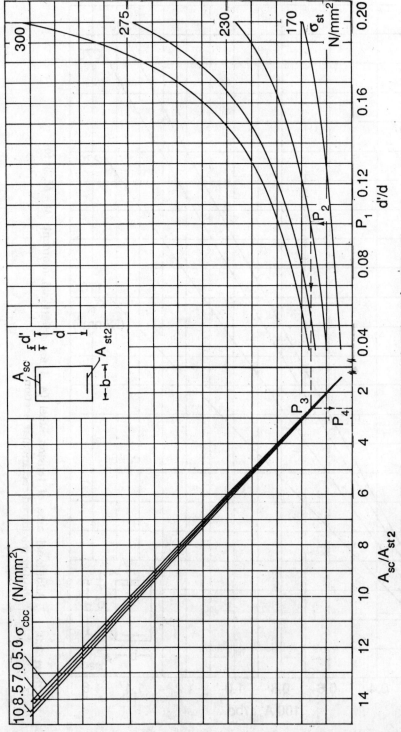

Chart 3.4 Area of compression steel (A_{sc}) for doubly reinforced rectangular section

3.2.2 Design of Flanged Beam Section

The design of flanged section for a given moment, M and cross-sectional dimensional dimensions may result in either singly reinforced balanced or under reinforced or doubly reinforced section. The neutral axis may lie in the flange or outside the flange. This can be ascertained by comparing the depth of the flange, D_f with the depth of the neutral axis of the balanced section, $K_{bal} d$.

If $D_f \geq K_{bal} d$, then the neutral axis always lies within the flange and the design of the section reduces to that of rectangular section of width b_f of flange of the beam.

If $D_f < K_{bal} d$, then the neutral axis may lie in the flange. This can be ascertained by comparing the given moment, M with the moment of resistance, M_1 of the section when the neutral axis coincides with the bottom of the flange ($Kd = D_f$). The moment, M_1 is obtained by:

$$M_1 = 0.5 \, \sigma_{st} \, b_f \, D_f^2 \, \frac{d - (D_f/3)}{m(d - D_f)} \tag{3.9}$$

where σ_{st} = permissible stress in steel

d = effective depth of section

m = modular ratio

If $M \leq M_1$, then the neutral axis lies in the flange ($Kd \leq D_f$) and the design of the section reduces to that of rectangular section of width b_f of flange of the section.

If $M > M_1$, then the neutral axis lies in the web and the design of the section may result in either singly reinforced or doubly reinforced section which can be ascertained by comparing the given moment, M with the allowable moment of resistance, M_{bal} of the section. The allowable moment resistance, M_{bal} by neglecting compression in the web is obtained by:

$$M_{bal} = \sigma_{cbc} \, b_f \, D_f \left(1 - \frac{D_f}{2K_{bal} d}\right) \left(d - \frac{D_f(3K_{bal} d - 2D_f)}{6K_{bal} d - 3D_f}\right) \tag{3.10}$$

I $M = M_{bal}$, it is a single reinforced balanced section and the corresponding area of steel, A_{st1} is given by:

$$A_{st1} = \frac{M_{bal}}{\sigma_{st} \, j \, d} \tag{3.11}$$

where

$$jd = d - \frac{D_f(3K_{bal} d - 2D_f)}{(6K_{bal} d - 3D_f)} \tag{3.12}$$

If $M < M_{bal}$, it is a singly under reinforced section and the area of steel, A_{st} is obtained by,

$$A_{st} = \frac{M}{\sigma_{st} \, j \, d} \tag{3.13}$$

where

$$jd = d - \frac{D_f(3Kd - 2D_f)}{6Kd - 3D_f} \tag{3.14}$$

$$K = \frac{Q + 3d \, D_f - 2D_f^2}{Q + 6d^2 - 3d \, D_f} \tag{3.15}$$

22 Handbook of Reinforced Concrete Design

$$Q = \frac{6 M m d}{\sigma_{st} b_f D_f}$$

If $M > M_{bal}$, it is a doubly reinforced section and the area of tension steel, A_{st} and compression steel, A_{sc} are given by:

$$A_{st} = A_{st1} + A_{st2}$$

$$= \frac{M_{bal}}{\sigma_{st} j d} + \frac{M - M_{bal}}{\sigma_{st} (d - d')} \tag{3.16}$$

$$A_{sc} = \frac{M - M_{bal}}{\sigma_{cbc} (1.5m - 1)(1 - d'/Kd)(d - d')} \tag{3.17}$$

where

$$jd = d - \frac{D_f (3 K_{bal} d - 2 D_f)}{(6 K_{bal} d - 3 D_f)} \tag{3.18}$$

d' = effective cover to compression reinforcement

Based on the above calculations, following charts have been prepared to facilitate design.

Chart 3.5: To determine the area of steel, A_{st} for singly reinforced T section when the neutral axis lies within the flange. It also gives the value of $M_1/b_f d^2$ at the point of intersection of curves for D_f/d and σ_{st} for that value of D_f/d and σ_{st}.

Charts 3.6 to 3.9: To determine the allowable moment of resistance, M_{bal} and the corresponding area of steel, A_{st1}.

Chart 3.10: To determine the area of steel, A_{st} for singly reinforced T section when neutral axis lies within the web. It also gives the value of $M_{bal}/b_f d^2$ on the curve for different values of σ_{st}.

Additional area of tension steel A_{st2} and compression steel A_{sc} in doubly reinforced flanged section for resisting moment $M - M_{bal}$ can be determined using Charts 3.3 and 3.4 respectively.

Example 3.2: Calculate the area of steel required for a T-beam section of 1500 mm width of flange, 300 mm width of web, 100 mm depth of flange and 500 mm effective depth when subjected to moments (i) 150 kN.m, (ii) 250 kN.m and (iii) 350 kN.m. Consider concrete of grade M20 and steel of grade Fe 415.

Solution

The permissible stresses for concrete of grade M20 and steel of grade Fe 415 are,

$$\sigma_{cbc} = 7 \text{ N/mm}^2$$

$$\sigma_{st} = 230 \text{ N/mm}^2$$

Case (i) $M = 150$ kN.m
Compute $b_w/b_f = 300/1500 = 0.2$

$$D_f/d = 100/500 = 0.2$$

$$\frac{M}{\sigma_{cbc}\, b_f\, d^2} = \frac{150 \times 10^6}{7 \times 1500 \times 500 \times 500} = 0.5714$$

Consider that the neutral axis lies within the flange. This is ascertained from Chart 3.5, where point P_2 corresponding to the value of $M/\sigma_{cbc}\, b_f\, d^2 = 0.05714$ lies within the line drawn for $D_f/d = 0.2$. So, the neutral axis lies within the flange and point P_4 gives area of steel as,

$$100 A_{st}/b_f\, d = 0.1875$$

$$A_{st} = 0.1875 \times 1500 \times 500 / 100 = 1406.25 \text{ mm}^2$$

Case (ii) $M = 250$ kN.m

Compute $\quad \dfrac{M}{\sigma_{cbc}\, b_f\, d^2} = \dfrac{250 \times 10^6}{7 \times 1500 \times 500 \times 500} = 0.09524$

Consider that the neutral axis lies outside the flange. This is ascertained from Chart 3.5 where the point corresponding to the value of $M/\sigma_{cbc}\, b_f\, d^2 = 0.09524$ lies outside the line drawn for $D_f/d = 0.2$. So, the neutral axis lies within the web. To ascertain the design to be singly or doubly reinforced, allowable moment of resistance of singly reinforced balanced section is determined from Chart 3.7. Corresponding to the values of $D_f/d = 0.2$ and $b_w/b_f = 0.2$, point P_3 gives,

$$\frac{M_{bal}}{\sigma_{cbc}\, b_f\, d^2} = 0.122$$

Since $\dfrac{M_{bal}}{\sigma_{bal}\, b_f\, d^2} > \dfrac{M}{\sigma_{cbc}\, b_f\, d^2}$, it is an under reinforced section.

The design of section is made with the help of Chart 3.10 where point P_4 gives,

$$100 A_{st}/b_f\, d = 0.31$$

$$A_{st} = 0.31 \times 1500 \times 500 / 100 = 2325 \text{ mm}^2$$

Case (iii) $M = 350$ kN.m

Compute $\quad \dfrac{M}{\sigma_{cbc}\, b_f\, d^2} = \dfrac{350 \times 10^6}{7 \times 1500 \times 500 \times 500} = 0.1333$

The value of $\dfrac{M_{bal}}{\sigma_{cbc}\, b_f\, d^2}$ determined in case (ii) is,

$$\frac{M_{bal}}{\sigma_{cbc}\, b_f\, d^2} = 0.122$$

Since $\dfrac{M}{\sigma_{cbc}\, b_f\, d^2} > \dfrac{M_{bal}}{\sigma_{cbc}\, b_f\, d^2}$, it is a doubly reinforced section.

The area of steel for singly reinforced balanced section is determined from Chart 3.7 where point P_5 gives,

$$100 A_{st1} / b_f d = 0.406$$

$$A_{st1} = 0.406 \times 1500 \times 500 / 100$$

$$= 3045 \text{ mm}^2$$

The area of additional tension steel A_{st2} and compression steel A_{sc} for resisting additional moment $M - M_{bal}$ are determined by Charts 3.3 and 3.4 respectively as follows.

Consider $\qquad\qquad\qquad d' = 50$ mm

Compute $\qquad\qquad\qquad d'/d = 50/500$

$$= 0.1$$

$$\frac{M - M_{bal}}{b_f d^2 (1 - d'/d)} = \frac{M - M_{bal}}{\sigma_{cbc} b_f d^2} \times \frac{\sigma_{cbc}}{(1 - d'/d)}$$

$$= (0.1333 - 0.122) \frac{7}{0.9}$$

$$= 0.08789$$

The area of additional tension steel A_{st2} determined from Char' 3.3 is given by,

$$100 A_{st2} / b_f d = 0.045$$

$$A_{st2} = 0.045 \times 1500 \times 500 / 100$$

$$= 337.5 \text{ mm}^2$$

Total area of tension steel,

$$A_{st} = A_{st1} + A_{st2}$$

$$= 3045 + 337.5 = 3382.5 \text{ mm}^2$$

The area of compression steel A_{sc} determined from Chart 3.4 is given by,

$$A_{sc} / A_{st2} = 2.75$$

$$A_{st} = 2.75 \times A_{st2}$$

$$= 2.75 \times 337.5$$

$$= 928.1 \text{ mm}^2$$

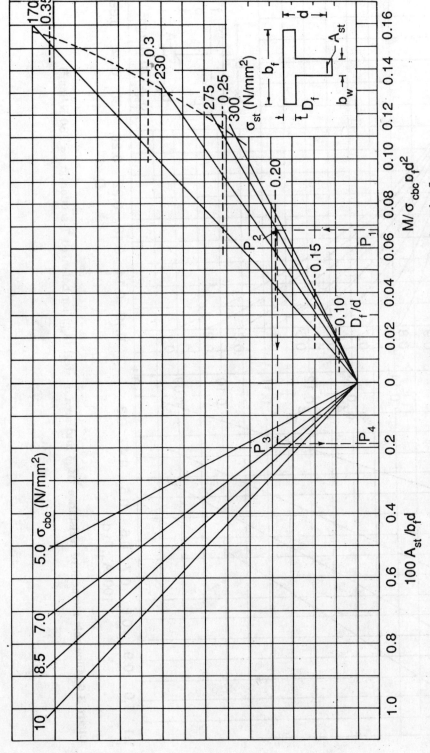

Chart 3.5 Area of steel (A_{st}) for under reinforced T-section when neutral axis lies within the flange

Handbook of Reinforced Concrete Design

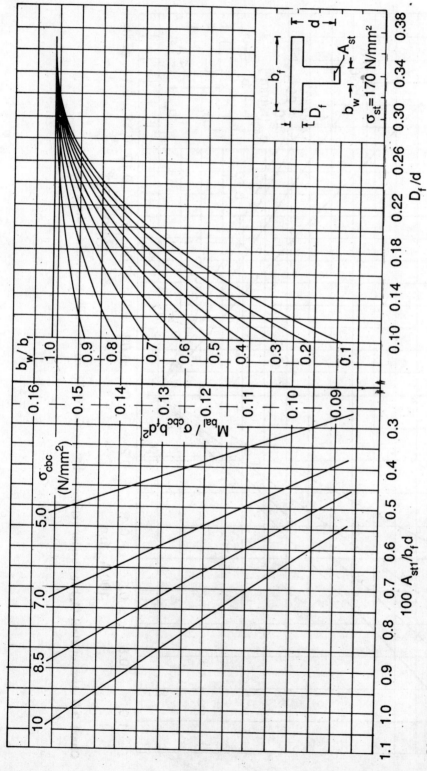

Chart 3.6 Allowable moment of resistance (M_{bal}) and area of steel (A_{st1}) for balanced T-section (σ_{st} = 130 N/mm²)

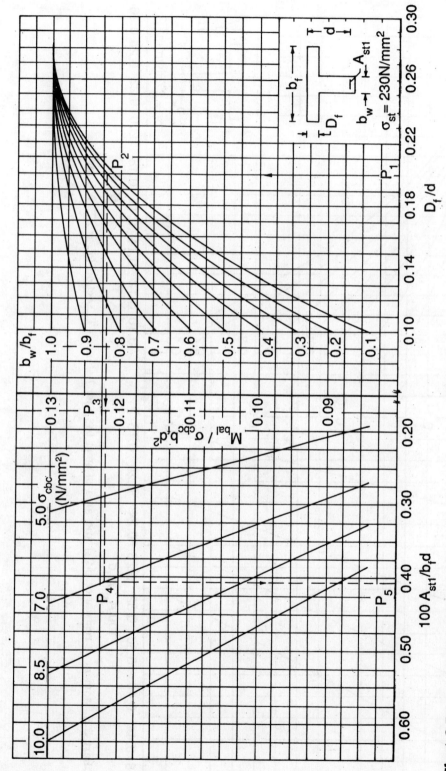

Chart 3.7 Moment of resistance (M_{bal}) and area of steel (A_{st1}) for balanced T- section ($\sigma_{st} = 230$ N/mm²)

28 Handbook of Reinforced Concrete Design

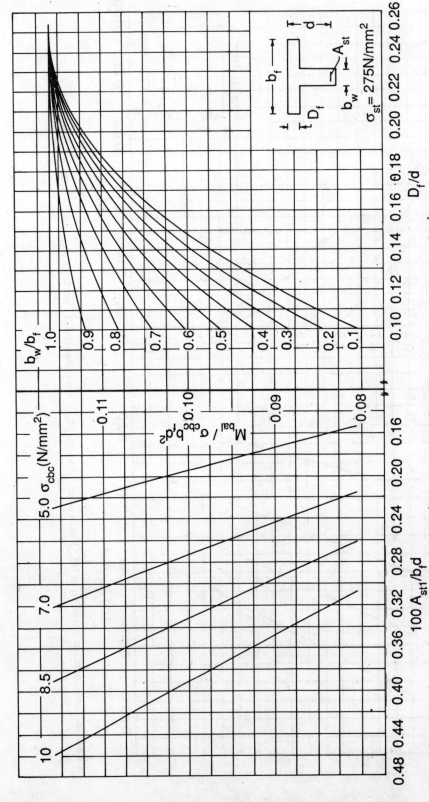

Chart 3.8 Moment of resistance (M_{bal}) and area of steel (A_{st1}) for balanced T- section ($\sigma_{st} = 275$ N/mm²)

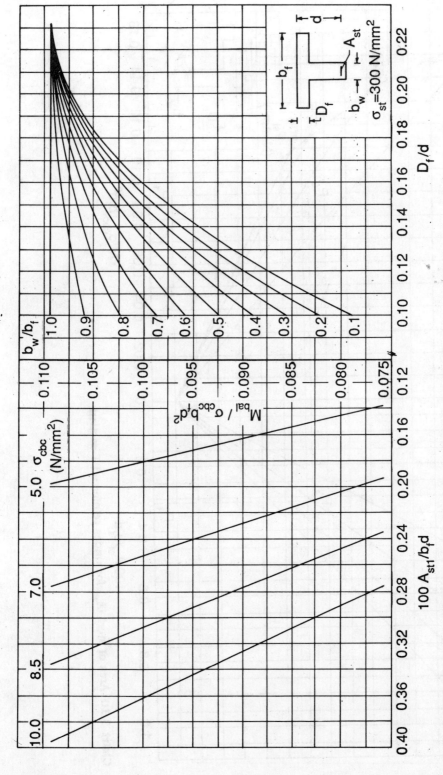

Chart 3.9 Moment of resistance (M_{bal}) and area of steel (A_{st1}) for balanced T-section ($\sigma_{st} = 300$ N/mm²)

30 Handbook of Reinforced Concrete Design

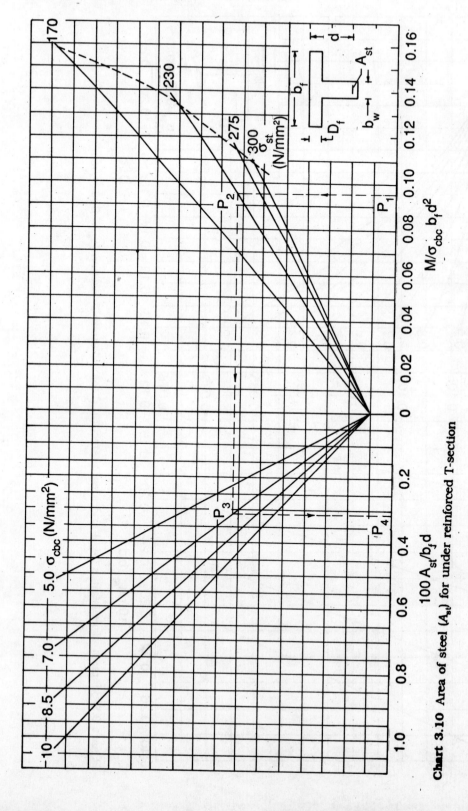

Chart 3.10 Area of steel (A_{st}) for under reinforced T-section

3.2.3 Design of Slab

As the thickness of slab is governed by the serviceability limit state of deflection than the limit state of collapse, the design of the slab results in an under reinforced section. The reinforcement for moment in the slab is expressed as diameter and spacing of bars than the diameter and number of bars as in the case of beams. Therefore, the design Charts 3.11 to 3.22 have been developed to determine the spacing of bars (S_v) for different values of diameter of bars (ϕ) for resisting design moment, M.

Example 3.3: Calculate the area of steel required for a slab of 200 mm thickness subjected to a moment of 15 kN.m/m. Consider concrete of grade M15, steel grade Fe 415 and clear cover to reinforcement equal to 15 mm.

Solution

The permissible stresses of concrete of grade M15 and steel of grade Fe 415 are:

$$\sigma_{cbc} = 5 \text{ N/mm}^2$$

$$\sigma_{st} = 230 \text{ N/mm}^2$$

Compute $d = 200 -$ clear cover of $15 -$ assumed diameter of 12 mm of reinforcement bar$/2$

$$= 179 \text{ mm}$$

$$\frac{M}{1000\, d^2} = \frac{15 \times 10^6}{1000 \times 179 \times 179} = 0.468 \text{ N/mm}^2$$

Spacing of 12 mm ϕ bars is determined with the help of Chart 3.12 where point P_3 gives,

$$S_v\, d \times 10^{-4} = 5.1$$

$\therefore \qquad S_v = 5.1 \times 10^4 / 179 = 284.9 \text{ mm}$

For smaller spacing of reinforcement bars, consider $\phi = 10$ mm.

$\therefore \qquad d = 200 - 15 - 5 = 180 \text{ mm}$

and

$$\frac{M}{1000\, d^2} = \frac{15 \times 10^6}{1000 \times 180 \times 180} = 0.463 \text{ N/mm}^2$$

Spacing of 10 mm ϕ bars is determined with the help of Chart 3.12 as,

$$S_v\, d \times 10^{-4} = 3.6$$

$\therefore \qquad S_v = 3.6 \times 10^4 / 180 = 200.0 \text{ mm}$

Hence provide 10 mm ϕ bars at spacing of 200 mm c/c.

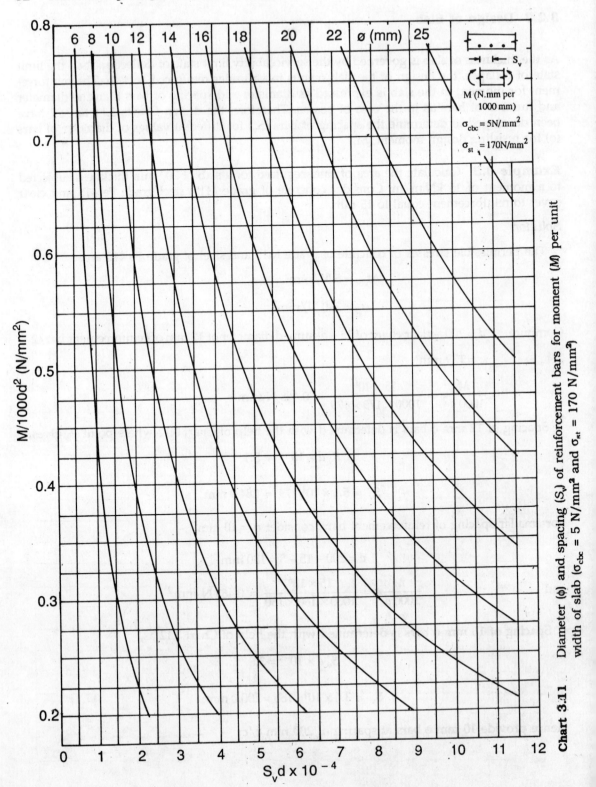

Chart 3.11 Diameter (ϕ) and spacing (S_v) of reinforcement bars for moment (M) per unit width of slab ($\sigma_{cbc} = 5$ N/mm² and $\sigma_{st} = 170$ N/mm²)

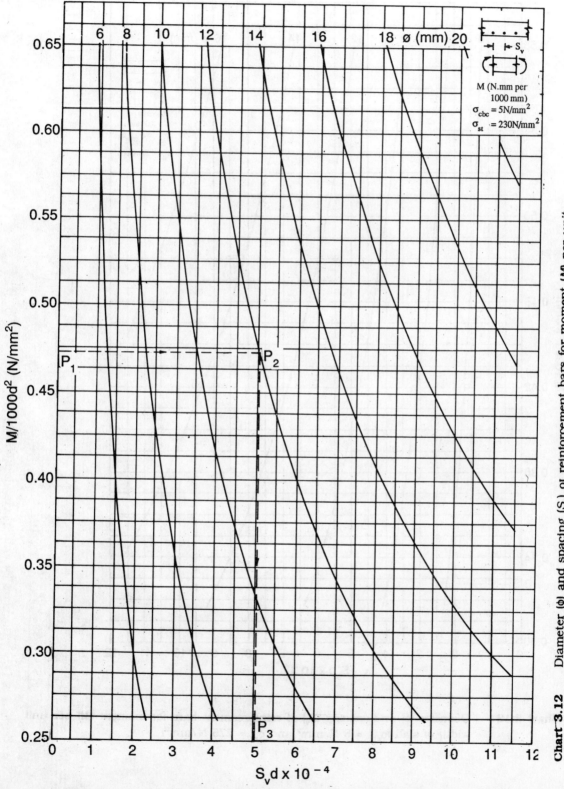

Chart 3.12 Diameter (ϕ) and spacing (S_v) of reinforcement bars for moment (M) per unit width of slab (σ_{cbc} = 5 N/mm² and σ_{st} = 230 N/mm²)

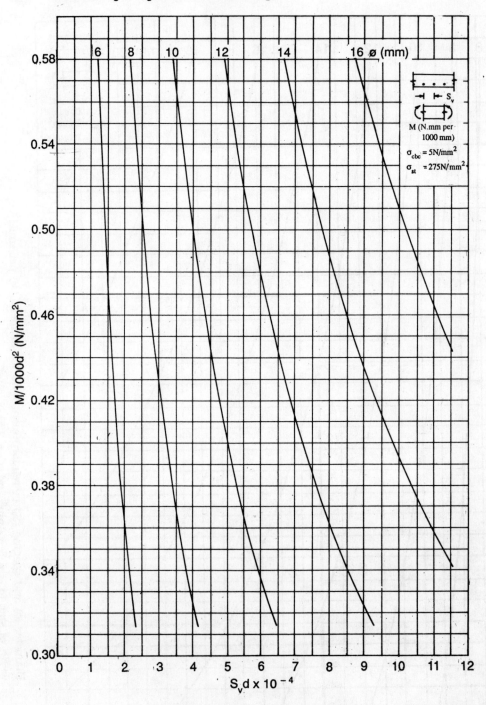

Chart 3.13 Diameter (ϕ) and spacing (S_v) of reinforcement bars for moment (M) per unit width of slab (σ_{cbc} = 5 N/mm² and σ_{st} = 275 N/mm²)

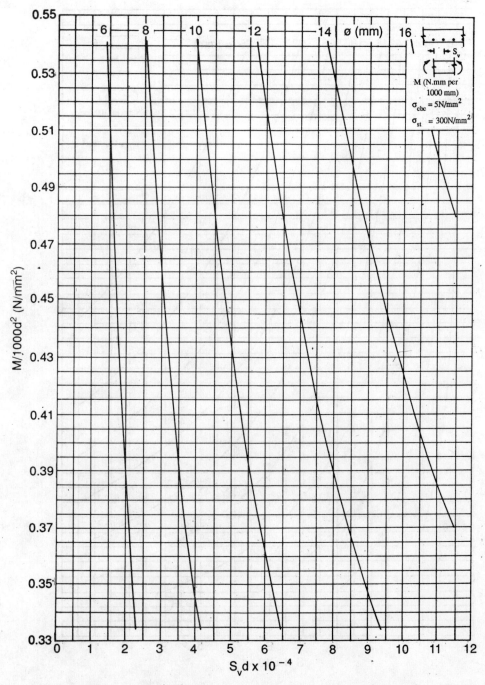

Chart 3.14 Diameter (φ) and spacing (S_v) of reinforcement bars for moment (M) per unit width of slab (σ_{cbc} = 5 N/mm² and σ_{st} = 300 N/mm²)

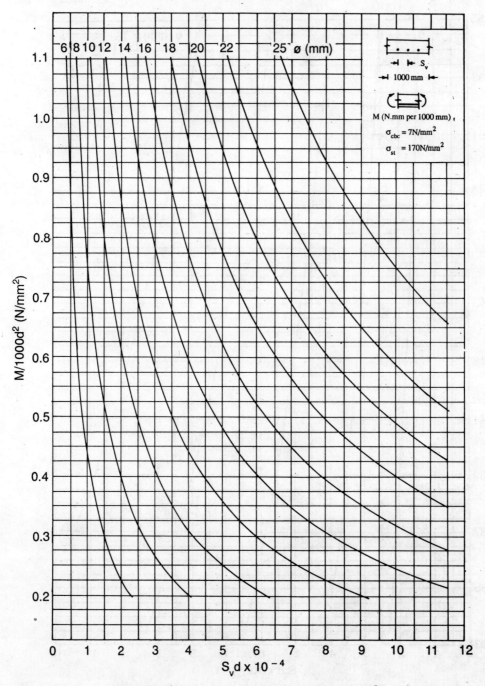

Chart 3.15 Diameter (ϕ) and spacing (S_v) of reinforcement bars for moment (M) per unit width of slab (σ_{cbc} = 7 N/mm² and σ_{st} = 170 N/mm²)

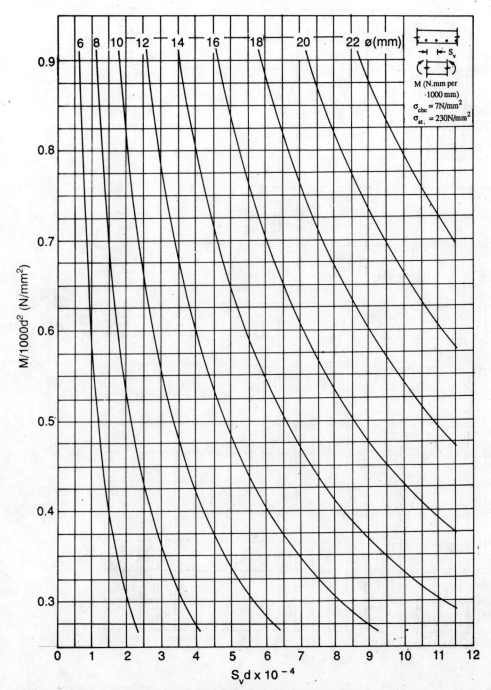

Chart 3.16 Diameter (φ) and spacing (S_v) of reinforcement bars for moment (M) per unit width of slab (σ_{cbc} = 7 N/mm² and σ_{st} = 230 N/mm²)

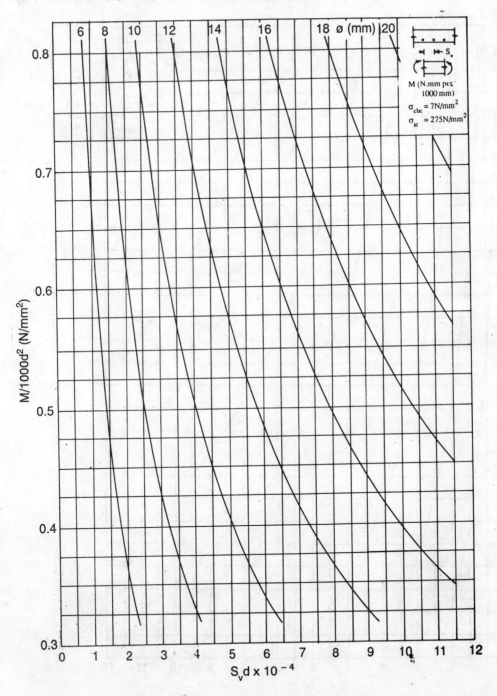

Chart 3.17 Diameter (ϕ) and spacing (S_v) of reinforcement bars for moment (M) per unit width of slab ($\sigma_{cbc} = 7$ N/mm² and $\sigma_{st} = 275$ N/mm²)

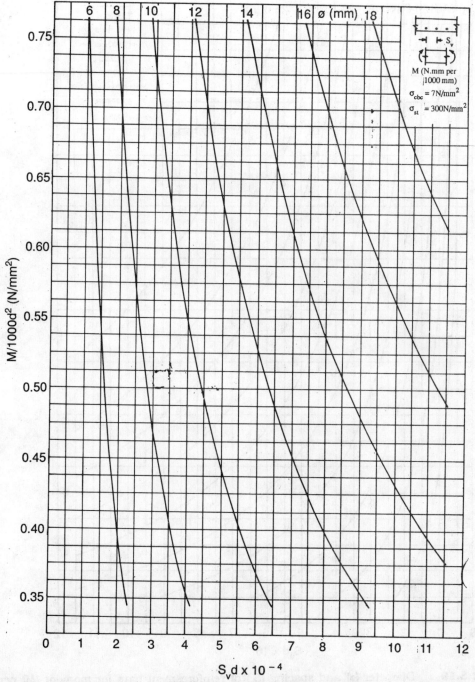

Chart 3.18 Diameter (φ) and spacing (S_v) of reinforcement bars for moment (M) per unit width of slab (σ_{cbc} = 7 N/mm² and σ_{st} = 300 N/mm²)

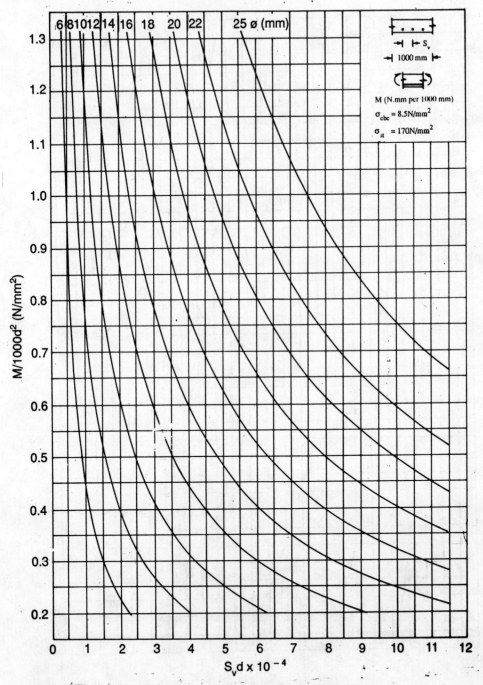

Chart 3.19 Diameter (φ) and spacing (S_v) of reinforcement bars for moment (M) per unit width of slab (σ_{cbc} = 8.5 N/mm² and σ_{st} = 170 N/mm²)

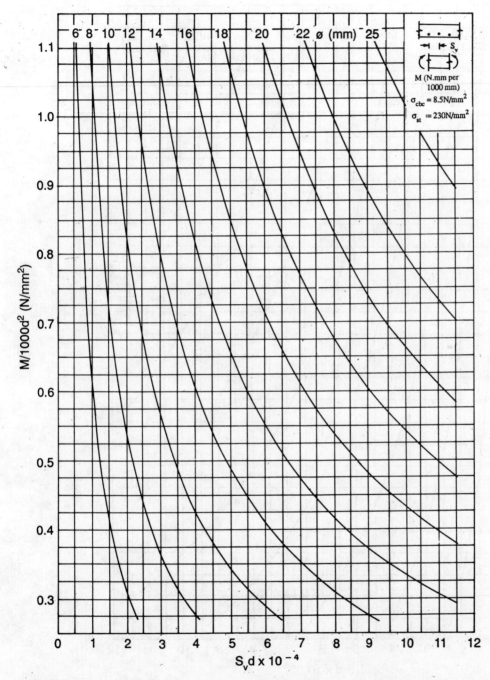

Chart 3.20 Diameter (φ) and spacing (S_v) of reinforcement bars for moment (M) per unit width of slab (σ_{cbc} = 8.5 N/mm² and σ_{st} = 230 N/mm²)

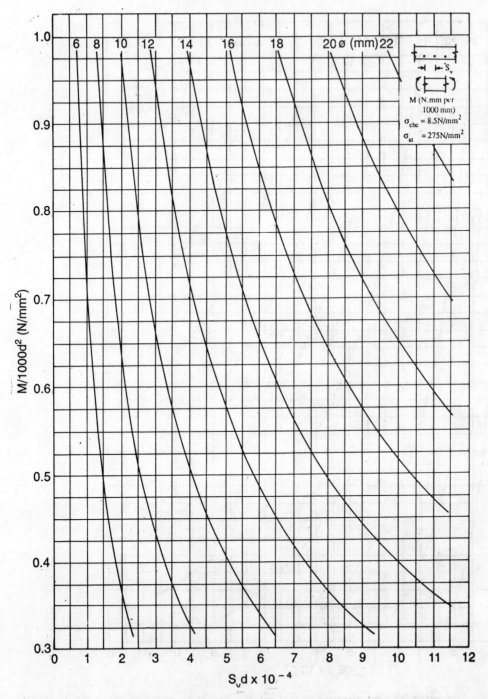

Chart 3.21 Diameter (φ) and spacing (S_v) of reinforcement bars for moment (M) per unit width of slab (σ_{cbc} = 8.5 N/mm² and σ_{st} = 275 N/mm²)

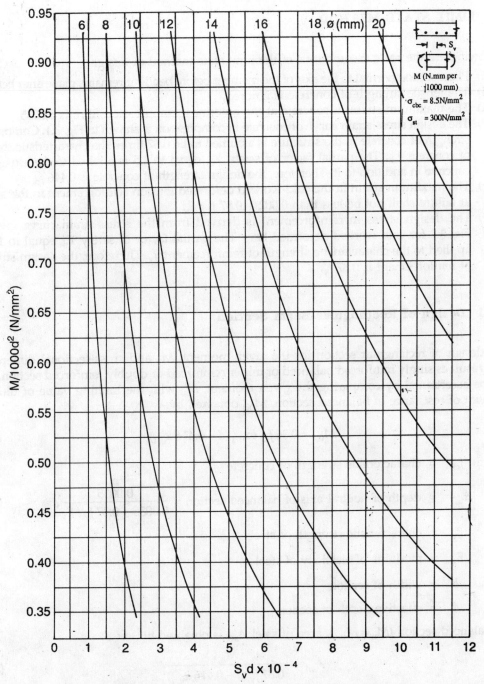

Chart 3.22 Diameter (φ) and spacing (S_v) of reinforcement bars for moment (M) per unit width of slab (σ_{cbc} = 8.5 N/mm² and σ_{st} = 300 N/mm²)

3.3 LIMIT STATE METHOD

Following are the basic assumptions of the limit state method of design:
  (i) Plane section normal to the axis of the member before bending remains plane after bending.
  (ii) The tensile strength of concrete is ignored.
  (iii) The maximum strain in concrete at the outermost compression fibre is 0.0035.
  (iv) The design stress-strain curve of concrete in compression is shown in Fig. 3.1. Compressive strength of concrete in the structure is assumed to be 0.67 times the characteristic strength of concrete, f_{ck}. The partial factor of safety Y_m equal to 1.5 is applied to the strength of concrete in addition to it. Therefore, the design strength of concrete is $0.446 f_{ck}$.
  (v) The maximum strain in the cold-worked deformed tension reinforcement in the section at failure shall not be less than $0.002 + 0.87 f_y / E_s$.
  (vi) The design stress in reinforcement is derived from the stress-strain curve given in Fig. 3.2 for cold-worked deformed bars. The partial factor of safety Y_m equal to 1.15 is applied to the characteristic strength of reinforcement, f_y. Therefore, the design strength of reinforcement is $0.87 f_y$.

3.3.1 Design of Rectangular Beam Section

The design of rectangular section for the given moment, M_u and cross-sectional dimensions may result as singly reinforced balanced or under reinforced or doubly reinforced section. This can be ascertained by comparing the given moment, M_u with the limiting value of ultimate moment of resistance of balanced section, $M_{u,1}$ obtained by:

$$M_{u,1} = 0.36 f_{ck} b x_{u,1} (d - 0.416 x_{u,1}) \qquad (3.19)$$

where f_{ck} = characteristic strength of concrete

$x_{u,1}$ = depth of neutral axis of balanced section = $\dfrac{0.0035}{0.0055 + 0.87 f_y / E_s} d \qquad (3.20)$

f_y = characteristic strength of steel

E_s = modulus of elasticity of steel

b = width of section

d = effective depth of section

For balanced section ($M_u = M_{u,1}$), area of steel A_{st1} is obtained by:

$$A_{st1} = \dfrac{M_{u,1}}{0.87 f_y (d - 0.416 x_{u,1})} \qquad (3.21)$$

For under reinforced section ($M_u < M_{u,1}$), area of steel A_{st} is obtained by:

$$A_{st} = \dfrac{M_u}{0.87 f_y (d - 0.416 x_u)} \qquad (3.22)$$

Flexure 45

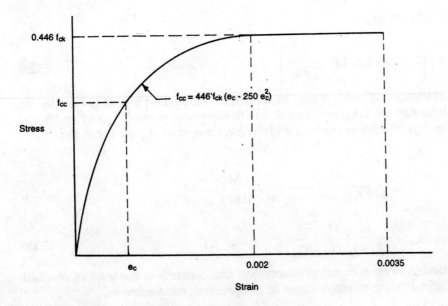

Fig. 3.1 Design stress-strain curve for concrete

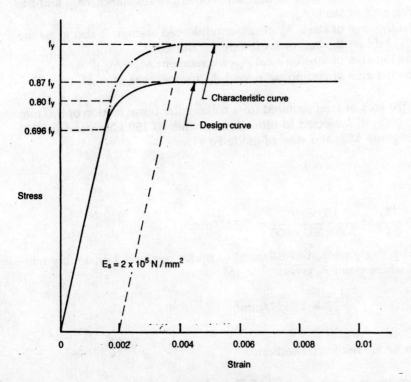

Fig. 3.2 Design stress-strain curve for steel

where x_u = depth of neutral axis)

$$= 1.202 + \left(1.4446 - 6.67735 \frac{M_u}{f_{ck} b d^2}\right)^{0.5} d \qquad (3.23)$$

The doubly reinforced section ($M_u > M_{u,1}$) may be considered equivalent to a singly reinforced balanced section with tension steel A_{st1} and a section with compression steel A_{sc} and additional tension steel A_{st2}. The area of tension steel A_{st} and compression steel A_{sc} are given by:

$$A_{st} = A_{st1} + A_{st2}$$

$$= \frac{M_{u,1}}{0.87 f_y (d - 0.416 x_{u,1})} + \frac{M_u - M_{u,1}}{0.87 f_y (d - d')} \qquad (3.24)$$

and

$$A_{sc} = \frac{M_u - M_{u,1}}{(f_{sc} - f_{cc})(d - d')} \qquad (3.25)$$

where f_{sc} and f_{cc} are design stresses in compression steel and concrete at the level of centroid of compression steel and d' is the effective cover to compression reinforcement.

Based on above calculations, the following charts have been prepared to facilitate design.

Chart 3.23: To determine the limiting value of ultimate moment of resistance, $M_{u,1}$ and the corresponding area of steel, A_{st1}

Chart 3.24: To determine the area of steel, A_{st} of under reinforced section. It also gives the value of $M_{u,1}/bd^2$ on the curves for different values of f_y

Chart 3.25: To determine the area of tension steel A_{st2} for moment $M_u - M_{u,1}$.

Chart 3.26: To determine the area of compression steel A_{sc} for moment $M_u - M_{u,1}$

Example 3.4: Calculate the area of steel required for a rectangular beam section of 300 mm width and 500 mm effective depth subjected to ultimate moments (i) 150 kN.m and (ii) 300 kN.m. Consider concrete of grade M20 and steel of grade Fe 415.

Solution

Case (i) $M_u = 150$ kN.m

Compute

$$\frac{M_u}{b d^2} = \frac{150 \times 10^6}{300 \times 500 \times 500} = 2.0 \, N/mm^2$$

To ascertain the section to be singly reinforced balanced or under reinforced or doubly reinforced, consider Chart 3.23 where point P_3 gives,

$$\frac{M_{u,1}}{b d^2} = 2.76 \, N/mm^2$$

Since $\dfrac{M_{u,1}}{b d^2} > \dfrac{M_u}{b d^2}$, it is an under reinforced section.

The area of steel A_{st} is determined with the help of Chart 3.24 as follows,

Compute
$$\frac{M_u}{f_{ck}\,b\,d^2} = \frac{150 \times 10^6}{20 \times 300 \times 500 \times 500} = 0.1$$

Corresponding to the above value of $M_u / f_{ck}\,b\,d^2$, point P_4 on Chart 3.24 gives,
$$100\,A_{st}/bd = 0.64$$

∴
$$A_{st} = 0.64\,bd/100 = 0.64 \times 300 \times 500/100 = 960 \text{ mm}^2$$

Case (ii) $M_u = 300$ kN.m

Compute
$$\frac{M_u}{b\,d^2} = \frac{300 \times 10^6}{300 \times 500 \times 500} = 4.0 \text{ N/mm}^2$$

The value of $M_{u,1}/bd^2$ determined in case (i) is,
$$M_{u,1}/b\,d^2 = 2.76 \text{ N/mm}^2$$

Since $\dfrac{M_u}{b\,d^2} > \dfrac{M_{u,1}}{b\,d^2}$, it is a doubly reinforced section.

The area of steel A_{st1} for balanced section is determined from Chart 3.23 where point P_5 gives,
$$100 A_{st1}/b\,d = 0.96$$

∴
$$A_{st1} = 0.96\,b\,d/100 = 0.96 \times 300 \times 500/100 = 1440 \text{ mm}^2$$

Consider effective cover to compression reinforcement, $d' = 50$ mm

Compute
$$d'/d = 50/500 = 0.1$$

$$\frac{M_u - M_{u,1}}{b\,d^2\,(1 - d'/d)} = \frac{4.0 - 2.76}{0.9} = 1.378 \text{ N/mm}^2$$

Area of additional tension steel A_{st2} is determined by Chart 3.25 where point P_3 gives,
$$100 A_{st2}/bd = 0.38$$

∴
$$A_{st2} = 0.38\,bd/100 = 0.38 \times 300 \times 500/100 = 570 \text{ mm}^2$$

Total area of tension steel,
$$A_{st} = A_{st1} + A_{st2} = 1440 + 570 = 2010 \text{ mm}^2$$

Area of compression steel A_{sc} is determined by Chart 3.26 where point P_4 gives,
$$A_{st}/A_{st2} = 1.043$$

∴
$$A_{sc} = 1.043\,A_{st2} = 1.043 \times 570 = 594.5 \text{ mm}^2$$

48 Handbook of Reinforced Concrete Design

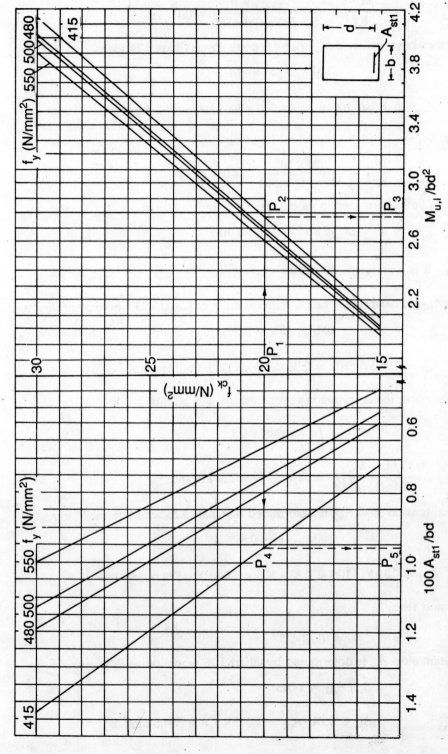

Chart 3.23 Ultimate moment of resistance ($M_{u,1}$) and area of steel (A_{st1}) for balanced rectangular section

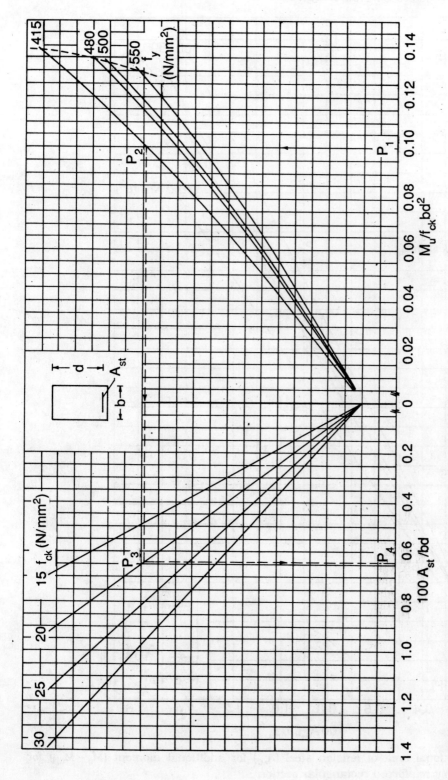

Chart 3.24 Area of steel (A_{st}) for under reinforced rectangular section

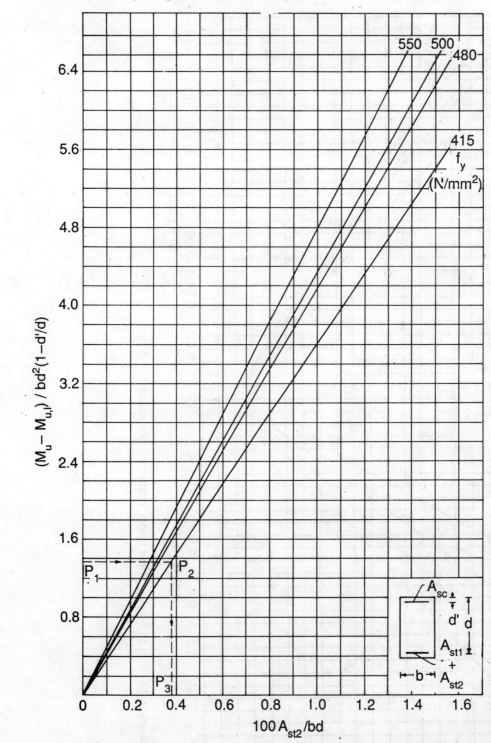

Chart 3.25 Additional area of tension steel (A_{st2}) for additional moment ($M_u - M_{u,1}$) for doubly reinforced rectangular section

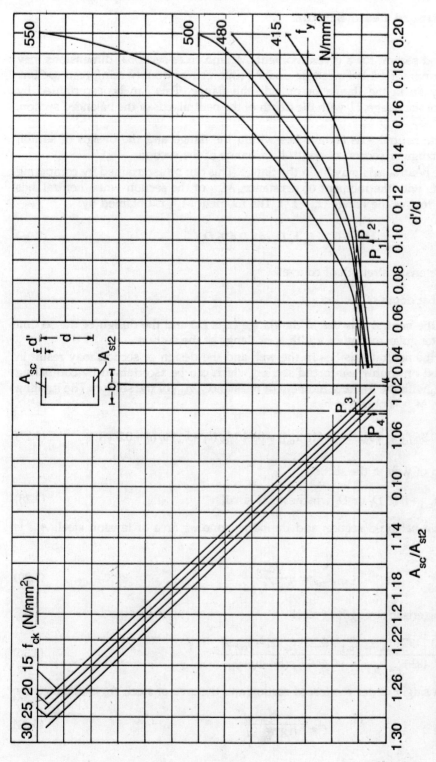

Chart 3.26 Area of compression steel (A_{sc}) for doubly reinforced rectangular section

3.3.2 Design of Flanged Beam Section

The design of a flanged section for a given moment, M_u and cross-sectional dimensions may result in either singly reinforced balanced or under reinforced or doubly reinforced section. The neutral axis may lie in the flange or outside the flange. This can be ascertained by comparing the depth of the flange, D_f with the depth of the neutral axis of the balanced section, $x_{u,1}$.

If $D_f \geq x_{u,1}$, then the neutral axis always lies within the flange and the design of section reduces to that of rectangular section of width b_f of flange of the section.

If $D_f < x_{u,1}$, then the neutral axis may lie in the flange. This can be ascertained by comparing the given moment, M_u with the moment of resistance, $M_{u,1}$ of the section when neutral axis coincides with the bottom of the flange ($x_u = D_f$). The moment $M_{u,1}$ is obtained by:

$$M_{u1} = 0.36 f_{ck} b_f D_f (d - 0.416 D_f) \tag{3.26}$$

where f_{ck} = characteristic strength of concrete

d = effective depth of section

If $M_u \leq M_{u1}$, then the neutral axis lies in the flange ($x_u \leq D_f$) and the design of the section reduces to that of rectangular section of width b_f of flange of the section.

If $M_u > M_{u1}$, then the neutral axis lies in the web and the design of section may result in either singly reinforced or doubly reinforced section which can be ascertained by comparing the given moment, M_u with the limiting moment of resistance, $M_{u,1}$ of the section. The limiting moment of resistance, $M_{u,1}$ is obtained by:

$$M_{u,1} = 0.36 f_{ck} x_{u,1} b_w (d - 0.416 x_{u,1}) + 0.446 f_{ck} (b_f - b_w) y_f (d - 0.5 y_f) \tag{3.27}$$

where b_w = width of web of the section

$$y_f = 0.15 x_{u,1} + 0.65 D_f \text{ or } D_f \text{ whichever is small} \tag{3.28}$$

If $M_u = M_{u,1}$, it is a balanced section and the corresponding area of tension steel, A_{st1} is given by:

$$A_{st1} = \frac{M_{u,1}}{0.87 f_y j d} \tag{3.29}$$

where f_y = characteristic strength of steel

$$jd = d - \frac{0.14976 f_{ck} b_w x_{u,1}^2 + 0.223 f_{ck} (b_f - b_w) y_f^2}{0.36 f_{ck} b_w x_{u,1} + 0.446 f_{ck} (b_f - b_w) y_f} \tag{3.30}$$

If $M_u < M_{u,1}$, it is a singly under reinforced section and the area of steel, A_{st} is obtained by:

$$A_{st} = \frac{M_u}{0.87 f_y j d} \tag{3.31}$$

where $jd = d - \dfrac{0.14976 f_{ck} b_w x_u^2 + 0.223 f_{ck} (b_f - b_w) y_f^2}{0.36 f_{ck} b_w x_u + 0.446 f_{ck} (b_f - b_w) y_f}$ (3.32)

$$x_u = \dfrac{-b + (b^2 - 4ac)^{0.5}}{2a}$$ (3.33)

For $y_f = 0.15 x_u + 0.65 D_f$,

$a = -0.14976 f_{ck} b_w - 0.0050175 f_{ck} (b_f - b_w)$

$b = 0.36 f_{ck} b_w d + 0.0669 f_{ck} (b_f - b_w)(d - 0.65 D_f)$

$c = 0.2899 f_{ck} D_f (b_f - b_w)(d - 0.325 D_f) - M_u$

For $y_f = D_f$,

$a = -0.14976 f_{ck} b_w$

$b = 0.36 f_{ck} b_w d$

$c = 0.446 f_{ck} D_f (b_f - b_w)(d - 0.5 D_f) - M_u$

If $M_u > M_{u,1}$, it is a doubly reinforced section and the area of tension steel, A_{st} and compression steel, A_{sc} are given by:

$$A_{st} = A_{st1} + A_{st2}$$
$$= \dfrac{M_{u,1}}{0.87 f_y jd} + \dfrac{M_u - M_{u,1}}{0.87 f_y (d - d')}$$ (3.34)

$$A_{sc} = \dfrac{M_u - M_{u,1}}{(f_{sc} - f_{cc})(d - d')}$$ (3.35)

where jd is given by Eq. 3.30, f_{sc} and f_{cc} are design stresses in compression steel and concrete at the level of centroid of compression steel respectively and d' is the effective cover to compression reinforcement.

Based on the above calculations, the following charts have been prepared to facilitate design.

Chart 3.27: To determine the area of steel, A_{st} for singly reinforced T section when the neutral axis lies within the flange. It also gives the value of $M_{u,1}/b_f d^2$ at the point of intersection of curves for D_f/d and f_y for that value of D_f/d and f_y.

Charts 3.28 to 3.31: To determine the limiting moment of resistance, $M_{u,1}$ and the corresponding area of steel, A_{st1}.

Chart 3.32: To determine area of steel, A_{st} for singly reinforced T section when neutral axis lies within the web. It also gives the value of $M_{u,1}/b_f d^2$ on the curve for different values of f_y.

Additional area of tension steel A_{st2} and compression steel A_{sc} in doubly reinforced flanged section for resisting moment $M_u - M_{u,1}$ can be determined using Charts 3.25 and 3.26 respectively.

Example 3.5: Calculate the area of steel required for a T-beam section of 1500 mm width of flange, 300 mm width of web, 100 mm depth of flange and 500 mm effective depth when subjected to ultimate moments (i) 450 kN.m, (ii) 600 kN.m and (iii) 750 kN.m. Consider concrete of grade M20 and steel of grade Fe 415.

Solution

Case (i) $M_u = 450$ kN.m

Compute

$$b_w / b_f = 300 / 1500 = 0.2$$

$$D_f / d = 100 / 500 = 0.2$$

$$\frac{M_u}{f_{ck} b_f d^2} = \frac{450 \times 10^6}{20 \times 1500 \times 500 \times 500} = 0.06$$

Consider that the neutral axis lies within the flange. This is ascertained from Chart 3.27 where point P_2 corresponding to the value of $M_u / f_{ck} b_f d^2 = 0.06$ lies within the line drawn for $D_f / d = 0.2$. So the neutral axis lies within the flange and point P_4 gives area of steel as,

$$100 A_{st} / b_f d = 0.365$$

$$A_{st} = 0.365 \times 1500 \times 500 / 100 = 2737.5 \text{ mm}^2$$

Case (ii) $M_u = 600$ kN.m

Compute

$$\frac{M_u}{f_{ck} b_f d^2} = \frac{600 \times 10^6}{20 \times 1500 \times 500 \times 500} = 0.08$$

Consider that the neutral axis lies outside the flange. This is ascertained from Chart 3.27 where the point corresponding to the value of $M_u / f_{ck} b_f d^2 = 0.08$ lies above the line drawn for $D_f / d = 0.2$. So, the neutral axis lies within the web. To ascertain the design to be singly reinforced or doubly reinforced, ultimate moment of resistance of singly reinforced balanced section is determined from Chart 3.28. Corresponding to the values of $D_f / d = 0.2$ and $b_w / b_f = 0.2$, point P_3 gives,

$$\frac{M_{u,1}}{f_{ck} b_f d^2} = 0.0917$$

Since $\dfrac{M_{u,1}}{f_{ck} b_f d^2} > \dfrac{M_u}{f_{ck} b_f d^2}$, it is an under reinforced section.

The design of section is made with the help of Chart 3.32 where point P_4 gives,

$$100 A_{st} / b_f d = 0.51$$

$$A_{st} = 0.51 \times 1500 \times 500 / 100 = 3825 \text{ mm}^2$$

Case (iii) $M_u = 750$ kN.m

Compute
$$\frac{M_u}{f_{ck}\, b_f\, d^2} = \frac{750 \times 10^6}{20 \times 1500 \times 500 \times 500} = 0.1$$

The value of $\dfrac{M_{u,1}}{f_{ck}\, b_f\, d^2}$ determined in case (ii) is,

$$\frac{M_{u,1}}{f_{ck}\, b_f\, d^2} = 0.0917$$

Since $\dfrac{M_u}{f_{ck}\, b_f\, d^2} > \dfrac{M_{u,1}}{f_{ck}\, b_f\, d^2}$, it is a doubly reinforced section.

The area of steel for singly reinforced balanced section is determined from Chart 3.28 where point P_5 gives,

$$100\, A_{st1} / b_f\, d = 0.59$$

$$A_{st1} = 0.59 \times 1500 \times 500 / 100 = 4425 \text{ mm}^2$$

The area of additional tension steel A_{st2} and compression steel A_{sc} for resisting additional moment, $M_u - M_{u,1}$ are determined by Charts 3.25 and 3.26 respectively as follows.

Consider $\qquad d' = 50$ mm

Compute $\qquad d'/d = 50/500 = 0.1$

$$\frac{M_u - M_{u,1}}{b_f\, d^2 (d - d')} = \frac{M_u - M_{u,1}}{f_{ck}\, b_f\, d^2} \times \frac{f_{ck}}{(1 - d'/d)}$$

$$= (0.1 - 0.0917)\frac{20}{1 - 0.01} = 0.1844$$

The area of additional tension steel A_{st2} determined from Chart 3.25 is given by,

$$100\, A_{st2} / b_{fd} = 0.05$$

$$A_{st2} = 0.05 \times 1500 \times 500 / 100 = 375 \text{ mm}^2$$

Total area of tension steel,

$$A_{st} = A_{st1} + A_{st2} = 4425 + 375 = 4800 \text{ mm}^2$$

The area of compression steel A_{sc} determined by Chart 3.26 is given by,

$$A_{sc} / A_{st2} = 1.043$$

∴ $\qquad A_{sc} = 1.043 \times A_{st2} = 1.043 \times 375 = 391.13 \text{ mm}^2$

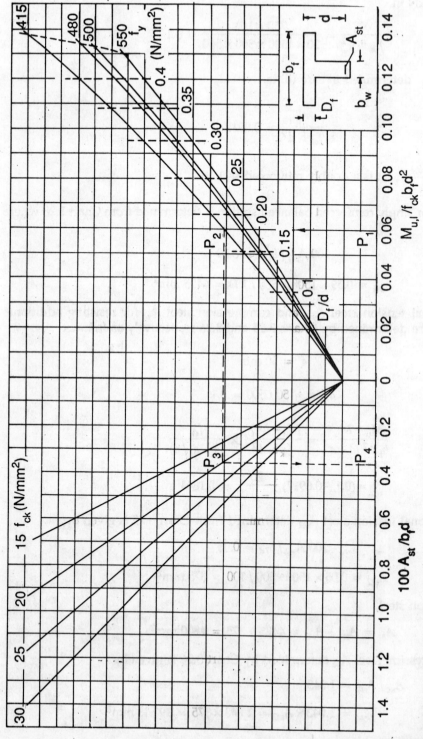

Chart 3.27 Area of steel (A_{st}) for under reinforced T-section when neutral axis lies within the flange

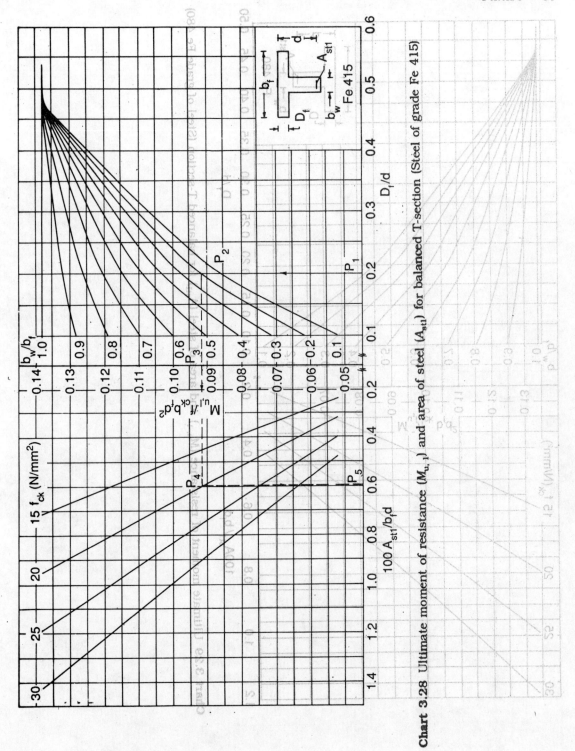

Chart 3.28 Ultimate moment of resistance ($M_{u,1}$) and area of steel (A_{st1}) for balanced T-section (Steel of grade Fe 415)

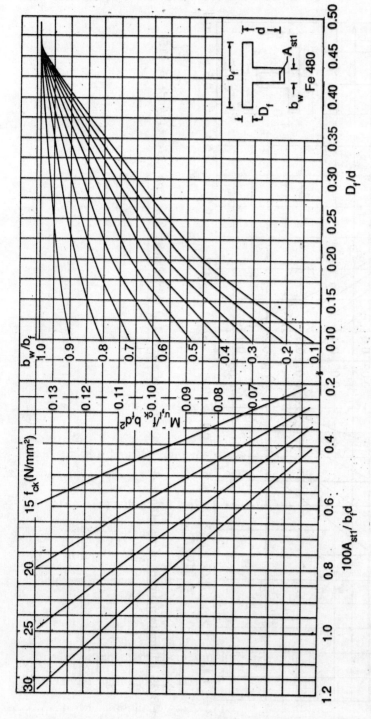

Chart 3.29 Ultimate moment of resistance ($M_{u,1}$) and area of steel (A_{st1}) for balanced T-section (Steel of grade Fe 480)

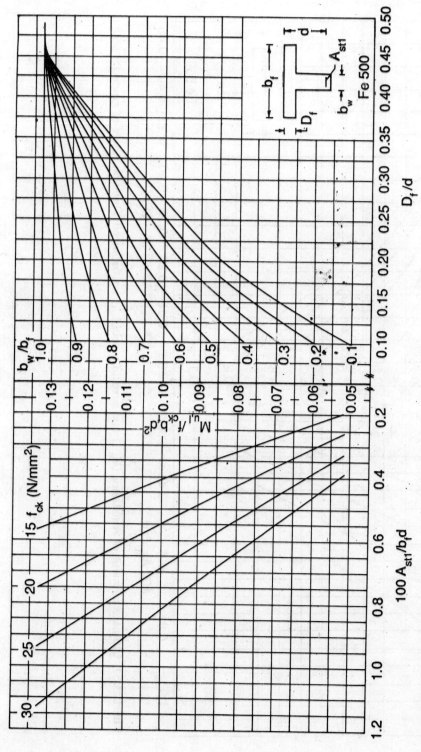

Chart 3.30 Ultimate moment of resistance ($M_{u,1}$) and area of steel (A_{st1}) for balanced T-section (Steel of grade Fe 500)

60 Handbook of Reinforced Concrete Design

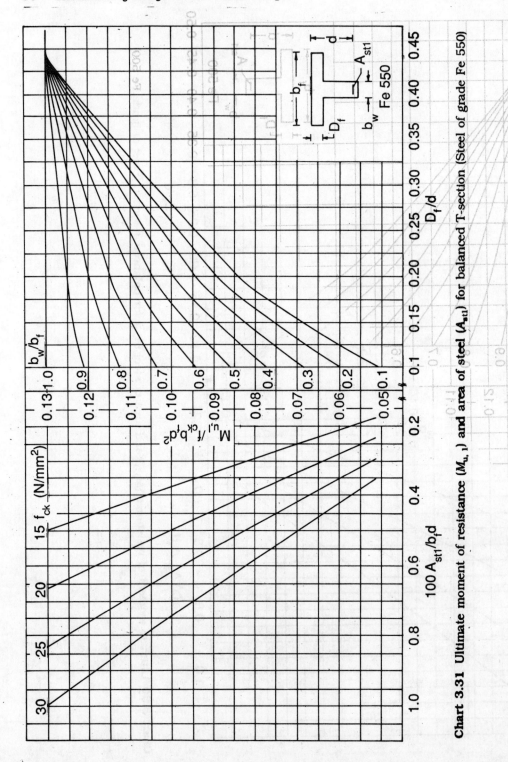

Chart 3.31 Ultimate moment of resistance ($M_{u,1}$) and area of steel (A_{st1}) for balanced T-section (Steel of grade Fe 550)

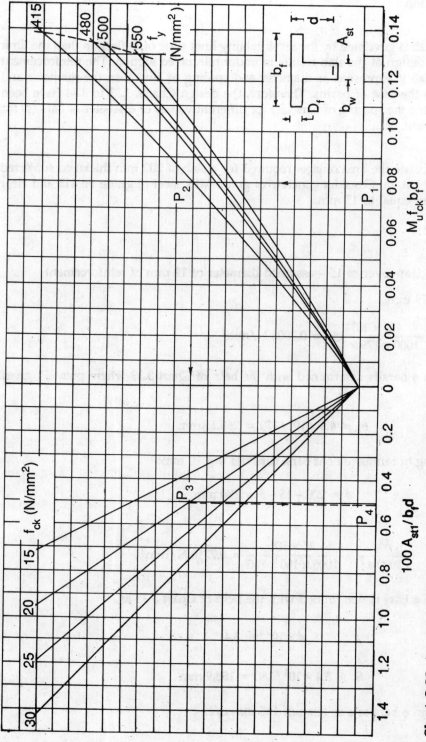

Chart 3.32 Area of steel (A_{st}) for under reinforced T-section

3.3.3 Design of Slab

As the thickness of slab is governed by the serviceability limit state of deflection than the limit state of collapse, the design of the slab results in under reinforced section. The reinforcement for moment in the slab is expressed as diameter and spacing of bars than the diameter and number of bars as in the case of beams. Therefore, the design Charts 3.33 to 3.44 have been developed to determine the spacing of bars (S_v) for different values of diameter of bars (ϕ) for resisting design moment, M_u.

Example 3.6: Calculate the area of steel required for a slab of 200 mm thickness subjected to a moment of 25 kN.m/m. Consider concrete of grade M15, steel of grade Fe 415 and clear cover to reinforcement equal to 15 mm.

Solution

Compute d = 200 – clear cover of 15 – assumed diameter of 12 mm of reinforcement

$$\text{bar}/2 = 179 \text{ mm}.$$

$$\frac{M_u}{1000\, d^2} = \frac{25 \times 10^6}{1000 \times 179 \times 179} = 0.7803 \text{ N/mm}^2$$

Spacing of 12 mm ϕ bars is determined with the help of Chart 3.33 where point P_3 gives, $S_v\, d \times 10^{-4} = 4.8$

$$\therefore \quad S_v = 4.8 \times 10^4 / 179 = 268.15 \text{ mm}$$

For smaller spacing of reinforcement bars, consider ϕ = 10 mm.

$$\therefore \quad d = 200 - 15 - 5 = 180 \text{ mm}$$

and

$$\frac{M_u}{1000\, d^2} = \frac{25 \times 10^6}{1000 \times 180 \times 180} = 0.7716 \text{ N/mm}^2$$

Spacing of 10 mm ϕ bars is determined with the help of Chart 3.33 as,

$$S_v\, d \times 10^{-4} = 3.4$$

$$\therefore \quad S_v = 3.4 \times 10^4 / 180 = 188.9 \text{ mm}$$

Hence provide 10 mm ϕ bars at a spacing of 185 mm c/c.

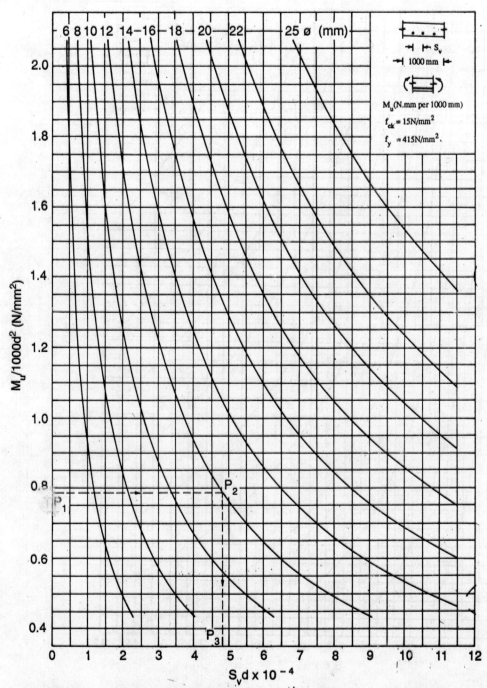

Chart 3.33 Diameter (φ) and spacing (S_v) of reinforcement bars for moment (M_u) per unit width of slab (f_{ck} = 15 N/mm² and f_y = 415 N/mm²)

Flexure 63

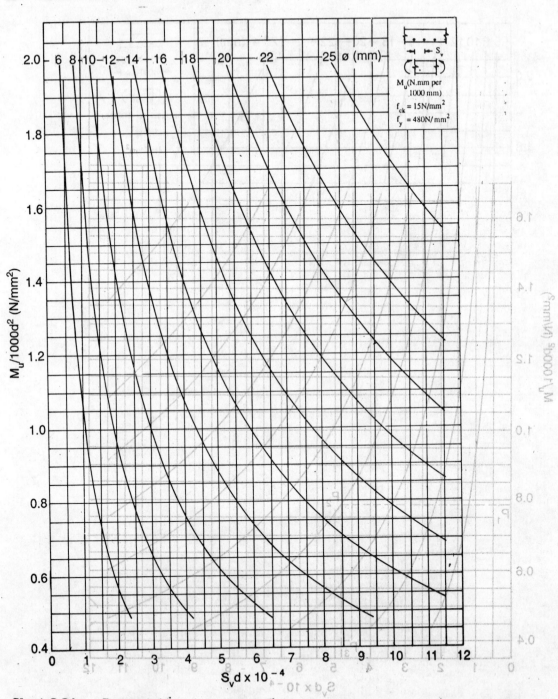

Chart 3.34 Diameter (ϕ) and spacing (S_v) of reinforcement bars for moment (M_u) per unit width of slab (f_{ck} = 15 N/mm² and f_y = 480 N/mm²)

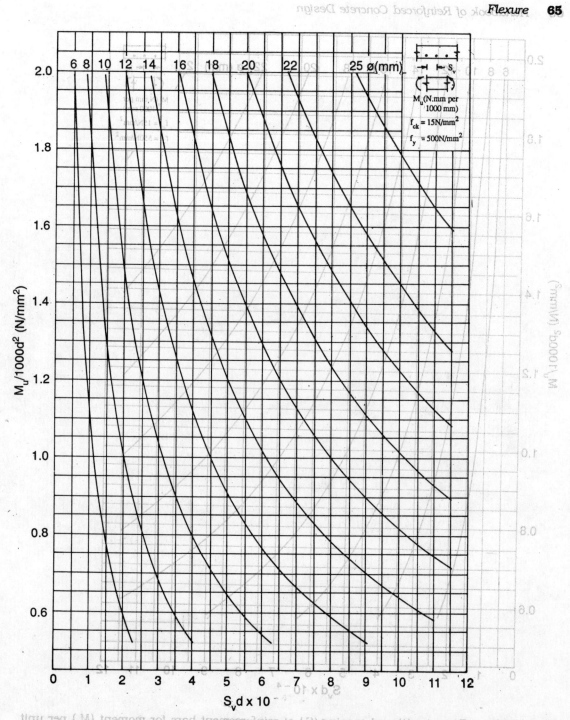

Chart 3.35 Diameter (ϕ) and spacing (S_v) of reinforcement bars for moment (M_u) per unit width of slab (f_{ck} = 15 N/mm² and f_y = 500 N/mm²)

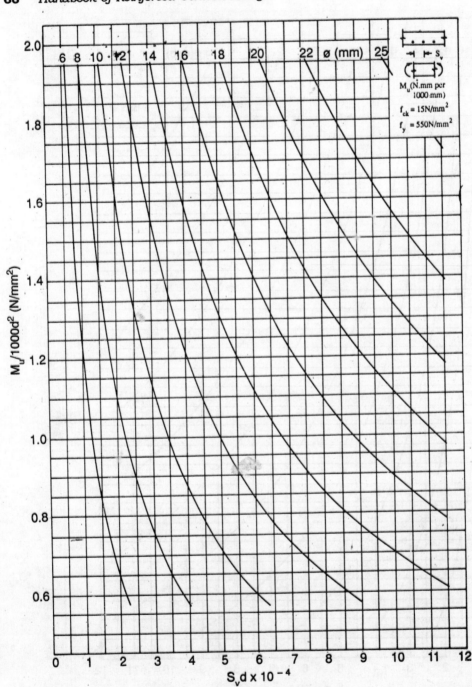

Chart 3.36 Diameter (ϕ) and spacing (S_v) of reinforcement bars for moment (M_u) per unit width of slab (f_{ck} = 15 N/mm² and f_y = 550 N/mm²)

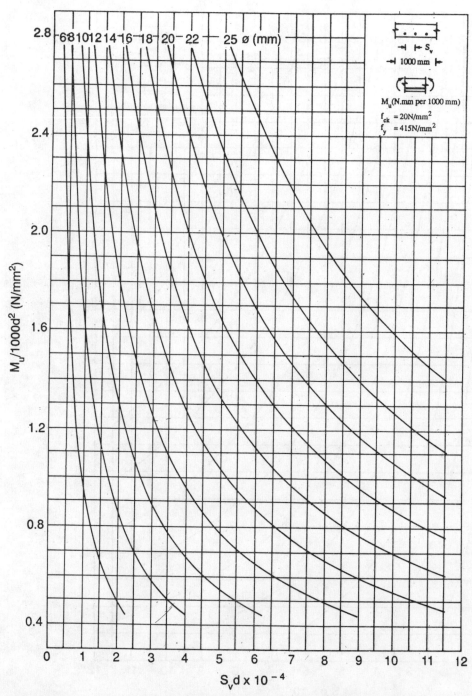

Chart 3.37 Diameter (ϕ) and spacing (S_v) of reinforcement bars for moment (M_u) per unit width of slab (f_{ck} = 20 N/mm² and f_y = 415 N/mm²)

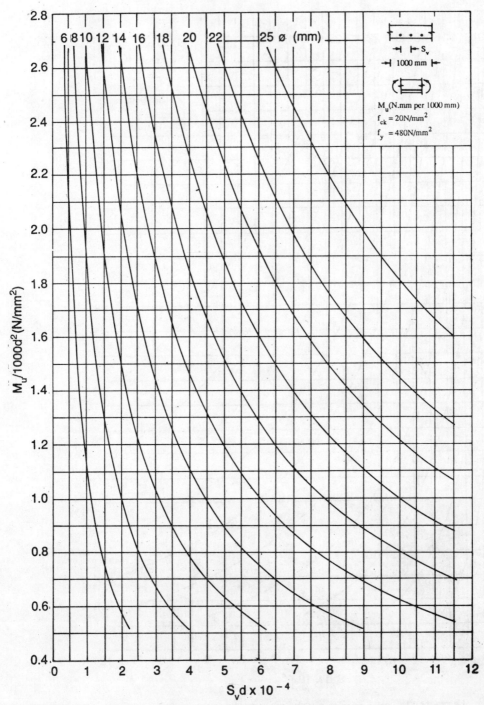

Chart 3.38 Diameter (φ) and spacing (S_v) of reinforcement bars for moment (M_u) per unit width of slab (f_{ck} = 20 N/mm² and f_y = 480 N/mm²)

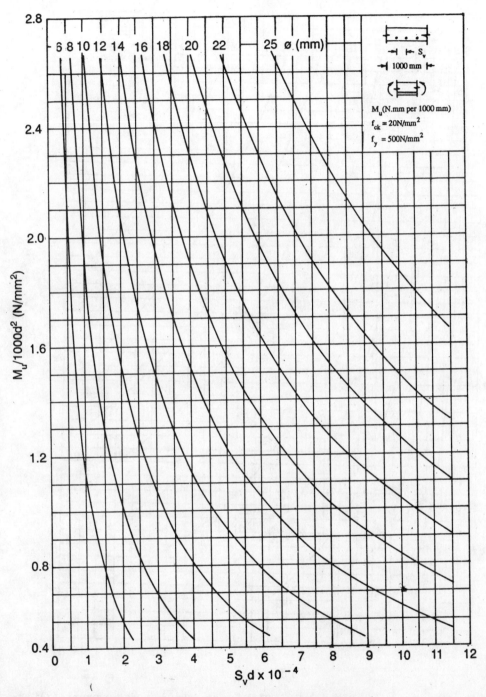

Chart 3.39 Diameter (φ) and spacing (S_v) of reinforcement bars for moment (M_u) per unit width of slab (f_{ck} = 20 N/mm² and f_y = 500 N/mm²)

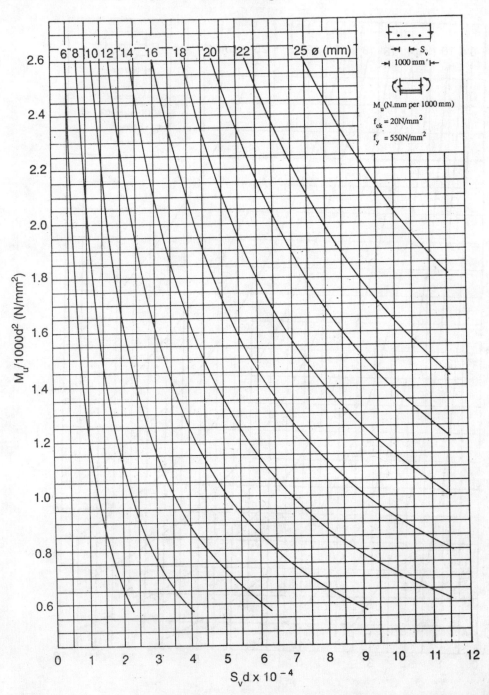

Chart 3.40 Diameter (ϕ) and spacing (S_v) of reinforcement bars for moment (M_u) per unit width of slab (f_{ck} = 20 N/mm² and f_y = 550 N/mm²)

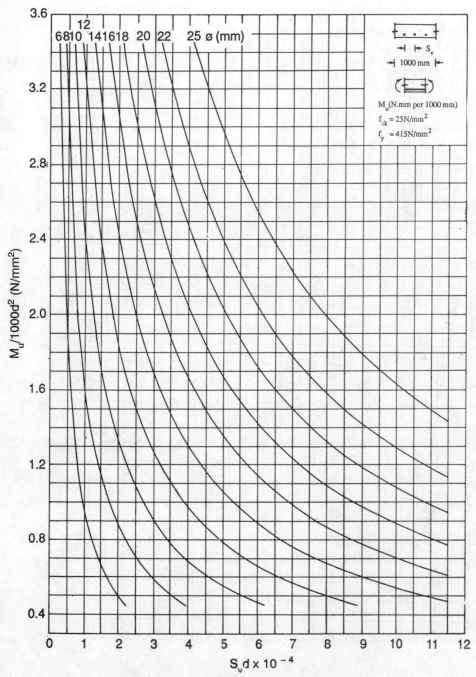

Chart 3.41 Diameter (φ) and spacing (S_v) of reinforcement bars for moment (M_u) per unit width of slab (f_{ck} = 25 N/mm² and f_y = 415 N/mm²)

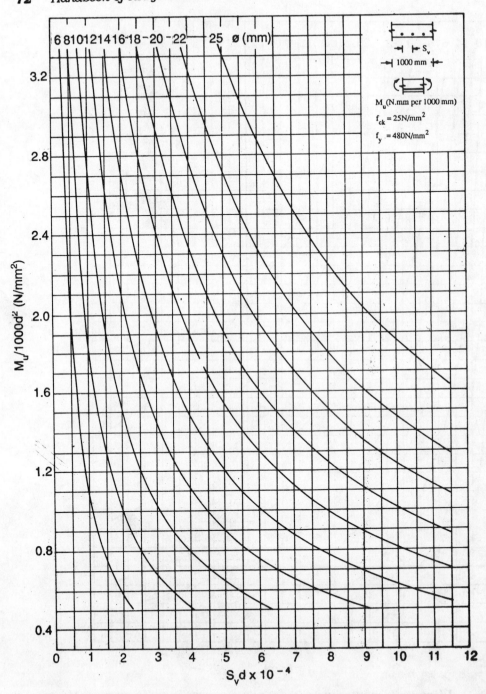

Chart 3.42 Diameter (φ) and spacing (S_v) of reinforcement bars for moment (M_u) per unit width of slab (f_{ck} = 25 N/mm² and f_y = 480 N/mm²)

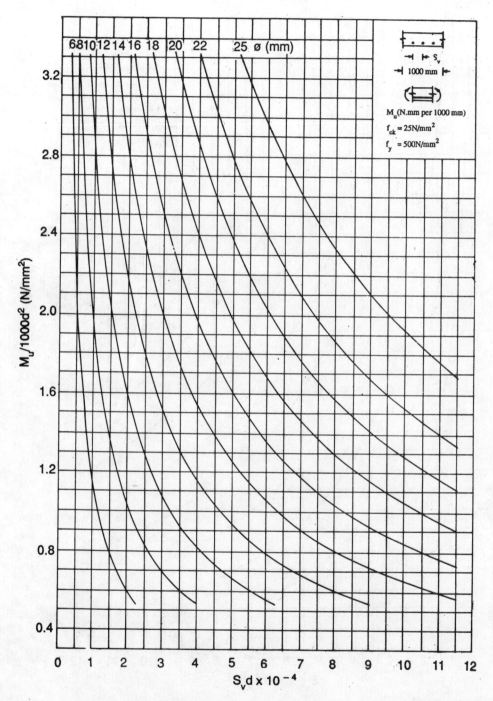

Chart 3.43 Diameter (φ) and spacing (S_v) of reinforcement bars for moment (M_u) per unit width of slab (f_{ck} = 25 N/mm² and f_y = 500 N/mm²)

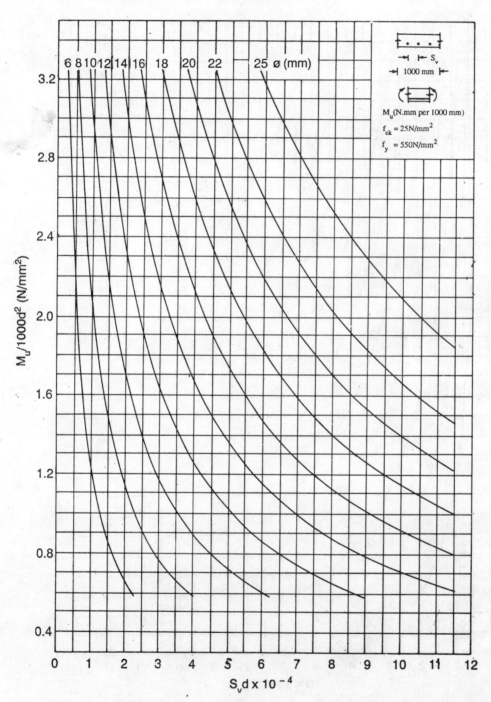

Chart 3.44 Diameter (φ) and spacing (S_v) of reinforcement bars for moment (M_u) per unit width of slab (f_{ck} = 25 N/mm² and f_y = 550 N/mm²)

Chapter 4

Shear

4.1 DESIGN METHODS

IS : 456 – 1978 recommends both working stress and limit state methods of design for shear. Although emphasis has been laid on the limit state method, working stress method is also used for the design of some structures such as water retaining and bridge structures. Therefore, both the methods are discussed below.

4.2 WORKING STRESS METHOD

The area and spacing of shear reinforcement required to resist shear force can be obtained as follows.

For inclined stirrups or a series of bars bent up at different cross-sections,

$$\frac{A_s}{S_v} = \frac{V_s}{\sigma_{sv} d (\sin i + \cos i)} \qquad (4.1)$$

For single bar or a group of parallel bars all bent up at the same cross-section,

$$A_{sv} = \frac{V_s}{\sigma_{sv} \sin i} \qquad (4.2)$$

For vertical stirrups,

$$\frac{A_{sv}}{S_v} = \frac{V_s}{\sigma_{sv} d} \qquad (4.3)$$

where A_{sv} = total cross-sectional area of stirrup legs or bent up bars within a distance S_v

S_v = spacing of stirrups or bent up bars along the length of the member

σ_{sv} = permissible tensile stress in shear reinforcement which shall not be taken greater than 230 N/mm²

i = angle beteeen the inclined stirrups or bent up bars and the axis of the member

d = effective depth of beam

V_s = shear force to be resisted by shear reinforcement at working state

   = $V - V_c$

V = shear force at working load

V_c = permissible shear strength of concrete

   = $\tau_c\,b\,d$

τ_c = permissible shear strength of concrete which can be obtained from Chart 4.1

b = width of beam

If $V \leq V_c$, minimum shear reinforcement is provided which is given by,

$$\frac{A_{sv}}{b\,S_v} = \frac{0.4}{f_y} \tag{4.4}$$

where f_y is characteristic strength of steel.

In no circumstances the nominal shear stress $\tau_v\,(=V/bd)$ shall exceed the maximum permissible shear stress $\tau_{c,max}$ which is tabulated in Chart 4.1.

Based on the above equations, the design Charts 4.2 to 4.7 have been prepared to facilitate the design.

Example 4.1: Determine the shear reinforcement required to resist a shear force of 150 kN at service state for a rectangular beam section of 250 mm width and 450 mm overall depth reinforced with 4 bars of 20 mm diameter. Consider concrete of grade M15, steel of grade Fe 415 and clear cover to reinforcement equal to 25 mm.

Solution

Area of tension steel, $A_{st}\,(4 \times 20\,\text{mm}\,\phi) = 1257\,\text{mm}^2$

Effective depth of beam, $d = 450 - 25 - 20/2 = 415\,\text{mm}$

Per cent area of tension steel, $p = 100\,A_{st}/bd = 100 \times 1257/(250 \times 415) = 1.2116$

Nominal shear stress, $\tau_v = V_u/bd = 150 \times 1000/(250 \times 415) = 1.4458\,\text{N/mm}^2$

Consider shear reinforcement of 10 mm ϕ 2 legged vertical stirrups. Its spacing is determined with the help of Chart 4.3 where point P_4 gives,

$$S_v\,b = 3.4 \times 10^4$$

$\therefore \qquad S_v = 3.4 \times 10^4/250 = 136\,\text{mm} \approx 135\,\text{mm c/c}$

Hence provide 10 mm ϕ two legged vertical stirrups at spacing of 135 mm c/c.

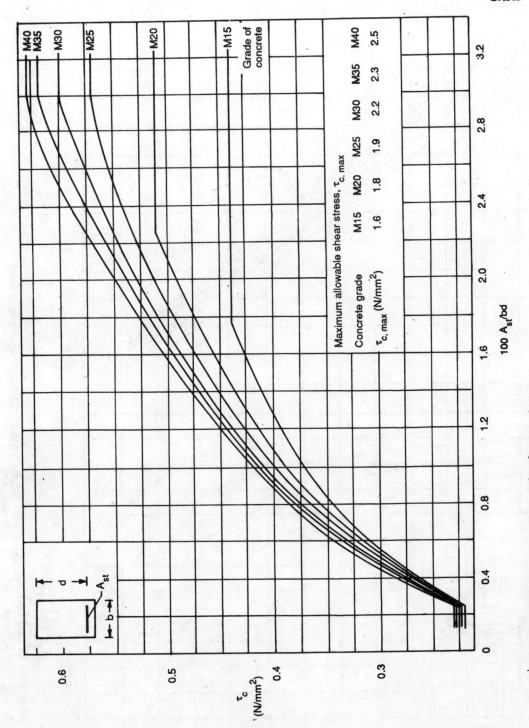

Chart 4.1 Allowable shear strength of concrete

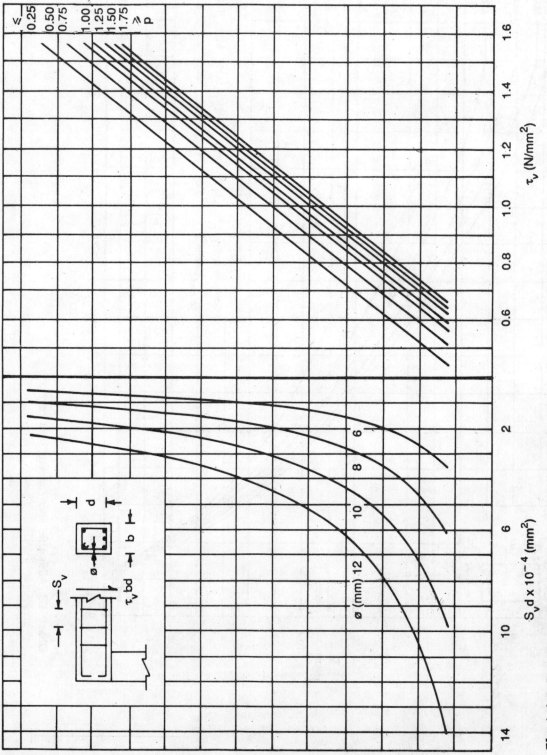

Chart 4.2 Diameter (ϕ) and spacing (S_v) of two legged vertical stirrups for shear stress, τ_v (σ_{cbc} = 5 N/mm² and σ_{st} = 140 N/mm²)

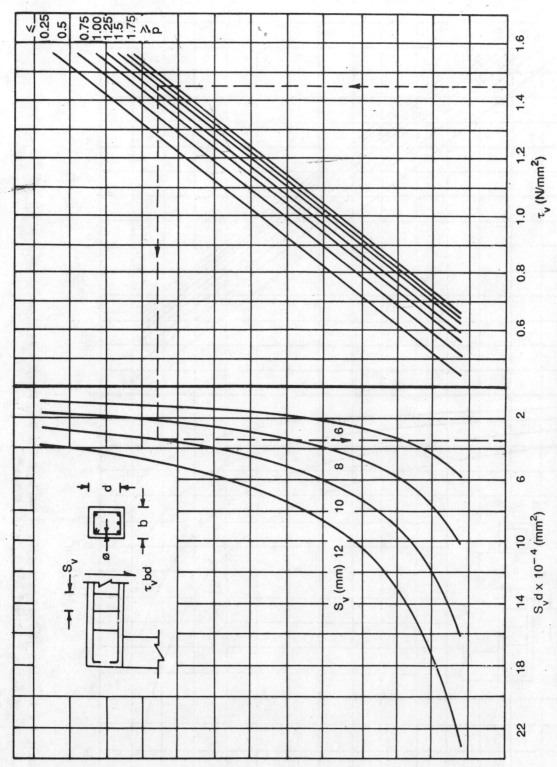

Chart 4.3 Diameter (φ) and spacing (S_v) of two legged vertical stirrups for shear stress, τ_v (σ_{cbc} = 5 N/mm² and σ_{st} = 230 N/mm²)

80 Handbook of Reinforced Concrete Design

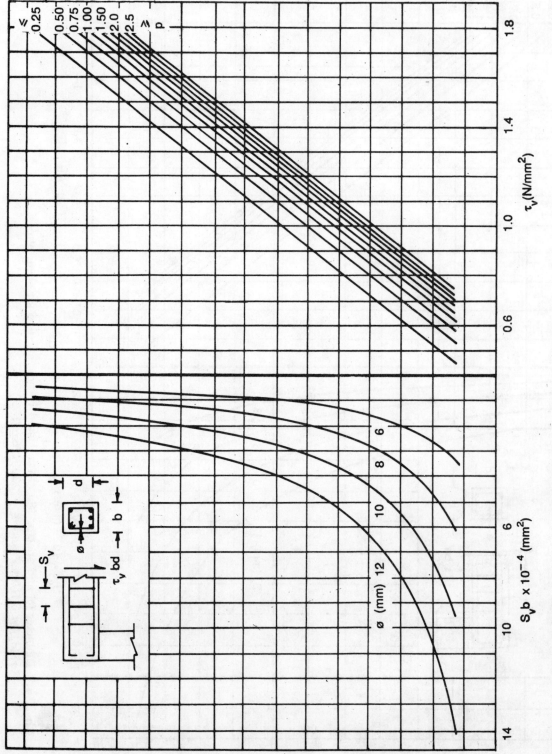

Chart 4.4 Diameter (ϕ) and spacing (S_v) of two legged vertical stirrups for shear stress, τ_v ($\sigma_{cbc} = 7$ N/mm² and $\sigma_{st} = 140$ N/mm²)

Shear 81

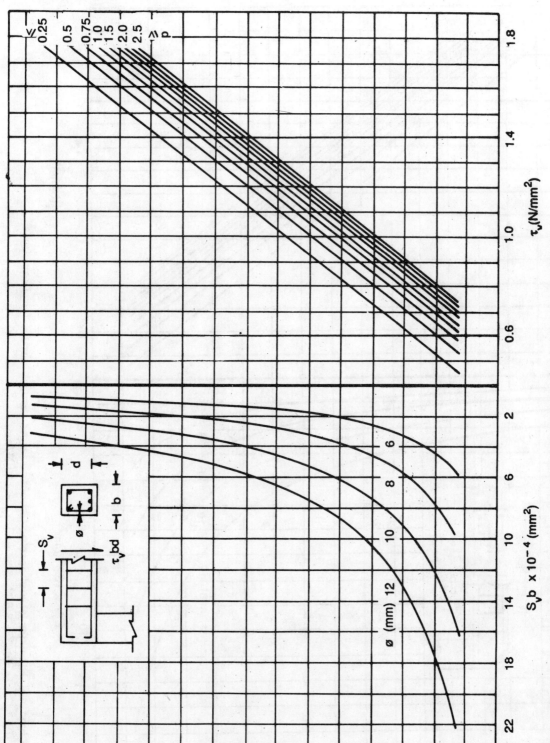

Chart 4.5 Diameter (φ) and spacing (S_v) of two legged vertical stirrups for shear stress, τ_v (σ_{cbc} = 7 N/mm² and σ_{st} = 230 N/mm²)

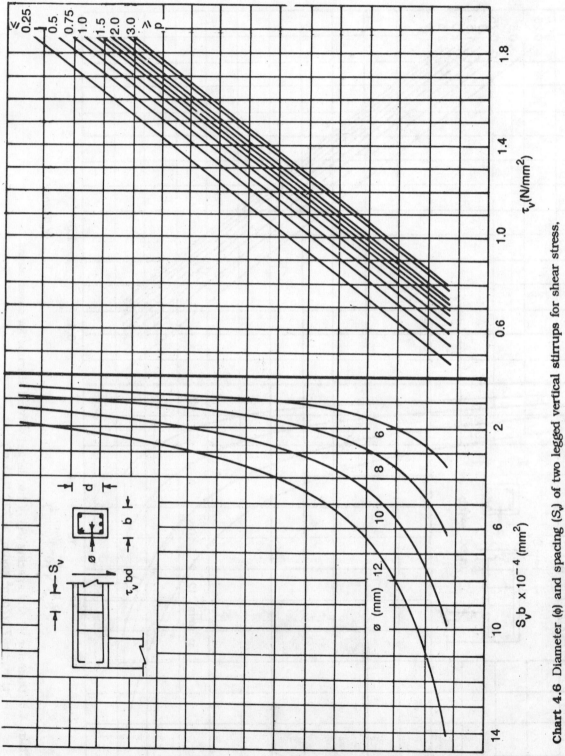

Chart 4.6 Diameter (ϕ) and spacing (S_v) of two legged vertical stirrups for shear stress, τ_v (σ_{cbc} = 8.5 N/mm² and σ_{st} = 140 N/mm²)

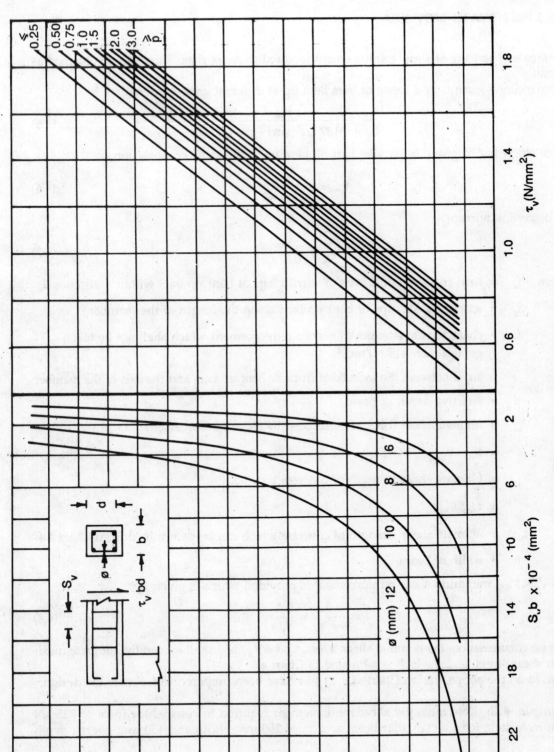

Chart 4.7 Diameter (ø) and spacing (S_v) of two legged vertical stirrups for shear stress, τ_v (σ_{cbc} = 8.5 N/mm² and σ_{st} = 230 N/mm²)

4.3 LIMIT STATE METHOD

The area and spacing of shear reinforcement required to resist shear force can be obtained as follows.

For inclined stirrups or a series of bars bent up at different cross-sections,

$$\frac{A_{sv}}{S_v} = \frac{V_{us}}{0.87 f_y d (\sin i + \cos i)} \qquad (4.5)$$

For single bar or group of parallel bars all bent up at the same cross-section,

$$A_{sv} = \frac{V_{us}}{0.87 f_y \sin i} \qquad (4.6)$$

For vertical stirrups,

$$\frac{A_{sv}}{S_v} = \frac{V_{us}}{0.87 f_y} \qquad (4.7)$$

where A_{sv} = total cross sectional area of stirrup legs or bent up bars within a distance S_v

S_v = spacing of stirrups or bent up bars along the length of the member

f_y = characteristic strength of stirrup reinforcement which shall not be taken greater than 415 N/mm^2

i = angle between the inclined stirrups or bent up bars and the axis of the member

d = effective depth of beam

V_{us} = ultimate shear force to be resisted by shear reinforcement

= $V_u - V_{uc}$

V_{uc} = ultimate shear strength of concrete

= $\tau_{uc} bd$

τ_{uc} = ultimate shear strength of concrete which can be determined from Chart 4.4

b = width of beam.

If $V_u < V_{uc}$, minimum shear reinforcement is provided which is given by,

$$\frac{A_{sv}}{b S_v} = \frac{0.4}{f_y} \qquad (4.8)$$

In no circumstances the nominal shear stress, τ_{uc} ($= V_u / bd$) shall exceed the ultimate maximum shear stress $\tau_{uc, max}$ which is tabulated in Chart 4.8.

Based on the above, design Charts 4.9 to 4.14 have been prepared to facilitate the design.

Example 4.2: Determine the shear reinforcement required to resist shear force of 225 kN at ultimate state for a rectangular beam section of 250 mm width and 450 mm overall depth reinforced with 4 bars of 20 mm diameter. Consider concrete a grade M15, steel of grade Fe 415 and a clear cover to reinforcement equal to 25 mm.

Solution

Area of tension steel, A_{st} (4 × 20mm ϕ) = 1257 mm²

Effective depth of beam,

$$d = \text{Total depth of Beam} - \text{clear cover to reinforcement} - \text{diameter of bar}/2$$
$$= 450 - 25 - 20/2$$
$$= 415 \text{ mm}$$

Per cent area of tension steel,

$$p = 100 \, A_{st} / bd$$
$$= 100 \times 1257 / (250 \times 415)$$
$$= 1.2116$$

Nominal shear stress,

$$\tau_v = V_u / bd$$
$$= 225 \times 1000 / (250 \times 415)$$
$$= 2.169 \text{ N/mm}^2$$

Consider shear reinforcement of 10 mm ϕ 2 legged vertical stirrups whose spacing is determined with the help of Chart 4.10 where point P_4 gives,

$$S_v \, b = 3.6 \times 10^4$$

∴ $$S_v = 3.6 \times 10^4 / 250$$
$$= 144 \text{ mm}$$
$$\simeq 140 \text{ mm}$$

Hence, provide 10 mm ϕ 2-legged vertical stirrups at spacing of 140 mm c/c.

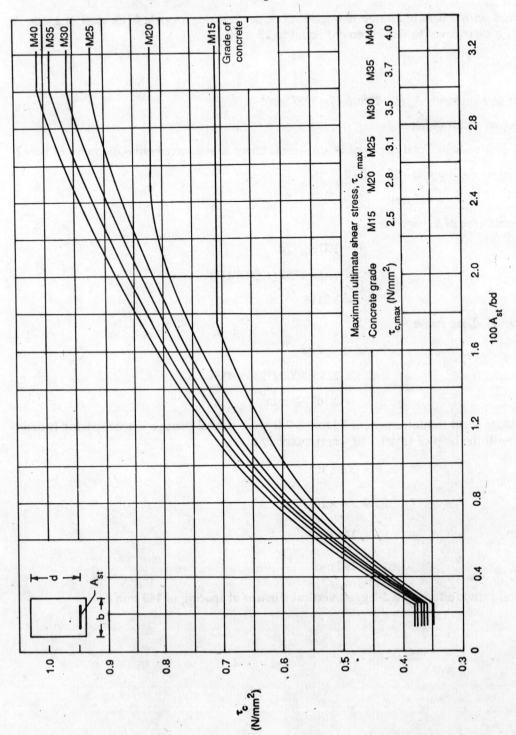

Chart 4.8 Ultimate shear strength of concrete

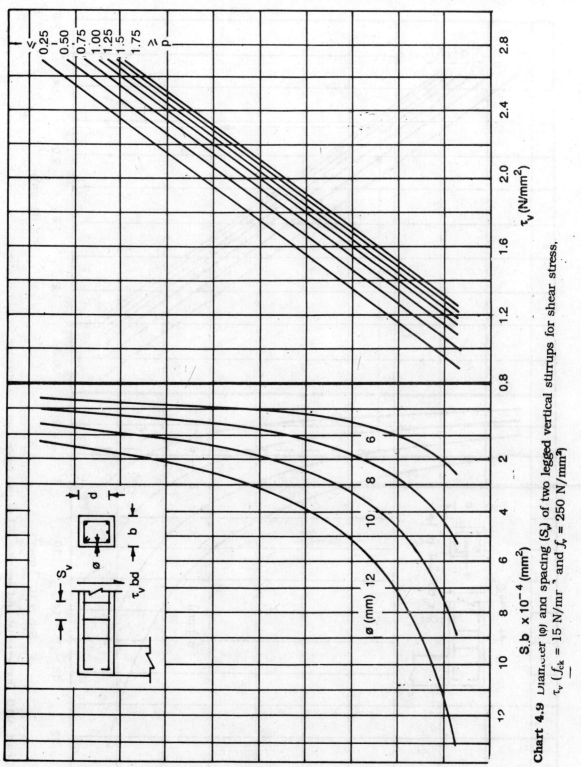

Chart 4.9 Diameter (ϕ) and spacing (S_v) of two legged vertical stirrups for shear stress, τ_v (f_{ck} = 15 N/mm² and f_y = 250 N/mm²)

88 Handbook of Reinforced Concrete Design

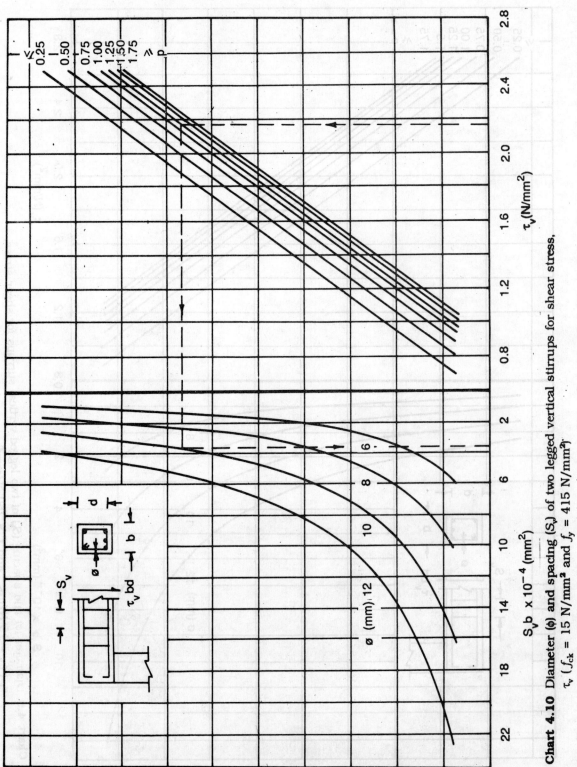

Chart 4.10 Diameter (ϕ) and spacing (S_v) of two legged vertical stirrups for shear stress, τ_v (f_{ck} = 15 N/mm² and f_y = 415 N/mm²).

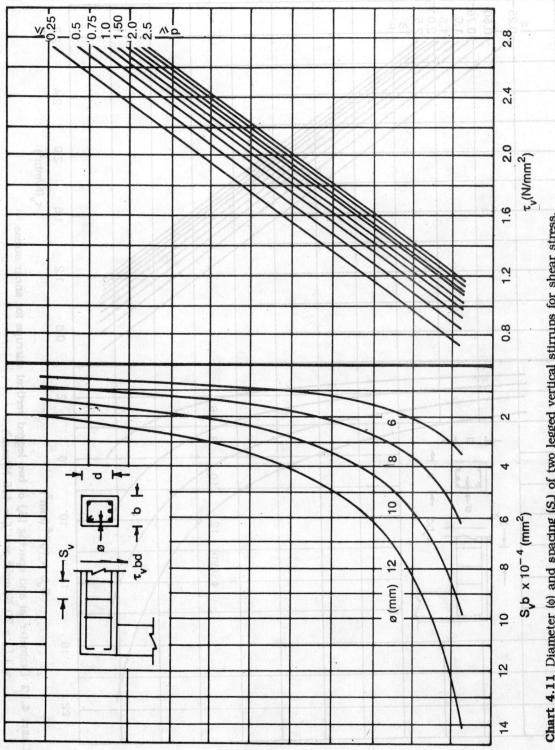

Chart 4.11 Diameter (ϕ) and spacing (S_v) of two legged vertical stirrups for shear stress, τ_v (f_{ck} = 20 N/mm² and f_y = 250 N/mm²)

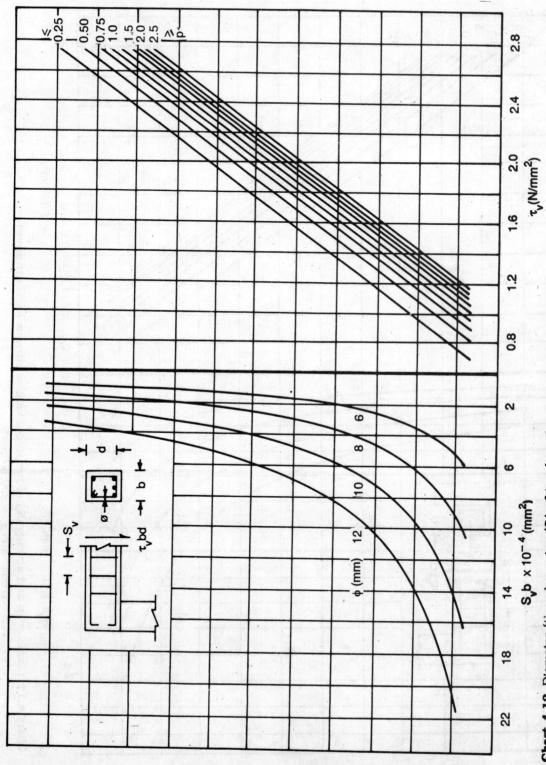

Chart 4.12 Diameter (φ) and spacing (S_v) of two legged vertical stirrups for shear stress, τ_v (f_{ck} = 20 N/mm² and f_y = 415 N/mm²)

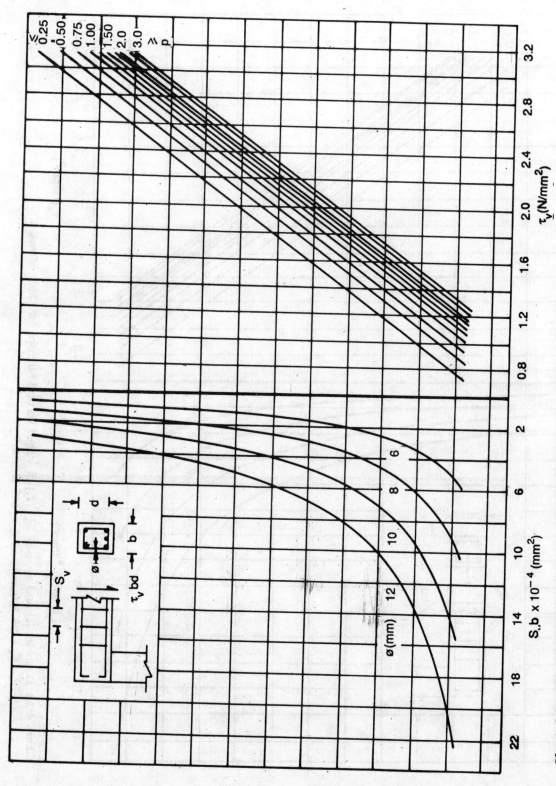

Chart 4.13 Diameter (ϕ) and spacing (S_v) of two legged vertical stirrups for shear stress, τ_v. ($f_{ck} = 25$ N/mm² and $f_y = 250$ N/mm²)

92 Handbook of Reinforced Concrete Design

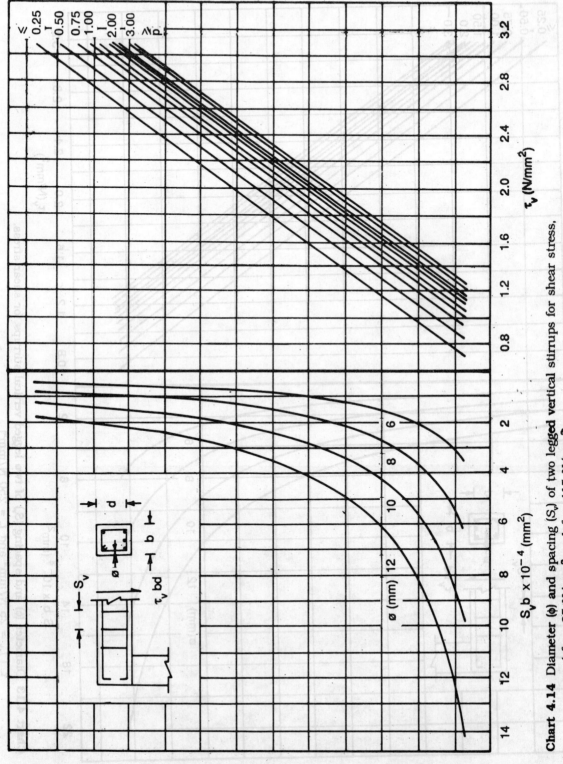

Chart 4.14 Diameter (φ) and spacing (S_v) of two legged vertical stirrups for shear stress, τ_v (f_{ck} = 25 N/mm² and f_y = 415 N/mm²)

Chapter 5

Columns

5.1 INTRODUCTION

Columns are important structural elements which support floors and the roof. They are compression members and their failure may endanger the whole structure. They may have shapes such as circular, rectangular, ele (L), tee (T) and cross (+) as shown in Fig. 5.1. Concentrically loaded columns are subjected to pure axial load. However, such columns rarely occur in practice. They are generally subjected to moments along with the axial load. If moments act about one axis, they are classified as uniaxially eccentrically loaded columns. If moments act about both the axes, they are called biaxially eccentrically loaded columns (Fig. 5.2).

Columns are also classified as short and slender columns depending on their length, lateral dimensions and support conditions. Columns with effective lengths about both the axes of bending less than or equal to 12 times the corresponding lateral dimensions are called *short*

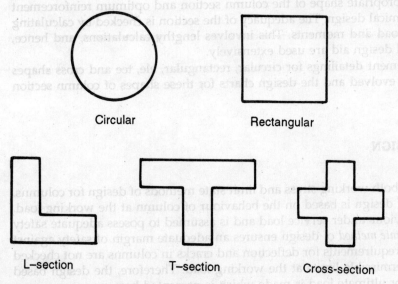

Fig. 5.1 Shapes of column section

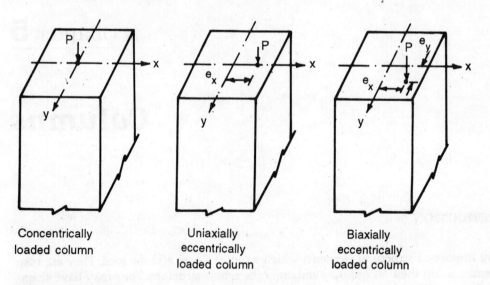

Fig. 5.2 Classification of columns based on types of loading

columns. Their load carrying capacity is governed by the strength of materials. On the other hand, columns with effective length about an axis of bending greater than 12 times the corresponding lateral dimension are called *long or slender columns.* Their load carrying capacity is influenced by the slenderness effect, which produces additional moments because of the transverse deformations. Such columns may fail either due to the failure of materials or by buckling.

The method of design of the column section consists of choosing its cross-section and reinforcement distribution. Appropriate shape of the column section and optimum reinforcement distribution result in economical design. The adequacy of the section is checked by calculating its capacity to resist axial load and moments. This involves lengthy calculations, and hence, computer-aided design and design aid are used extensively.

In this chapter, reinforcement detailings for circular, rectangular, ele, tee and cross shapes column sections have been evolved and the design charts for these shapes of column section have been presented.

5.2 METHODS OF DESIGN

IS : 456 – 1978 recommends both working stress and limit state methods of design for columns. The *working stress method* of design is based on the behaviour of column at the working load. It ensures satisfactory behaviour under service load and is assumed to posess adequate safety against collapse. The *limit state method* of design ensures an adequate margin of safety against collapse. The serviceability requirements for deflection and cracks in columns are not checked as these are within their permissible limits at the working load. Therefore, the design based on the limit state method for ultimate load is made which is presented here.

Design of column is made for the maximum value of axial load and moments acting along the length of the column. In case of the concentrically loaded column, design is made for minimum eccentricity of load, e_{min}, which may arise due to support or load conditions or tolerances.

$$e_{min} = \frac{\text{unsupported length of colum}}{500} + \frac{\text{lateral dimension}}{30} \not< 20 \text{ mm} \qquad (5.1)$$

where the unsupported length of a compression member shall be taken as the clear distance between the end restraints except that:
(i) In flat slab construction, it shall be the clear distance between the floor and the lower extremity of the capital, drop panel or the slab, whichever is the least.
(ii) In beam slab construction, it shall be the clear distance between the floor and the underside of the shallower beam framing into the column in each direction at the next higher floor level.
(iii) In columns restrained laterally by struts, it shall be the clear distance between consecutive struts in each vertical plane, provided that, to be an adequate support, two such struts shall meet the column at approximately the same level, and the angle between the vertical planes through the struts shall not vary more than 30° from a right angle. Such struts shall be of adequate dimensions and have sufficient anchorage to restrain the member against lateral deflection.
(iv) In columns restrained laterally by struts or beams, with brackets used at the junction, it shall be the clear distance between the floor and the lower edge of the bracket, provided that the bracket width equals that of the beam strut and is at least half of that of the column.

Long columns with effective length to lateral dimension greater than 12 are subjected to buckling and hence additional moment due to its lateral deflection is induced. Therefore IS : 456-1978 has recommended limits to very slender columns by the stipulation of unsupported length as follows:

Column with both ends restrained:

$$\text{Unsupported length} < 60 \times \text{least lateral dimension of column} \qquad (5.2a)$$

Column with one end restrained:

$$\text{Unsupported length} < 100 \, B^2 / D \qquad (5.2b)$$

where D = depth of section measured in the plane under consideration
        B = width of section

IS : 456 – 1978 has also stipulated the theoretical and recommended values of effective length of column for idealised support conditions (Table 5.1). However, the reinforced concrete columns are generally part of a large frame which may be a *braced frame*, where joint translation is prevented by rigid bracings of shear walls, or an *unbraced frame*, where joint translation is not prevented and buckling stability is dependent on the stiffness of the beams and columns. The effective length of column in such frames depends on the end restraints provided by the

Table 5.1 Effective length of compression members

Degree of end restraint of compression member	Symbol	Theoretical value of effective length	Recommended value of effective length
Effectively held in position and restrained against rotation at both ends		$0.5\,l$	$0.65\,l$
Effectively held in position at both ends, and restrained against rotation at one end		$0.7\,l$	$0.8\,l$
Effectively held in position at both ends, but not restrained against rotation		$1.0\,l$	$1.0\,l$
Effectively held in position and restrained against rotation at one end, and restrained against rotation but not held in position at the other end		$1.1\,l$	$1.2\,l$
Effectively held in position and restrained against rotation at one end, and partially restrained against rotation but not held in position at the other end		–	$1.5\,l$
Effectively held in position but not restrained against rotation at one end, and restrained against rotation but not held in position at the other end		$2.0\,l$	$2.0\,l$
Effectively held in position and restrained against rotation at one end but not held in position nor restrained against rotation at the other end		$2.0\,l$	$2.0\,l$

adjoining members. It is the function of the end restraint factors K_t and K_b at the top and bottom of the member respectively, defined as:

$$K_t = \frac{K_{ct}}{(K_{ct} + K_{bt})} \quad (5.3a)$$

$$K_b = \frac{K_{cb}}{(K_{cb} + K_{bb})} \quad (5.3b)$$

where,

K_{ct}, K_{cb} = summation of the flexural stiffness of columns framing into the top and bottom joints respectively

K_{bt}, K_{bb} = summation of the flexural stiffness of beams framing into the top and bottom joints respectively

A plot of the ratio of effective length (l_{ef}) to unsupported length l for different values of K_t and K_b is given in Figs. 5.3(a) and 5.3(b) for columns in braced and unbraced frames respectively. The unbraced compression members, at any given level or storey, subjected to lateral

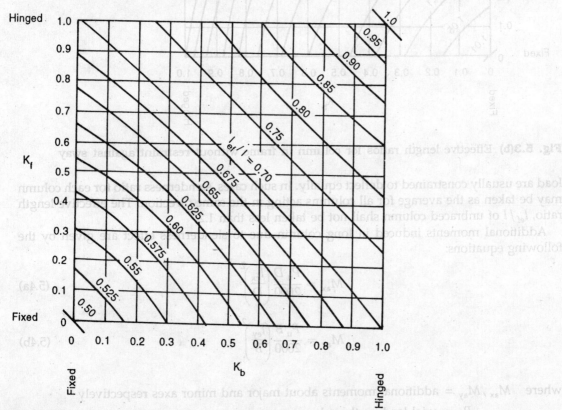

Fig. 5.3(a) Effective length ratios for column in frame with no sway

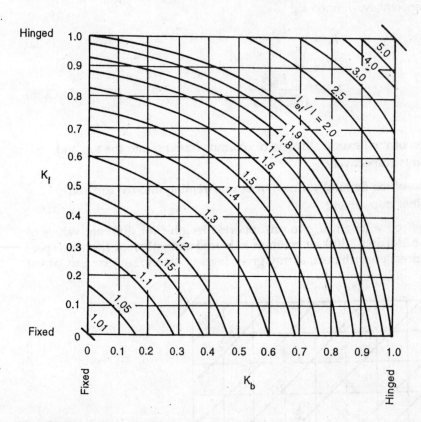

Fig. 5.3(b) Effective length ratios for column in frame without restraint against sway

load are usually constrained to deflect equally. In such cases, slenderness ratio for each column may be taken as the average for all columns acting in the same direction. The effective length ratio, l_{ef}/l of unbraced column shall not be taken less than 1.2.

Additional moments induced in long column due to slenderness effect are given by the following equations:

$$M_{ax} = \frac{P_u D}{2000}\left(\frac{l_{ex}}{D}\right)^2 \qquad (5.4a)$$

$$M_{ay} = \frac{P_u B}{2000}\left(\frac{l_{ey}}{B}\right)^2 \qquad (5.4b)$$

where M_{ax}, M_{ay} = additional moments about major and minor axes respectively

P_u = axial load on the column

l_{ex} = effective length of column with respect to X-X axis
D = depth of section at right angle to the X-X axis
l_{ey} = effective length of column with respect to Y-Y axis
B = width of the section at right angle to the Y-Y axis

The value of additional moments based on above expressions for balanced failure condition is modified by multiplying with a factor K given below for any possible mode of failure of column.

$$K = \frac{P_{uz} - P_u}{P_{uz} - P_{ub}} \leq 1.0 \tag{5.5}$$

where P_u = ultimate axial load on column
P_{uz} = ultimate concentric axial load capacity of the column section
P_{ub} = $0.446 f_{ck} A_c + 0.75 f_y A_s$
= ultimate axial load corresponding to the condition of the maximum compressive strain of 0.0035 in extreme compression fibre of concrete and tensile strain of 0.002 in outermost layer of tension steel

The values of P_{ub} for different sections of circular, rectangular, ele, tee and cross shapes of column are given in their respective sections.

The value of K equal to 1.0 corresponds to the balanced failure, and greater than 1.0 corresponds to the tension failure. Therefore, the maximum value of K is limited to 1.0.

The design moment is obtained by superposing the additional moment to the maximum initial moment. The maximum initial moment is obtained by a suitable combination of the end moments. In the case of braced column without any transverse loads occurring on its height, the additional moment may be added to the initial moment equal to $0.4 M_{u1} + 0.6 M_{u2}$, where M_{u2} is the higher end moment and M_{u1} is the smaller end moment, which is taken negative if the column bends in double curvature. In no case the initial moment shall be taken smaller than $0.4 M_{u2}$, nor the total moment considering initial and additional moments shall be taken less than M_{u2}. In the case of single curvature, $0.4 M_{u1} + 0.6 M_{u2} + M_a$ may be smaller than M_{u2}, for which the column could have been normally designed even without the consideration of slenderness effects. Therefore, the value of $M_{u1} + 0.6 M_{u2} + M_a$ shall not be taken less than M_{u2}. In the case of double curvature, initial moment may be very small and so a minimum of $0.4 M_{u2}$ is recommended. However, the total moment, i.e. the sum of initial moment and the additional moment shall not be taken less than M_{u2}. For unbraced columns, additional moments shall be added to the end moments.

After this, the design of the column section is made by choosing its cross-section and reinforcement distribution, and its adequacy is checked by calculating its capacity to resist the axial load and moments, based on the following assumptions:

(i) Plane section normal to the axis of the column before deformation remains plane after deformation.
(ii) The tensile strength of concrete is ignored.

(iii) The failure of concrete is governed by the maximum strain criteria. For the entire section in compression, the transition of strain from 0.002 for pure axial load condition to 0.0035 for pure bending condition governs the failure of column section, as shown in Fig. 5.4. The strain distribution lines of two extreme conditions are of uniform strain of 0.002 for the pure axial load case and when the strain at the least compressed edge is 0.0 and strain at the highly compressed edge is 0.0035 for neutral axis lying at the edge of the least compressed edge. The strain distribution lines for these two cases intersect each other at a depth of $3D/7$ from the highly compressed edge. Thus, the maximum strain at the most compressed edge of the section shall be taken as $0.0035 - 0.75$ times the strain in the least compressed edge of the section.

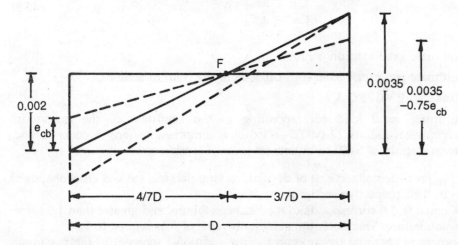

Fig. 5.4 Limiting strain diagram for column section

(iv) The design stress-strain curve of concrete in compression is shown in Fig. 5.5. Compressive strength of concrete in the structure is assumed to be 0.67 times the characteristic strength of concrete. The partial factor of safety y_m, equal to 1.5, is applied to the strength of concrete in addition to it. Therefore, the design strength of concrete is $0.446 f_{ck}$.

(v) The design stress in reinforcement is derived from the stress-strain curve given in Fig. 5.6 for the cold-worked deformed bar. The partial factor of safety y_m, equal to 1.15, is applied to the strength of the reinforcement. Therefore, the design strength of reinforcement is $f_y/1.15$ i.e., $0.87 f_y$. The design stress-strain curve is linear up to a design stress of $0.8 \times 0.87 f_y$, i.e. $0.696 f_y$. Thereafter, it is non-linear and attains a design stress of $0.87 f_y$ at a strain of $0.002 + 0.87 f_y/E_s$.

The ultimate load capacity of the concentrically loaded column section reinforced with high strength deformed bars is given by:

$$P_{uz} = 0.446 f_{ck} + 0.75 f_y A_s \qquad (5.6)$$

where A_c = area of concrete

Columns 101

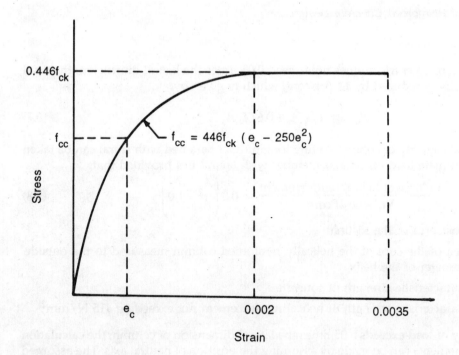

Fig. 5.5 Design stress-strain curve for concrete

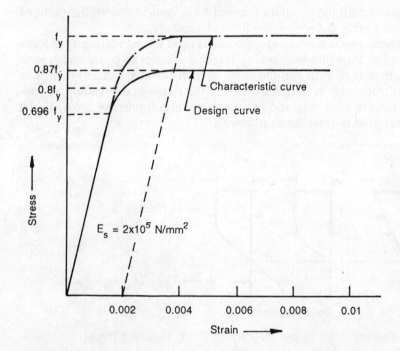

Fig. 5.6 Design stress-strain curve for cold worked deformed steel

102 *Handbook of Reinforced Concrete Design*

A_s = area of steel

When the minimum eccentricity does not exceed 0.05 times the lateral dimension, the axial load carrying capacity is reduced by 11 per cent, which is given by:

$$P_u = 0.4 f_{ck} A_c + 0.67 f_y A_s \qquad (5.7)$$

The ultimate load capacity of column having longitudinal bars tied with spiral can be taken as 1.05 times the ultimate load for similar member with lateral ties provided that,

$$\frac{\text{Volume of helical reinforcement}}{\text{Volume of core}} < 0.3 \left(\frac{A_g}{A_c} - 1.0\right) \frac{f_{ck}}{f_y} \qquad (5.8)$$

where A_g = gross area of the section

A_c = area of the core of the helically reinforced column measured to the outside diameter of the helix

f_{ck} = characteristic strength of concrete

f_y = characteristic strength of helical reinforcement not exceeding 415 N/mm²

If the eccentricity of load exceeds 0.05 times the lateral dimension of column, the calculation of axial load and moments can be made by assuming the position of neutral axis. The assumed position of neutral axis should satisfy the requirement of resultant internal force acting at the eccentricities of external load. If it is not satisfied, then the assumed position of neutral axis is altered and the method is repeated till the resultant internal force coincides with the point of application of the external load within the acceptable limit of accuracy.

For assumed position of neutral axis, a numerical approach is used for calculating axial force and moments where compression zone of concrete is split into a number of equal width strip as shown in Fig. 5.7. The length of each strip and the strain at its centroid are determined. The stress-strain at the centroid of each strip is determined from the stress-strain relation of concrete. Then, the compressive force in each strip and moments due to compressive force in each strip about the two orthogonal axes is obtained as follows.

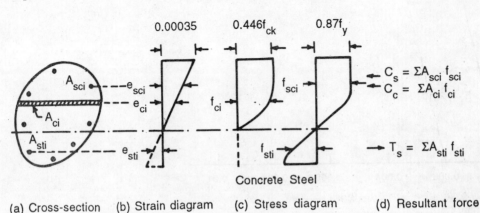

(a) Cross-section   (b) Strain diagram   (c) Stress diagram   (d) Resultant force

Fig. 5.7 Section under combined axial load and biaxial moments

Axial force in strip i,

$$P_{ci} = l_i B_i f_{ci} \tag{5.9}$$

Moment due to axial force P_{ci} in strip i about X-X axis,

$$M_{uxci} = P_{ci} y_{ci} \tag{5.10a}$$

Moment due to axial force P_{ci} in strip i about Y-Y axis,

$$M_{uyci} = P_{ci} x_{ci} \tag{5.10b}$$

where l_i = length of centre line of strip i
B_i = width of strip i
f_{ci} = stress at centre of strip i
x_{ci} = distance of mid point of centre line of strip i from Y-Y axis
y_{ci} = distance of mid point of centre line of strip i from X-X axis

Forces and moments due to reinforcement bars in the section are calculated by determining strain in the reinforcement bars and then stress from their stress-strain curve as follows.

Axial force in reinforcement bar i,

$$P_{si} = A_{si} f_{si} \tag{5.11}$$

Moment due to axial force P_{si} in reinforcement bar i about X-X axis,

$$M_{uxsi} = P_{si} y_{si} \tag{5.12a}$$

Moment due to axial force P_{si} in reinforcement bar i about Y-Y axis,

$$M_{uysi} = P_{si} x_{si} \tag{5.12b}$$

where A_{si} = area of reinforcement bar i
f_{si} = stress in reinforcement bar i
x_{si} = distance of reinforcement bar i from Y-Y axis
Y_{si} = distance of reinforcement bar i from X-X axis

The total axial force and moments are obtained by summing axial forces and moments due to concrete strips and reinforcement bars as follows.

Total axial force,

$$P_u = \sum P_{uci} + \sum P_{usi} \tag{5.13}$$

Total moment about X-X axis,

$$M_{ux} = \sum M_{uxci} + \sum M_{uxsi} \qquad (5.14a)$$

Total moment about Y-Y axis,

$$M_{uy} = \sum M_{uyci} + \sum M_{uysi} \qquad (5.14b)$$

The section is safe if the design ultimate load is within the ultimate load capacity, otherwise it is unsafe. Accordingly, the assumed section and the reinforcement area are successively corrected until the strength of the section approaches the required value.

The design of column section as discussed above involves lengthy calculations. Therefore, design is made by using interaction curves. For column section subjected to axial load and uniaxial moment, a typical interaction curve is shown in Fig. 5.8. For column section subjected to axial load and biaxial moments, a typical interaction surface is shown in Fig. 5.9. It consists of a series of interaction curves obtained by varying the inclination of the neutral axis. Any typical point b on the interaction surface represents the failure load, P_{ub} and moments, M_{uxb} and M_{uyb}. A typical horizontal section taken through the interaction surface gives the interaction curve for ultimate moments M_{ux} and M_{uy} at ultimate axial load P_u (Fig. 5.10). It is a constant load contour of the interaction surface and its shape depends on the geometry of the section, strength of materials, area of steel and its arrangement and the value of the axial load.

The design of section based on the actual interaction surface requires consideration of a large number of variables to cover all the possible design cases that give rise to a large number of charts. As the equation of such curves are complex and cannot be obtained easily, the design of section is made based on the simplifying approximations for the shape of interaction surface.

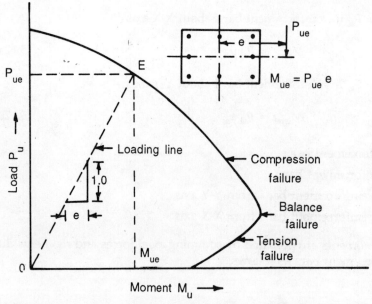

Fig. 5.8 $P_u - M_u$ interaction curve

Columns 105

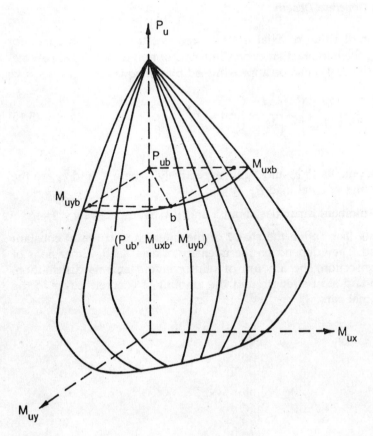

Fig. 5.9 $P_u - M_{ux} - M_{uy}$ interaction surface

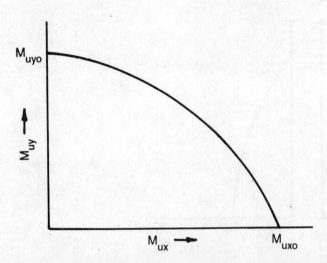

Fig. 5.10 Biaxial moments interaction curve at constant load

The interaction surface at constant value of axial force P_u reduces to interaction curve for ultimate moments M_{ux} and M_{uy}. Such interaction curves in terms of non-dimensional parameters M_{ux}/M_{uxo} and M_{uy}/M_{uyo} (Fig. 5.11) can be approximated by the equation:

$$\left(\frac{M_{ux}}{M_{uxo}}\right)^m + \left(\frac{M_{uy}}{M_{uyo}}\right)^n = 1 \tag{5.15}$$

where M_{ux}, M_{uy} = biaxial moments $P_u e_y$ and $P_u e_x$ respectively, where e_x and e_y are the eccentricities of axial load P_u

M_{uxo}, M_{uyo} = uniaxial moment capacities about x and y axes respectively

m, n = exponents that define the shape of the interaction curve at constant axial load. They depend on the intensity of axial load, dimensions of the cross-section, the amount of reinforcement and its distribution, concrete and steel strength and the amount of concrete cover to the longitudinal bars.

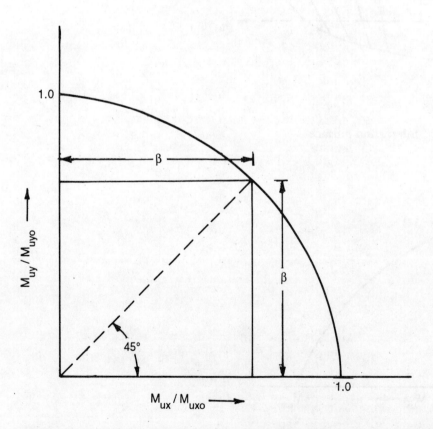

Fig. 5.11 Non-dimensional biaxial moments interaction curve at constant load

The interaction curve defined by Eq. 5.15 and plotted in Fig. 5.11 can be considered symmetrical about the vertical line bisecting the two ordinate planes and can be expressed as,

$$\left(\frac{M_{ux}}{M_{uxo}}\right)^{\alpha_n} + \left(\frac{M_{uy}}{M_{uyo}}\right)^{\alpha_n} = 1 \quad (5.16)$$

where the value of α_n depends on a large number of design parameters such as cross-sectional dimensions, amount of reinforcement and its distribution, strength of concrete and steel and the cover to longitudinal reinforcement. However, the IS code has considered a single important parameter P_u/P_{uz}, that governs the value of α_n. It is given by,

$$\alpha_n = 0.667 + 1.667\, P_u/P_{uz} \geq 1.0 \quad (5.17)$$

$$\leq 2.0$$

The interaction curves for different values of P_u/P_{uz} are shown in Fig. 5.12. The design of column section subjected to axial load and biaxial moments with the help of interaction curves for axial load and uniaxial moment is simple as described below.

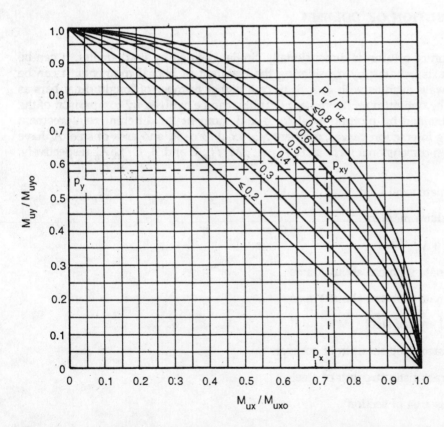

Fig. 5.12 Interaction curves for biaxial moments for different values of P_u/P_{uz}

(i) Assume the cross-section dimensions and area of reinforcement and its distribution.
(ii) Determine concentric load capacity P_{uz}. Compute the value of P_u/P_{uz}.
(iii) Determine uniaxial moment capacities M_{uxo} and M_{uyo} combined with given axial load P_u with the use of appropriate interaction curve for the section subjected to combined axial load and uni-axial moment.
(iv) Then the adequacy of the column section is checked either by interaction equation (Eq. 5.16) or with the use of Fig. 5.12. For checking the adequacy of the section with the use of Fig. 5.12, the values of M_{ux}/M_{uxo} and M_{uy}/M_{uyo} are plotted as P_x and P_y respectively (Fig. 5.12). Draw vertical and horizontal lines from points P_x and P_y to intersect at point P_{xy}. If it is within the interaction curve for the computed value of P_u/P_{uz}, then the section is safe, otherwise it is unsafe. Accordingly, the assumed section and reinforcement area are successively corrected until the strength of section approaches the required value.

The interaction curves for a very wide range of cross-sectional dimensions and reinforcement detailing for the circular, rectangular, ele, tee, and cross shapes of column section are made. These are presented in the following sections.

5.3 CIRCULAR SECTION OF COLUMN

As the circular section of column is dimensionally similar defined by its diameter, it can be designed for single axis bending by considering the resultant of biaxial moments. It can be reinforced in many ways such as with 4, 6, 8, 10 or a higher number of longitudinal bars as shown in Fig. 5.13. The optimum reinforcement detailing can be evolved by comparison of the interaction curves obtained by plotting axial force and moment for different reinforcement detailing. For plotting interaction curves, moment M_u, axial force P_u and area of steel A_s have been expressed in non-dimensional form as $P_u/f_{ck}A_g$, $M_u/f_{ck}Z$ and $A_s f_y/f_{ck}A_g$ respectively, where:

P_u = Ultimate axial load

M_u = Resultant moment

  = $(M_{ux}^2 + M_{uy}^2)^{0.5}$

M_{ux} = Ultimate moment about X-axis

M_{uy} = Ultimate moment about Y-axis

A_s = Total area of steel

f_{ck} = Characteristic strength of concrete

f_y = Characteristic strength of steel

A_g = Gross area of section

  = $0.25\pi D^2$

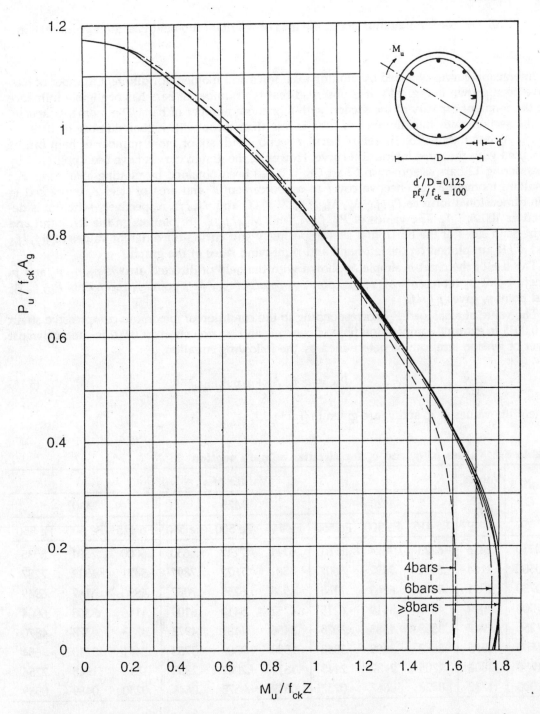

Fig. 5.13 Interaction cruves for different number of reinforcement bars for circular section of column

Z = Section modulus about the axis of resultant moment based on gross section

$\quad = \pi D^3 / 32$

Interaction curves obtained by plotting axial force and moment for different number of bars have been shown in Fig. 5.13. It is observed that the number of bars has negligible influence on the moment capacity of the section with 8 or more number of bars. The moment capacity of the section with four number of bars is less than those with higher number of bars for $P_u / f_{ck} A_g$ less than 0.775. Therefore circular section with six or more number of bars can be designed with the same interaction curve. However, the design curves have been prepared by considering 12 bars as shown in Chart Cr. 1. It has been obtained by plotting axial force P_u, resultant moment M_{uo}, effective cover to reinforcement d' and area of steel A_s expressed in non-dimensional form as $P_u / f_{ck} A_g$, $M_{uo} / f_{ck} Z$, d' / D and $p f_y / f_{ck}$ respectively where p is defined as $100 A_s / A_g$. The values of $P_u / f_{ck} A_g$ and $M_{uo} / f_{ck} Z$ are plotted on the left-hand and right-hand sides of the horizontal axis respectively and curves for different values of $p f_y / f_{ck}$ and d' / D are plotted on the left-hand and right-hand sides of the graph.

The use of the chart is simple as shown with the help of directed arrows p_1, p_2, p_3 and p_4 where point p_1 represents $M_u / f_{ck} Z$, point p_2 represents d' / D, point p_3 represents $P_u / f_{ck} A_g$ and point p_4 gives $p f_y / f_{ck}$.

The value of axial load P_{ub} corresponding to the condition of maximum compressive strain of 0.0035 in extreme compression fibre of concrete and tension strain of 0.002 in the outermost layer of tension steel can be determined by the following equation,

$$P_{ub} = q_c f_{ck} A_g + q_s p A_g \qquad (5.18)$$

where the values of q_s and q_c are given in Table 5.2.

Table 5.2 Values of q_c and q_s for circular column section

$\dfrac{d'}{D}$	Values of q_c	Values of q_s								
		M20			M25			M30		
		Fe 415	Fe 500	Fe 550	Fe 415	Fe 500	Fe 550	Fe 415	Fe 500	Fe 550
.0250	.2236	.6223	.7824	.8761	.6111	.7712	.8650	.6000	.7601	.8538
.0500	.2164	.5691	.7132	.8007	.5581	.7022	.7897	.5471	.6912	.7787
.0750	.2091	.5046	.6363	.7165	.4938	.6255	.7057	.4830	.6147	.6949
.1000	.2019	.4357	.5519	.6215	.4252	.5413	.6109	.4146	.5308	.6004
.1250	.1947	.3558	.4583	.5075	.3456	.4481	.4973	.3353	.4378	.4870
.1500	.1875	.2689	.3408	.3852	.2590	.3309	.3753	.2491	.3210	.3654
.1750	.1803	.1705	.2137	.2445	.1611	.2042	.2350	.1516	.1948	.2256
.2000	.1732	.0428	.0667	.0737	.0339	.0578	.0648	.0250	.0489	.0559

The use of design charts is explained with the help of the following numerical examples.

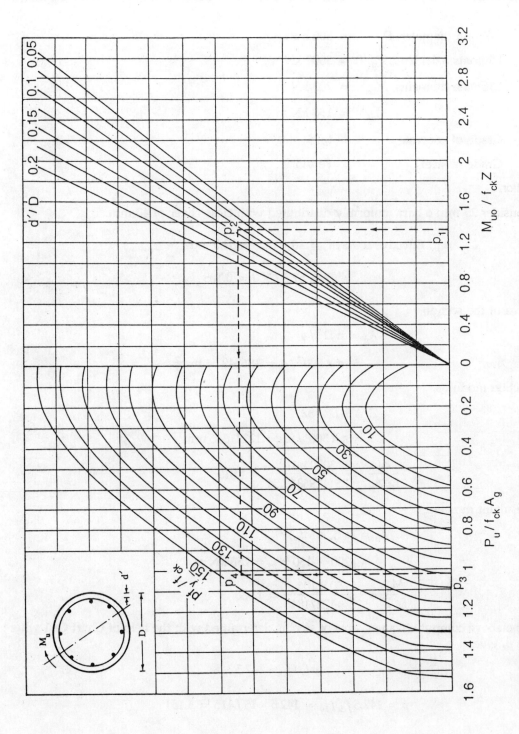

Chart Cr. 1

112 Handbook of Reinforced Concrete Design

Example 5.1: Design a biaxially eccentrically loaded circular column section for the following data.

Column diameter, D	=	500 mm
Ultimate axial load, P_u	=	3000 kN
Ultimate moments, M_{ux}	=	200 kN.m
M_{uy}	=	150 kN.m
Grade of concrete	:	M25
Grade of steel	:	Fe 415

Solution

Consider 25 mm ϕ bars uniformly distributed with clear cover of 40 mm.

∴ Effective cover, $d' = 40 + 12.5 = 52.5$ mm

∴ $d'/D = 52.5/500 = 0.105$

Area of the section,
$$A_g = \pi D^2 / 4$$
$$= \pi \times 500^2 / 4 = 196349.54 \text{ mm}^2$$

Section modulus,
$$Z = \pi D^3 / 32$$
$$= \pi \times 500^3 / 32 = 12271846 \text{ mm}^3$$
$$\frac{P_u}{f_{ck} A_g} = \frac{3000 \times 10^3}{25 \times 196349.54} = 1.019$$

Resultant moment,
$$M_u = (M_{ux}^2 + M_{uy}^2)^{0.5}$$
$$= (200^2 + 150^2)^{0.5} = 250 \text{ kN.m}$$

∴ $$\frac{M_u}{f_{ck} Z} = \frac{250 \times 10^6}{25 \times 12271846} = 1.358$$

For the above computed values, area of steel is determined with the help of Chart Cr.1 where point p_4 gives,

$$p f_y / f_{ck} = 142.5$$

∴ $p = 142.5 f_{ck} / f_y = 142.5 \times 15 / 415 = 5.151$

$$A_s = pA_g/100$$
$$= 5.151 \times 196349.54/100 = 10113.965 \text{ mm}^2$$

Provide 21 bars of 25 mm ϕ (A_s = 10307 mm²). This may cause congestion of reinforcement which may be avoided by providing 18 bars of 28 mm ϕ (A_s = 11083 mm²) as shown in Fig. Ex. 5.1.

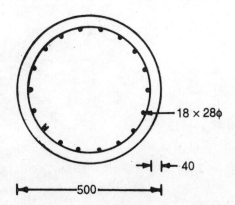

Fig. Ex. 5.1 Reinforcement detailing

Example 5.2: Design a biaxially eccentrically loaded braced circular column deforming in single curvature for the following data.

Ultimate axial load, P_u	= 3000 kN

Ultimate moment about x-axis:

at bottom, M_{ux1}	= 200 kN.m
at top, M_{ux2}	= 150 kN.m

Ultimate moment about y-axis:

at bottom, M_{uy1}	= 150 kN.m
at top, M_{uy2}	= 125 kN.m
Unsupported length of column, l	= 9 m
Effective length about x-axis, l_{ex}	= 6.5 m
Effective length about y-axis, l_{ey}	= 8 m
Column diameter, D	= 500 mm
Grade of concrete	: M25
Grade of steel	: Fe 415

Solution

Moment due to minimum eccentricity:

$$e_{min} = e_{x,\,min} = e_{y,\,min} = \frac{l}{500} + \frac{D}{30} \text{ or 20 mm whichever is more}$$

$$= \frac{9000}{500} + \frac{500}{30} \text{ or 20 mm whichever is more}$$

$$= 34.67 \text{ mm}$$

$$\therefore \quad M_{uxe} = M_{uye} = P_u\, e_{min}$$

$$= 3000 \times 34.67 \text{ kN.mm}$$

$$= 104.01 \text{ kN.m} < M_{ux} \text{ and } M_{uy}$$

Check for short/long column:

$$l_{ex}/D = 6500/500 = 13 > 12$$

$$l_{ey}/D = 8000/500 = 15 > 12$$

Hence it is a slender column about both axes of bending.

Additional moment due to slenderness effect,

$$M_{ax} = P_u\, e_y$$

$$= P_u \frac{D}{2000}\left(\frac{l_{ey}}{D}\right)^2$$

$$= 3000 \times \frac{500}{2000} \times 15^2 \text{ kN.mm} = 168.75 \text{ kN.m}$$

and

$$M_{ay} = P_u\, e_x$$

$$= P_u \frac{D}{2000}\left(\frac{l_{ex}}{D}\right)^2$$

$$= 3000 \times \frac{500}{2000} \times 13^2 \text{ kN.mm} = 126.75 \text{ kN.m}$$

The additional moments M_{ax} and M_{ay} are reduced by multiplying with a factor K given by,

$$K = \frac{P_{uz} - P_u}{P_{uz} - P_{ub}}$$

where P_{uz} and P_{ub} are determined for the assumed area of longitudinal reinforcement and its distribution. Consider percentage of reinforcement, $p = 5.5$. It is considered uniformly distributed with clear cover of 40 mm and tied with lateral ties. Consider 28 mm ϕ bars.

$\therefore$ \qquad Effective cover, $d' = 40 + 14 = 54$ mm

Compute, $\quad P_{uz} = 0.446 f_{ck} A_g + (0.75 f_y - 0.446 f_{ck}) p A_g / 100$

$$= 0.446 \times 25 \times \frac{\pi}{4} \times 500^2 + (0.75 \times 415 - 0.446 \times 25) \times 5.5 \times \frac{\pi}{4} \times \frac{500^2}{100} \text{ N}$$

$$= 5430.17 \text{ kN}$$

and $\quad P_{ub} = q_c f_{ck} A_g + q_s p A_g$

where for $d'/D = 54/500 = 0.108$,

$$q_c = 0.2019 - \frac{0.2019 - 0.1947}{0.125 - 0.1} \times (0.108 - 0.1) = 0.1996$$

and $\quad q_s = 0.4252 - \dfrac{0.4252 - 0.3456}{0.125 - 0.1} \times (0.108 - 0.1) = 0.3997$

$\therefore \quad P_{ub} = q_c f_{ck} \dfrac{\pi}{4} D^2 + q_s p \dfrac{\pi}{4} D^2$

$$= 0.1996 \times 25 \times \frac{\pi}{4} \times 500^2 + 0.3997 \times 5.5 \times \frac{\pi}{4} \times 500^2 \text{ N} = 1411.43 \text{ kN}$$

and $\quad K = \dfrac{P_{uz} - P_u}{P_{uz} - P_{ub}}$

$$= \frac{5430.17 - 3000.0}{5430.17 - 1411.43} = 0.6047$$

Reduced additional moments are,

$$M'_{ax} = K M_{ax} = 0.6047 \times 168.75 = 102.043 \text{ kN.m}$$

and $\quad M'_{ay} = K M_{ay} = 0.6047 \times 126.75 = 76.646$ kN.m

Total ultimate moments are,

$$M_{ux} = M'_{ux} + M'_{ax}$$

where $\quad M'_{ux} = 0.6 M_{ux,\,max} + 0.4 M_{ux,\,min}$

$$= 0.6 \times 200 + 0.4 \times 150$$

$$= 180.0 \text{ kN.m} > 0.4 M_{ux,\,max}$$

$\therefore \quad M_{ux} = 180.0 + 102.043 = 282.043$ kN.m

and $M_{uy} = M'_{uy} + M'_{ay}$

where $M'_{uy} = 0.6 M_{uy,\,max} + 0.4 M_{uy,\,min} = 0.6 \times 150 + 0.4 \times 125 = 140.0$ kN.m $> 0.4 M_{uy,\,max}$

∴ $M_{uy} = 140.0 + 76.646 = 216.646$ kN.m

The area of steel is determined with the use of Chart Cr. 1 as follows.

Compute $\dfrac{P_u}{f_{ck} A_g} = \dfrac{3000 \times 10^3}{25 \times \pi \times 500^2 / 4} = 0.611$

Resultant biaxial moments

$$M_u = (M_{ux}^2 + M_{uy}^2)^{0.5} = (282.043^2 + 216.646^2)^{0.5} = 355.646 \text{ kN.m}$$

$$\dfrac{M_u}{f_{ck} Z} = \dfrac{355.646 \times 10^6}{25 \times \pi \times 500^3 / 32} = 1.16$$

For the above computed values, area of steel is determined with the help of Chart Cr. 1 as follows.

$$pf_y / f_{ck} = 86.5$$

∴ $p = 86.5 \times 25 / 415 = 5.21$

As the assumed value of p is approximately equal to its computed value, the value of K as computed above is acceptable.

$$A_s = p A_g / 100 = 0.0521 \times \pi \times 500^2 / 4 = 10229.811 \text{ mm}^2$$

Provide 18 bars of 28 mm φ ($A_s = 11083$ mm^2) as shown in Fig. Ex. 5.2.

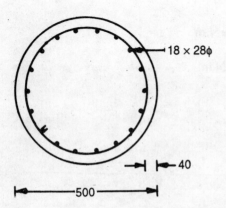

Fig. Ex. 5.2 Reinforcement detailing

5.4 RECTANGULAR SECTION OF COLUMNS

The rectangular section may have large variation in shape defined by the ratio of depth and width of the section. It may have many possible ways of reinforcement detailing. Different reinforcement detailing for a typical geometry of rectangular column section is shown in Fig. 5.14. For studying the effect of these reinforcement detailing on moment capacities, interaction curves for biaxial moments for a particular value of axial force have been shown in the figure. For interaction curves, moments M_{ux} and M_{uy}, axial force P_u and area of steel A_s have been expressed in non-dimensional form as $P_u/f_{ck}A_g$, $M_{ux}/f_{ck}Z_x$, $M_{uy}/f_{ck}Z_y$ and pf_y/f_{ck} respectively where,

P_u = Ultimate axial load

M_{ux} = Ultimate moment about X-axis

M_{uy} = Ultimate moment about Y-axis

A_s = Total area of steel

f_{ck} = Characteristic strength of concrete

f_y = Characteristic strength of steel

p = $100 A_s / A_g$

A_g = Gross area of section

= BD

Z_x = Section modulus about X-axis based on gross section

= $BD^2/6$

Z_y = Section modulus about Y-axis based on gross section

= $B^2D/6$

It may be observed that the same reinforcement detailing is not optimum for every combination of biaxial moments. This has also been observed for other values of axial force.

The effect of different reinforcement detailing on moment capacities for other geometry of rectangular column section has been shown in Fig. 5.15. The optimum reinforcement detailing for a particular combination of biaxial moments may be evolved by comparison of the interaction curves for different reinforcement detailing on the line passing through the origin and having inclination equal to the ratio of biaxial moments. The same reinforcement detailing is not optimum for every combination of biaxial moments. Therefore, some possible reinforcement detailing for such geometry of column section has been considered which are shown in Fig. 5.16. The effect of these reinforcement detailing on their moment capacity is also shown in the figure. In general, the most appropriate reinforcement detailing may be evolved based on maximum area under the interaction curve.

For column section of depth very large as compared to its width, the possible reinforcement detailing are shown in Fig. 5.17. The effect of these reinforcement detailing on moment

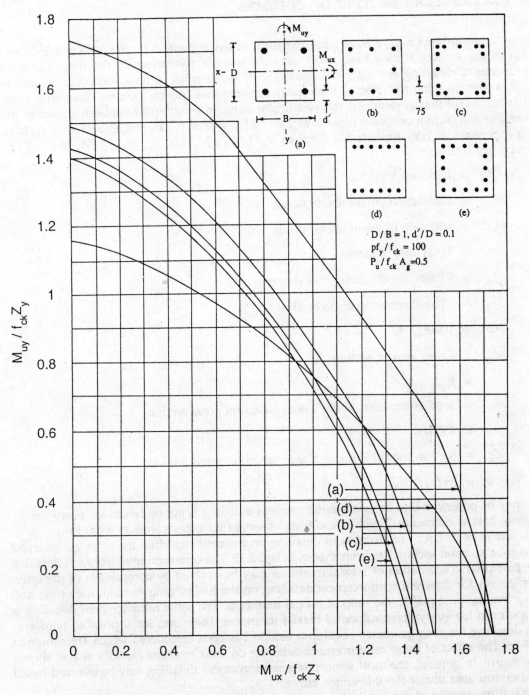

Fig. 5.14 Interaction curves for different types of reinforcement detailing for a typical rectangular section of column ($D/B = 1.0$)

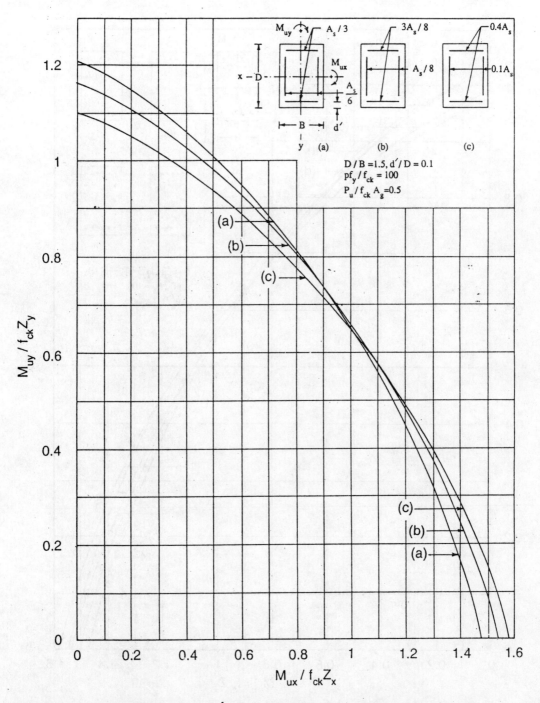

Fig. 5.15 Interaction curves for different types of reinforcement detailing for a typical rectangular section of column ($D/B = 1.5$)

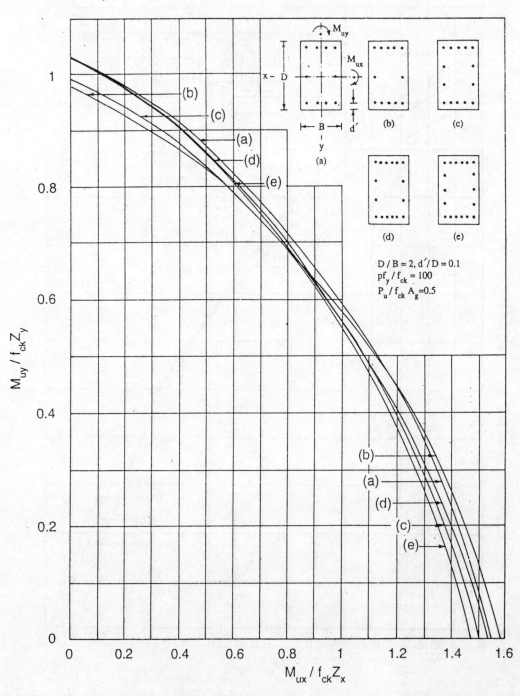

Fig. 5.16 Interaction curves for different types of reinforcement detailing for a typical rectangular section of column ($D/B = 2.0$)

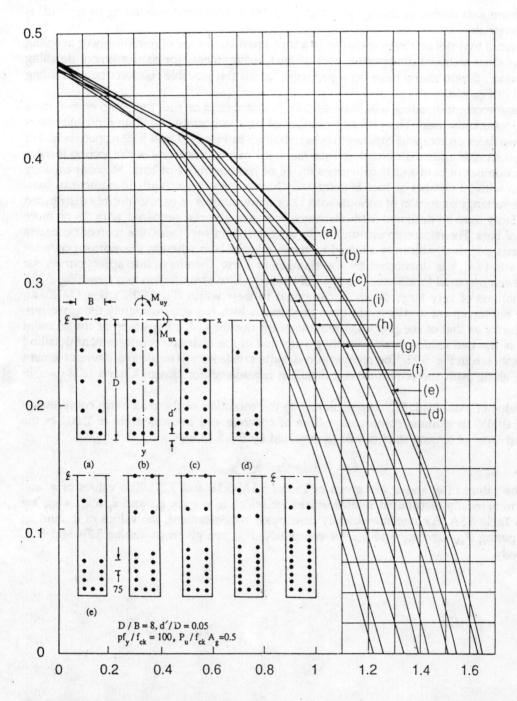

Fig. 5.17 Interaction curves for different types of reinforcement detailing for a typical rectangular section of column ($D/B = 8.0$)

capacities are also shown in the figure. In general, the reinforcement detailing of type (d) is most appropriate.

However, it may not be always possible to adopt a particular type of reinforcement detailing because of cross-sectional dimensions, area of steel, requirements for reinforcement detailing etc. Therefore, design charts have been prepared for several possible reinforcement detailing as shown in Fig 5.18.

The reinforcement detailing which are uniformly distributed on four faces and on two faces are most commonly used. The effect of number of bars uniformly distributed on four faces and on two faces on moment capacity has been shown in Figs 5.19 and 5.20 respectively. For reinforcement uniformly distributed on four faces, the moment capacity of the section is more with less number of bars and it converges for 16 or more number of bars. Moment capacity of column with 12 number of bars is marginally higher than that of with 16 number of bars. Therefore rectangular section of column with 12 or more number of bars uniformly distributed on four faces may be designed with the same interaction curve prepared with 16 or more number of bars. For reinforcement uniformly distributed on two faces, the moment capacity about y-axis (M_{uy} is significantly affected by the number of bars whereas the moment capacity about x-axis (M_{ux}) is uneffected by the number of bars. Therefore, interaction curves for M_{uy} has been prepared for different number of bars equally distributed on two faces (Fig. 5.21).

For columns of very large depth as compared to their width (Figs 5.18, cases 11–18), an alternate reinforcement detailing has been proposed which has approximately the same moment capacity as that of the proposed reinforcement detailing. A comparison of the moment capacity of the two reinforcement detailing with that of the alternate reinforcement detailing has been shown in Fig. 5.22. The comparison is quite satisfactory. The alternate reinforcement detailing along with the reinforcement detailing considered for design Charts is shown in Fig. 5.23.

The value of axial load, P_{ub} corresponding to the condition of the maximum compression strain of 0.0035 in extreme compression fibre of concrete and tension strain of 0.002 in the outermost layer of tension steel can be determined by Eq. 5.18 as,

$$P_{ub} = q_c f_{ck} A_g + q_s p A_g$$

where the values of q_c and q_s are given in Tables 5.3A, 5.3B and 5.3C. The values of q_c and q_s for symmetrically distributed reinforcement for which $q_c = q_{cx} = q_{cy}$ and $q_s = q_{sx} = q_{sy}$ are given in Table 5.3A. For unsymmetrically distributed reinforcement, the values of q_{cx} and q_{sx} for computing P_{ubx} and q_{cy} and q_{sy} for computing P_{uby} are given in Tables 5.3B and 5.3C respectively.

Columns 123

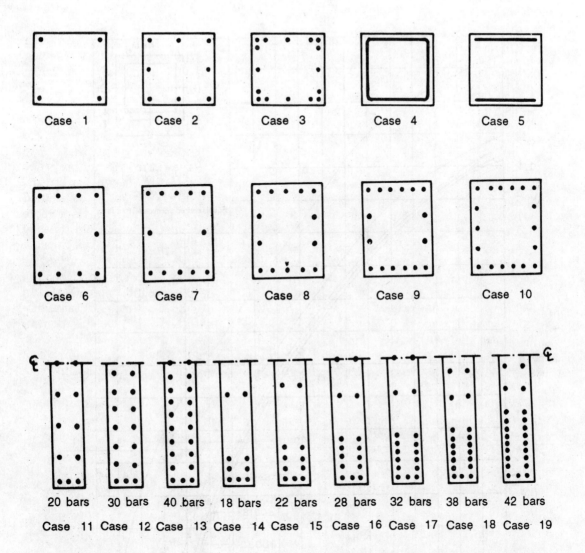

Fig. 5.18 Different types of reinforcement detailing for rectangular section of column

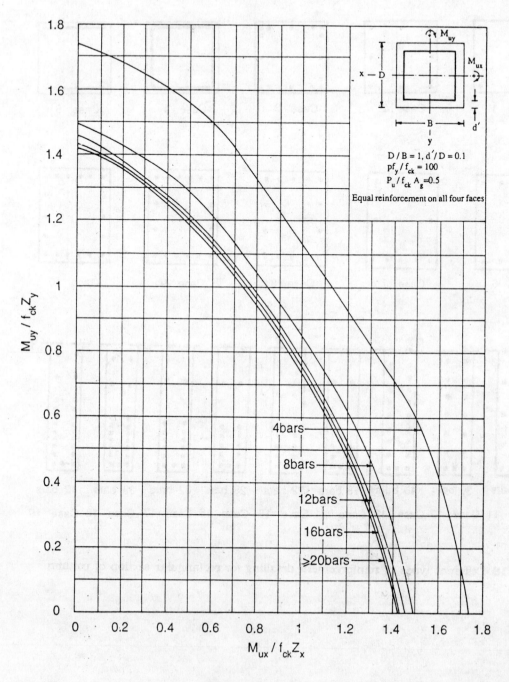

Fig. 5.19 Interaction curves for different number of reinforcement bars equally distributed on four faces for a typical rectangular section of column

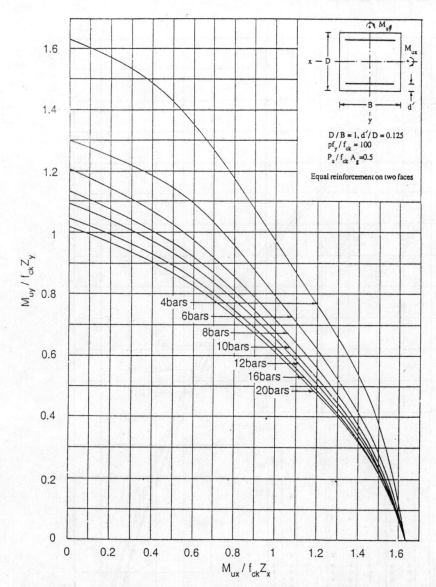

Fig. 5.20 Interaction curves for different number of reinforcement bars equally distributed on two opposite faces for a typical rectangular section of column

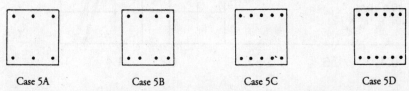

Fig. 5.21 Different types of reinforcement detailing for reinforcement bars equally distributed on two opposite faces of rectangular section of column

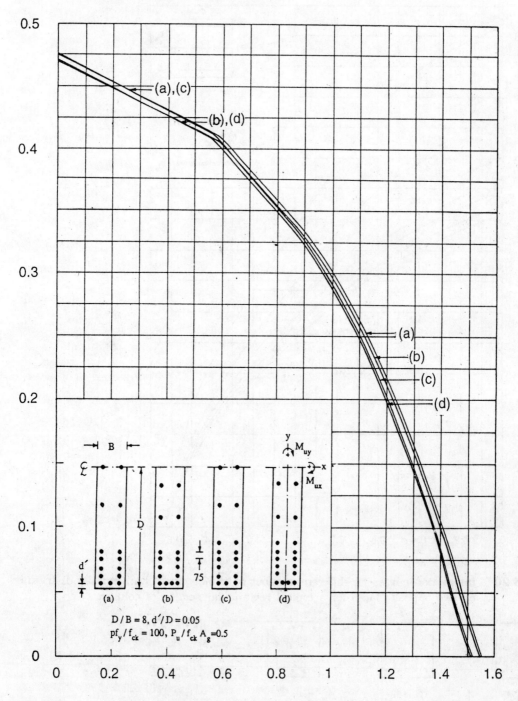

Fig. 5.22 Interaction curves for proposed and alternate reinforcement detailing for a typical rectangular section of column

Columns 127

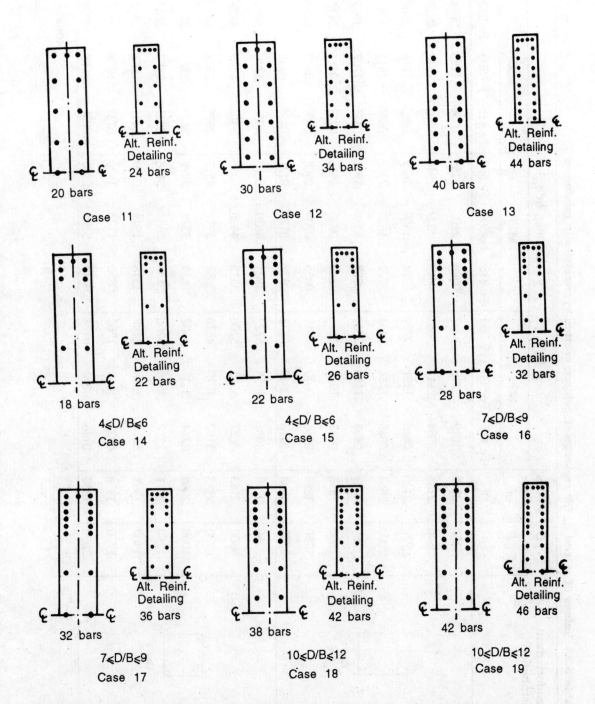

Fig. 5.23 Alternate reinforcement detailing for typical rectangular sections of column

Table 5.3A Values of q_c and q_s for rectangular column section symmetrically reinforced

Reinforcement distribution	d'_d/D or d'_b/B	Values of q_{cx} or q_{cy}	Values of q_{sy} or q_{sx}								
			M20			M25			M30		
			Fe 415	Fe 500	Fe 550	Fe 415	Fe 500	Fe 550	Fe 415	Fe 500	Fe 550
Case 1	.0250	.2240	.1019	.2218	.3267	.0908	.2107	.3155	.0796	.1995	.3044
	.0500	.2183	.0954	.2141	.3051	.0843	.2029	.2940	.0731	.1918	.2828
	.0750	.2125	.0886	.1991	.2790	.0774	.1880	.2679	.0663	.1768	.2567
	.1000	.2068	.0813	.1734	.2464	.0702	.1623	.2352	.0590	.1511	.2241
	.1250	.2011	.0698	.1430	.2062	.0587	.1319	.1951	.0475	.1207	.1839
	.1500	.1953	.0461	.1037	.1638	.0349	.0926	.1526	.0238	.0814	.1415
	.1750	.1896	.0166	.0620	.1019	.0055	.0509	.0908	−.0057	.0397	.0796
	.2000	.1838	−.0216	−.0009	.0233	−.0328	−.0120	.0121	−.0439	−.0232	.0010
Case 2	.0250	.2240	.4036	.4935	.5722	.3921	.4820	.5606	.3806	.4705	.5491
	.0500	.2183	.3627	.4518	.5200	.3515	.4405	.5088	.3403	.4293	.4976
	.0750	.2125	.3197	.4027	.4620	.3088	.3917	.4517	.2979	.3808	.4407
	.1000	.2068	.2744	.3435	.3982	.2639	.3329	.3876	.2533	.3223	.3771
	.1250	.2011	.2237	.2786	.3260	.2135	.2684	.3158	.2033	.2583	.3057
	.1500	.1953	.1614	.2046	.2497	.1516	.1949	.2399	.1419	.1851	.2302
	.1750	.1896	.0922	.1263	.1562	.0830	.1170	.1470	.0737	.1078	.1377
	.2000	.1838	.0136	.0292	.0474	.0049	.0205	.0386	−.0038	.0118	.0299

(contd.)

Columns 129

Table 5.3A (*Contd.*)

Reinforcement distribution	d'_d/D or d'_b/B	Values of q_{cx} or q_{cy}	Values of q_{sy} or q_{sx}									
			M20			M25			M30			
			Fe 415	Fe 500	Fe 550	Fe 415	Fe 500	Fe 550	Fe 415	Fe 500	Fe 550	
Case 3	.0250	.2240	.3595	.5052	.6097	.3481	.4938	.5984	.3368	.4825	.5871	
	.0500	.2183	.3366	.4788	.5702	.3254	.4677	.5590	.3142	.4565	.5478	
	.0750	.2125	.3110	.4435	.5240	.2999	.4325	.5130	.2889	.4214	.5019	
	.1000	.2068	.2840	.3971	.4654	.2731	.3862	.4546	.2622	.3753	.4437	
	.1250	.2011	.2484	.3386	.3821	.2378	.3280	.3715	.2272	.3173	.3609	
	.1500	.1953	.1992	.2500	.2875	.1889	.2397	.2772	.1786	.2294	.2669	
	.1750	.1896	.1237	.1521	.1770	.1138	.1422	.1671	.1039	.1323	.1572	
	.2000	.1838	.0238	.0368	.0519	.0144	.0274	.0425	.0051	.0181	.0332	
Case 4	.0250	.2240	.4718	.5993	.6864	.4606	.5881	.6752	.4494	.5769	.6640	
	.0500	.2183	.4327	.5498	.6258	.4217	.5387	.6147	.4106	.5277	.6037	
	.0750	.2125	.3843	.4925	.5596	.3734	.4816	.5488	.3625	.4707	.5379	
	.1000	.2068	.3307	.4243	.4853	.3200	.4136	.4746	.3093	.4029	.4639	
	.1250	.2011	.2719	.3499	.4011	.2615	.3394	.3906	.2510	.3289	.3801	
	.1500	.1953	.2000	.2633	.3013	.1898	.2531	.2911	.1796	.2429	.2809	
	.1750	.1896	.1207	.1616	.1849	.1108	.1517	.1750	.1009	.1418	.1651	
	.2000	.1838	.0276	.0398	.0538	.0182	.0303	.0444	.0087	.0208	.0349	

(Contd.)

Table 5.3B Values of q_{cx} and q_{sx} for rectangular column section unsymmetrically reinforced

Reinforcement distribution	$\dfrac{d'_d}{D}$	Values of q_{cx}	Values of q_{sx}								
			M20			M25			M30		
			Fe 415	Fe 500	Fe 550	Fe 415	Fe 500	Fe 550	Fe 415	Fe 500	Fe 550
Case 5	.0250	.2240	.1019	.2218	.3267	.0908	.2107	.3155	.0796	.1995	.3044
	.0500	.2183	.0954	.2141	.3051	.0843	.2029	.2940	.0731	.1918	.2828
	.0750	.2125	.0886	.1991	.2790	.0774	.1880	.2679	.0663	.1768	.2567
	.1000	.2068	.0813	.1734	.2464	.0702	.1623	.2352	.0590	.1511	.2241
	.1250	.2011	.0698	.1430	.2062	.0587	.1319	.1951	.0475	.1207	.1839
	.1500	.1953	.0461	.1037	.1638	.0349	.0926	.1526	.0238	.0814	.1415
	.1750	.1896	.0166	.0620	.1019	.0055	.0509	.0908	−.0057	.0397	.0796
	.2000	.1838	−.0216	−.0009	.0233	−.0328	−.0120	.0121	−.0439	−.0232	.0010
Case 6	.0250	.2240	.3433	.4392	.5231	.3318	.4277	.5116	.3204	.4163	.5002
	.0500	.2183	.3093	.4042	.4770	.2981	.3930	.4658	.2869	.3818	.4546
	.0750	.2125	.2735	.3620	.4259	.2625	.3510	.4149	.2516	.3400	.4039
	.1000	.2068	.2358	.3095	.3678	.2251	.2988	.3571	.2144	.2881	.3465
	.1250	.2011	.1929	.2515	.3021	.1825	.2411	.2917	.1722	.2308	.2813
	.1500	.1953	.1383	.1844	.2325	.1283	.1744	.2225	.1183	.1644	.2124
	.1750	.1896	.0771	.1134	.1454	.0675	.1038	.1357	.0579	.0942	.1261
	.2000	.1838	.0066	.0232	.0425	−.0026	.0140	.0333	−.0118	.0048	.0242

(Contd.)

Table 5.3B (*Contd.*)

Reinforcement distribution	$\dfrac{d'_d}{D}$	Values of q_{cx}	Values of q_{sx}									
			M20			M25			M30		
			Fe 415	Fe 500	Fe 550	Fe 415	Fe 500	Fe 550	Fe 415	Fe 500	Fe 550
Case 7	.0250	.2240	.3030	.4029	.4903	.2916	.3916	.4789	.2803	.3802	.4675
	.0500	.2183	.2736	.3725	.4484	.2624	.3613	.4372	.2512	.3501	.4260
	.0750	.2125	.2427	.3348	.4014	.2317	.3238	.3904	.2207	.3128	.3794
	.1000	.2068	.2101	.2868	.3476	.1993	.2760	.3368	.1885	.2653	.3261
	.1250	.2011	.1724	.2334	.2861	.1619	.2229	.2756	.1514	.2124	.2651
	.1500	.1953	.1229	.1710	.2210	.1127	.1608	.2108	.1025	.1506	.2006
	.1750	.1896	.0670	.1048	.1381	.0571	.0950	.1283	.0473	.0851	.1184
	.2000	.1838	.0019	.0192	.0393	−.0076	.0097	.0298	−.0172	.0002	.0203
Case 8	.0250	.2240	.4300	.5345	.6094	.4190	.5235	.5984	.4080	.5125	.5874
	.0500	.2183	.3981	.4869	.5519	.3872	.4760	.5410	.3762	.4651	.5301
	.0750	.2125	.3529	.4319	.4890	.3421	.4211	.4782	.3313	.4103	.4673
	.1000	.2068	.3011	.3668	.4190	.2904	.3561	.4082	.2797	.3454	.3975
	.1250	.2011	.2436	.2959	.3410	.2330	.2853	.3304	.2224	.2747	.3199
	.1500	.1953	.1745	.2157	.2586	.1641	.2053	.2482	.1537	.1949	.2378
	.1750	.1896	.0983	.1307	.1592	.0881	.1205	.1491	.0780	.1104	.1389
	.2000	.1838	.0125	.0273	.0446	.0026	.0174	.0347	−.0073	.0075	.0248

(*Contd.*)

Table 5.3B (*Contd.*)

Reinforcement distribution	d'_d/D	Values of q_{cx}	Values of q_{sx}									
			M20			M25			M30		
			Fe 415	Fe 500	Fe 550	Fe 415	Fe 500	Fe 550	Fe 415	Fe 500	Fe 550
Case 9	.0250	.2240	.3890	.4955	.5741	.3779	.4844	.5631	.3669	.4734	.5520
	.0500	.2183	.3603	.4528	.5211	.3493	.4419	.5101	.3383	.4309	.4992
	.0750	.2125	.3199	.4028	.4627	.3090	.3919	.4519	.2982	.3811	.4410
	.1000	.2068	.2736	.3427	.3974	.2628	.3319	.3866	.2521	.3211	.3758
	.1250	.2011	.2218	.2768	.3242	.2112	.2661	.3135	.2006	.2555	.3029
	.1500	.1953	.1584	.2017	.2467	.1479	.1912	.2362	.1375	.1807	.2257
	.1750	.1896	.0881	.1221	.1521	.0778	.1118	.1418	.0675	.1015	.1315
	.2000	.1838	.0082	.0238	.0419	−.0018	.0138	.0319	−.0119	.0037	.0218
Case 10	.0250	.2240	.4217	.5527	.6448	.4103	.5414	.6335	.3990	.5301	.6222
	.0500	.2183	.3882	.5131	.5919	.3770	.5019	.5807	.3659	.4907	.5696
	.0750	.2125	.3501	.4641	.5230	.3391	.4531	.5120	.3280	.4421	.5010
	.1000	.2068	.3099	.3984	.4470	.2991	.3875	.4362	.2883	.3767	.4254
	.1250	.2011	.2616	.3209	.3630	.2510	.3103	.3524	.2405	.2997	.3418
	.1500	.1953	.1957	.2342	.2742	.1854	.2239	.2639	.1751	.2136	.2536
	.1750	.1896	.1121	.1423	.1690	.1021	.1324	.1590	.0922	.1224	.1491
	.2000	.1838	.0188	.0326	.0487	.0092	.0231	.0392	−.0004	.0135	.0296

(Contd.)

Table 5.3B (*Contd.*)

Reinforcement distribution	$\frac{d'_d}{D}$	Values of q_{cx}	Values of q_{sx}								
			M20			M25			M30		
			Fe 415	Fe 500	Fe 550	Fe 415	Fe 500	Fe 550	Fe 415	Fe 500	Fe 550
20 bars Case 11	.0100	.2275	.7244	.8696	.9558	.7131	.8584	.9445	.7019	.8471	.9333
	.0300	.2229	.6752	.8155	.8978	.6641	.8045	.8868	.6531	.7934	.8758
	.0500	.2183	.6232	.7573	.8324	.6123	.7465	.8215	.6014	.7356	.8107
	.0700	.2137	.5669	.6941	.7565	.5562	.6834	.7458	.5456	.6727	.7351
	.0900	.2091	.5075	.6231	.6739	.4970	.6126	.6634	.4866	.6022	.6529
30 bars Case 12	.0100	.2275	.7724	.9284	1.0124	.7610	.9171	1.0010	.7497	.9057	.9897
	.0300	.2229	.7228	.8682	.9483	.7117	.8571	.9372	.7006	.8460	.9260
	.0500	.2183	.6708	.8035	.8763	.6599	.7926	.8654	.6491	.7817	.8546
	.0700	.2137	.6134	.7336	.7999	.6027	.7229	.7892	.5921	.7123	.7785
	.0900	.2091	.5502	.6559	.7136	.5398	.6454	.7032	.5294	.6350	.6927
40 bars Case 13	.0100	.2275	.7986	.9542	1.0412	.7873	.9429	1.0299	.7759	.9315	1.0185
	.0300	.2229	.7472	.8919	.9741	.7360	.8808	.9630	.7249	.8696	.9519
	.0500	.2183	.6924	.8264	.8987	.6815	.8155	.8878	.6706	.8046	.8769
	.0700	.2137	.6321	.7541	.8193	.6215	.7435	.8086	.6108	.7328	.7980
	.0900	.2091	.5655	.6726	.7327	.5551	.6622	.7223	.5447	.6518	.7119

(*Contd.*)

Table 5.3B (*Contd.*)

Reinforcement distribution	d'_d/D	Values of q_{cx}	Values of q_{sx}								
			M20			M25			M30		
			Fe 415	Fe 500	Fe 550	Fe 415	Fe 500	Fe 550	Fe 415	Fe 500	Fe 550
18 bars, Case 14	.0100	.2275	.4461	.5983	.7115	.4351	.5873	.7006	.4241	.5764	.6896
	.0300	.2229	.4197	.5674	.6771	.4088	.5565	.6662	.3979	.5456	.6553
	.0500	.2183	.3899	.5325	.6361	.3791	.5216	.6253	.3682	.5108	.6145
	.0700	.2137	.3587	.4950	.5923	.3480	.4842	.5816	.3372	.4735	.5709
	.0900	.2091	.3236	.4511	.5433	.3129	.4404	.5326	.3023	.4298	.5220
22 bars, Case 15	.0100	.2275	.4603	.6381	.7594	.4492	.6269	.7483	.4381	.6158	.7372
	.0300	.2229	.4393	.6111	.7291	.4283	.6001	.7181	.4173	.5891	.7071
	.0500	.2183	.4146	.5808	.6906	.4037	.5699	.6798	.3929	.5591	.6689
	.0700	.2137	.3880	.5467	.6488	.3773	.5360	.6380	.3665	.5252	.6273
	.0900	.2091	.3578	.5049	.5995	.3472	.4942	.5889	.3365	.4836	.5783
28 bars, Case 16	.0100	.2275	.4404	.6327	.7496	.4292	.6214	.7384	.4179	.6102	.7271
	.0300	.2229	.4229	.6018	.7151	.4117	.5906	.7039	.4005	.5794	.6927
	.0500	.2183	.4016	.5675	.6752	.3905	.5564	.6641	.3794	.5453	.6530
	.0700	.2137	.3776	.5297	.6309	.3666	.5188	.6199	.3556	.5078	.6089
	.0900	.2091	.3438	.4865	.5819	.3329	.4756	.5710	.3221	.4648	.5602

(Contd.)

Table 5.3B (Contd.)

Reinforcement distribution	$\frac{d'_d}{D}$	Values of q_{cx}	Values of q_{sx}								
			M20			M25			M30		
			Fe 415	Fe 500	Fe 550	Fe 415	Fe 500	Fe 550	Fe 415	Fe 500	Fe 550
Case 17 — 32 bars	.0100	.2275	.4628	.6596	.7816	.4515	.6484	.7703	.4403	.6371	.7591
	.0300	.2229	.4462	.6313	.7497	.4350	.6201	.7385	.4239	.6090	.7274
	.0500	.2183	.4268	.5999	.7107	.4158	.5889	.6997	.4047	.5778	.6886
	.0700	.2137	.3989	.5639	.6677	.3880	.5529	.6567	.3770	.5420	.6458
	.0900	.2091	.3686	.5220	.6178	.3577	.5112	.6070	.3469	.5003	.5962
Case 18 — 38 bars	.0100	.2275	.4447	.6404	.7693	.4334	.6292	.7580	.4222	.6179	.7467
	.0300	.2229	.4264	.6148	.7345	.4152	.6037	.7233	.4041	.5925	.7121
	.0500	.2183	.4062	.5840	.6950	.3951	.5729	.6839	.3841	.5619	.6728
	.0700	.2137	.3825	.5466	.6508	.3715	.5356	.6398	.3606	.5247	.6289
	.0900	.2091	.3545	.5033	.6015	.3437	.4925	.5907	.3328	.4817	.5799
Case 19 — 42 bars	.0100	.2275	.3970	.5921	.7243	.3858	.5809	.7130	.3745	.5696	.7018
	.0300	.2229	.3809	.5687	.6930	.3697	.5575	.6819	.3585	.5464	.6707
	.0500	.2183	.3621	.5417	.6556	.3510	.5306	.6445	.3399	.5195	.6334
	.0700	.2137	.3410	.5065	.6147	.3300	.4956	.6037	.3190	.4846	.5927
	.0900	.2091	.3151	.4664	.5683	.3042	.4555	.5574	.2934	.4447	.5466

(Contd.)

Table 5.3C Values of q_{cy} and q_{sy} for rectangular column section unsymmetrically reinforced

Reinforcement detailing	$\dfrac{d'_b}{B}$	Values of q_{cy}	Values of q_{sy}									
			M15			M20			M25		
			Fe 415	Fe 500	Fe 550	Fe 415	Fe 500	Fe 550	Fe 415	Fe 500	Fe 550
Case 1	.0250	.2240	.5042	.5841	.6540	.4925	.5725	.6424	.4809	.5608	.6307
	.0500	.2183	.4518	.5310	.5916	.4406	.5197	.5804	.4293	.5085	.5691
	.0750	.2125	.3968	.4705	.5238	.3860	.4597	.5129	.3751	.4488	.5021
	.1000	.2068	.3388	.4002	.4488	.3284	.3898	.4384	.3180	.3794	.4281
	.1250	.2011	.2750	.3238	.3659	.2651	.3139	.3561	.2553	.3041	.3462
	.1500	.1953	.1998	.2383	.2783	.1906	.2290	.2690	.1813	.2197	.2598
	.1750	.1896	.1174	.1477	.1743	.1088	.1391	.1657	.1002	.1305	.1571
	.2000	.1838	.0254	.0393	.0554	.0175	.0314	.0475	.0096	.0235	.0396
Case 2	.0250	.2240	.6760	.7691	.8215	.6651	.7582	.8106	.6542	.7473	.7997
	.0500	.2183	.6251	.6916	.7371	.6143	.6808	.7263	.6036	.6700	.7155
	.0750	.2125	.5512	.6065	.6465	.5406	.5959	.6359	.5300	.5853	.6253
	.1000	.2068	.4659	.5119	.5484	.4555	.5015	.5380	.4451	.4911	.5276
	.1250	.2011	.3739	.4105	.4421	.3637	.4003	.4319	.3536	.3902	.4218
	.1500	.1953	.2708	.2996	.3296	.2610	.2898	.3198	.2512	.2800	.3100
	.1750	.1896	.1596	.1823	.2022	.1501	.1728	.1928	.1407	.1634	.1834
	.2000	.1838	.0381	.0485	.0606	.0291	.0395	.0516	.0201	.0305	.0426

(Contd.)

Table 5.3C (*Contd.*)

Reinforcement detailing	$\frac{d'_b}{B}$	Values of q_{cy}	Values of q_{sy}									
			M15			M20			M25		
			Fe 415	Fe 500	Fe 550	Fe 415	Fe 500	Fe 550	Fe 415	Fe 500	Fe 550
Case 3	.0250	.2240	.6774	.8174	.8994	.6660	.8060	.8879	.6546	.7945	.8765
	.0500	.2183	.6224	.7523	.8214	.6112	.7411	.8102	.6000	.7299	.7990
	.0750	.2125	.5593	.6762	.7182	.5484	.6652	.7073	.5375	.6543	.6964
	.1000	.2068	.4928	.5783	.6075	.4822	.5677	.5969	.4717	.5572	.5864
	.1250	.2011	.4151	.4631	.4884	.4049	.4530	.4783	.3948	.4428	.4681
	.1500	.1953	.3154	.3385	.3625	.3058	.3289	.3529	.2962	.3193	.3433
	.1750	.1896	.1885	.2066	.2226	.1795	.1976	.2136	.1705	.1886	.2046
	.2000	.1838	.0511	.0594	.0690	.0428	.0511	.0608	.0345	.0428	.0525
Case 4	.0250	.2240	.7119	.8421	.9200	.7008	.8310	.9089	.6897	.8199	.8979
	.0500	.2183	.6444	.7689	.8401	.6337	.7582	.8293	.6229	.7474	.8186
	.0750	.2125	.5735	.6883	.7498	.5631	.6779	.7395	.5527	.6676	.7291
	.1000	.2068	.4976	.5972	.6501	.4875	.5871	.6400	.4774	.5770	.6299
	.1250	.2011	.4104	.4941	.5205	.4006	.4842	.5107	.3908	.4744	.5008
	.1500	.1953	.3142	.3644	.3844	.3047	.3549	.3749	.2952	.3454	.3655
	.1750	.1896	.2029	.2215	.2348	.1938	.2124	.2257	.1847	.2034	.2167
	.2000	.1838	.0569	.0638	.0718	.0483	.0552	.0633	.0398	.0467	.0548

(Contd.)

Table 5.3C (*Contd.*)

Reinforcement detailing	$\dfrac{d'_b}{B}$	Values of q_{cy}	Values of q_{sy}									
			M15			M20			M25		
			Fe 415	Fe 500	Fe 550	Fe 415	Fe 500	Fe 550	Fe 415	Fe 500	Fe 550
Case 6	.0250	.2240	.5612	.6596	.7226	.5502	.6487	.7116	.5393	.6378	.7007
	.0500	.2183	.5192	.5961	.6507	.5083	.5852	.6398	.4975	.5744	.6290
	.0750	.2125	.4587	.5250	.5730	.4480	.5143	.5623	.4373	.5036	.5516
	.1000	.2068	.3890	.4442	.4880	.3784	.4337	.4774	.3679	.4231	.4669
	.1250	.2011	.3130	.3570	.3949	.3027	.3466	.3846	.2924	.3363	.3742
	.1500	.1953	.2258	.2604	.2965	.2158	.2504	.2864	.2057	.2403	.2763
	.1750	.1896	.1310	.1582	.1822	.1212	.1484	.1724	.1114	.1387	.1626
	.2000	.1838	.0261	.0386	.0531	.0167	.0292	.0437	.0073	.0198	.0343
Case 7	.0250	.2240	.5815	.7181	.8039	.5701	.7067	.7925	.5587	.6954	.7812
	.0500	.2183	.5346	.6626	.7353	.5234	.6514	.7241	.5122	.6402	.7130
	.0750	.2125	.4808	.5967	.6450	.4699	.5857	.6341	.4589	.5747	.6231
	.1000	.2068	.4242	.5108	.5473	.4136	.5002	.5366	.4029	.4895	.5260
	.1250	.2011	.3575	.4098	.4414	.3472	.3995	.4311	.3369	.3892	.4208
	.1500	.1953	.2705	.2994	.3294	.2607	.2895	.3195	.2508	.2796	.3097
	.1750	.1896	.1598	.1825	.2025	.1505	.1732	.1931	.1411	.1638	.1838
	.2000	.1838	.0390	.0493	.0614	.0302	.0406	.0527	.0214	.0318	.0439

(Contd.)

Table 5.3C (*Contd.*)

Reinforcement detailing	d'_b/B	Values of q_{cy}	Values of q_{sy}									
			M15			M20			M25		
			Fe 415	Fe 500	Fe 550	Fe 415	Fe 500	Fe 550	Fe 415	Fe 500	Fe 550
Case 8	.0250	.2240	.5130	.6472	.7357	.5017	.6359	.7244	.4903	.6245	.7130
	.0500	.2183	.4719	.5985	.6739	.4607	.5873	.6627	.4495	.5761	.6515
	.0750	.2125	.4248	.5399	.5927	.4138	.5289	.5818	.4028	.5179	.5708
	.1000	.2068	.3752	.4626	.5043	.3645	.4519	.4936	.3588	.4412	.4829
	.1250	.2011	.3164	.3717	.4078	.3060	.3612	.3974	.2956	.3508	.3869
	.1500	.1953	.2385	.2714	.3057	.2284	.2614	.2957	.2184	.2513	.2856
	.1750	.1896	.1394	.1653	.1881	.1298	.1557	.1785	.1201	.1461	.1689
	.2000	.1838	.0303	.0422	.0560	.0212	.0331	.0469	.0121	.0240	.0378
Case 9	.0250	.2240	.5594	.6870	.7717	.5483	.6759	.7606	.5372	.6648	.7495
	.0500	.2183	.5072	.6302	.7063	.4963	.6194	.6995	.4855	.6085	.6846
	.0750	.2125	.4522	.5660	.6321	.4417	.5555	.6216	.4311	.5449	.6110
	.1000	.2068	.3935	.4912	.5491	.3832	.4809	.5388	.3728	.4705	.5284
	.1250	.2011	.3253	.4063	.4419	.3151	.3962	.4318	.3050	.3860	.4216
	.1500	.1953	.2471	.2992	.3293	.2372	.2893	.3194	.2273	.2794	.3095
	.1750	.1896	.1563	.1816	.2016	.1467	.1720	.1920	.1371	.1625	.1824
	.2000	.1838	.0372	.0476	.0597	.0280	.0384	.0505	.0189	.0292	.0413

(*Contd.*)

Table 5.3C (*Contd.*)

Reinforcement detailing	d'_b/B	Values of q_{cy}	Values of q_{sy}									
			M15			M20			M25		
			Fe 415	Fe 500	Fe 550	Fe 415	Fe 500	Fe 550	Fe 415	Fe 500	Fe 550
Case 10	.0250	.2240	.5086	.6353	.7222	.4974	.6242	.7111	.4863	.6131	.7000
	.0500	.2183	.4614	.5840	.6617	.4505	.5731	.6509	.4397	.5622	.6400
	.0750	.2125	.4118	.5252	.5929	.4012	.5146	.5823	.3906	.5040	.5717
	.1000	.2068	.3589	.4559	.5155	.3484	.4455	.5051	.3380	.4350	.4946
	.1250	.2011	.2969	.3771	.4158	.2866	.3668	.4055	.2764	.3565	.3952
	.1500	.1953	.2248	.2775	.3109	.2148	.2675	.3008	.2047	.2574	.2908
	.1750	.1896	.1408	.1683	.1905	.1310	.1586	.1808	.1213	.1488	.1710
	.2000	.1838	.0307	.0422	.0557	.0213	.0328	.0462	.0119	.0234	.0368
Case 11 (20 bars)	.0250	.2240	.2226	.3305	.4249	.2113	.3192	.4136	.2000	.3079	.4023
	.0500	.2183	.2023	.3092	.3911	.1912	.2980	.3799	.1800	.2868	.3687
	.0750	.2125	.1810	.2805	.3524	.1700	.2695	.3414	.1589	.2584	.3303
	.1000	.2068	.1586	.2414	.3071	.1477	.2305	.2962	.1367	.2196	.2853
	.1250	.2011	.1314	.1973	.2542	.1206	.1865	.2434	.1099	.1757	.2326
	.1500	.1953	.0922	.1441	.1981	.0816	.1335	.1875	.0710	.1229	.1770
	.1750	.1896	.0469	.0877	.1237	.0365	.0773	.1133	.0261	.0699	.1029

(Contd.)

Columns 141

Table 5.3C (*Contd.*)

Reinforcement detailing	$\dfrac{d'_b}{B}$	Values of q_{cy}	Values of q_{sy}									
			M15			M20			M25			
			Fe 415	Fe 500	Fe 550	Fe 415	Fe 500	Fe 550	Fe 415	Fe 500	Fe 550	
30 bars Case 12	.0250	.2240	.1824	.2943	.3921	.1711	.2830	.3809	.1599	.2718	.3696	
	.0500	.2183	.1667	.2775	.3624	.1555	.2663	.3512	.1444	.2551	.3401	
	.0750	.2125	.1502	.2534	.3280	.1391	.2423	.3169	.1280	.2312	.3058	
	.1000	.2068	.1328	.2188	.2869	.1218	.2078	.2759	.1108	.1968	.2649	
	.1250	.2011	.1109	.1792	.2382	.1000	.1683	.2273	.0891	.1574	.2164	
	.1500	.1953	.0768	.1306	.1867	.0660	.1199	.1759	.0553	.1091	.1651	
	.1750	.1896	.0368	.0791	.1164	.0261	.0685	.1058	.0155	.0579	.0951	
40 bars Case 13	.0250	.2240	.1622	.2761	.3758	.1510	.2649	.3645	.1398	.2537	.3533	
	.0500	.2183	.1489	.2616	.3481	.1377	.2505	.3369	.1266	.2393	.3258	
	.0750	.2125	.1348	.2398	.3157	.1237	.2287	.3046	.1126	.2176	.2935	
	.1000	.2068	.1200	.2074	.2767	.1089	.1964	.2657	.0979	.1854	.2547	
	.1250	.2011	.1006	.1702	.2302	.0896	.1592	.2192	.0787	.1482	.2083	
	.1500	.1953	.0691	.1239	.1809	.0583	.1130	.1701	.0474	.1022	.1592	
	.1750	.1896	.0317	.0749	.1128	.0210	.0641	.1020	.0102	.0533	.0913	

(*Contd.*)

Table 5.3C (*Contd.*)

Reinforcement detailing	d'_b/B	Values of q_{cy}	Values of q_{sy}								
			M15			M20			M25		
			Fe 415	Fe 500	Fe 550	Fe 415	Fe 500	Fe 550	Fe 415	Fe 500	Fe 550
18 bars Case 14	.0250	.2240	.2360	.3426	.4358	.2247	.3313	.4245	.2134	.3199	.4131
	.0500	.2183	.2142	.3197	.4006	.2030	.3085	.3894	.1919	.2974	.3782
	.0750	.2125	.1913	.2896	.3606	.1803	.2785	.3496	.1692	.2675	.3385
	.1000	.2068	.1672	.2490	.3138	.1563	.2381	.3030	.1454	.2272	.2921
	.1250	.2011	.1382	.2033	.2595	.1275	.1926	.2488	.1168	.1819	.2380
	.1500	.1953	.0973	.1486	.2019	.0868	.1380	.1914	.0763	.1275	.1809
	.1750	.1896	.0502	.0906	.1261	.0399	.0803	.1158	.0296	.0700	.1055
22 bars Case 15	.0250	.2240	.2116	.3206	.4159	.2003	.3093	.4046	.1891	.2980	.3934
	.0500	.2183	.1926	.3005	.3832	.1814	.2893	.3721	.1703	.2782	.3609
	.0750	.2125	.1726	.2731	.3458	.1616	.2621	.3347	.1505	.2510	.3236
	.1000	.2068	.1516	.2353	.3016	.1406	.2243	.2906	.1297	.2134	.2797
	.1250	.2011	.1258	.1923	.2498	.1150	.1815	.2390	.1042	.1707	.2282
	.1500	.1953	.0880	.1404	.1950	.0774	.1298	.1844	.0667	.1191	.1737
	.1750	.1896	.0441	.0854	.1217	.0337	.0749	.1112	.0232	.0645	.1008

(*Contd.*)

Table 5.3C (*Contd.*)

Reinforcement detailing	d'_b/B	Values of q_{cy}	Values of q_{sy}									
			M15			M20			M25		
			Fe 415	Fe 500	Fe 550	Fe 415	Fe 500	Fe 550	Fe 415	Fe 500	Fe 550
28 bars Case 16	.0250	.2240	.1881	.2994	.3968	.1769	.2882	.3855	.1656	.2769	.3743
	.0500	.2183	.1718	.2820	.3665	.1606	.2708	.3553	.1495	.2597	.3442
	.0750	.2125	.1546	.2573	.3315	.1435	.2462	.3204	.1325	.2351	.3093
	.1000	.2068	.1365	.2220	.2898	.1255	.2110	.2788	.1145	.2000	.2678
	.1250	.2011	.1138	.1818	.2405	.1029	.1709	.2296	.0920	.1600	.2187
	.1500	.1953	.0790	.1325	.1883	.0683	.1218	.1776	.0575	.1111	.1668
	.1750	.1896	.0382	.0804	.1175	.0276	.0698	.1068	.0170	.0591	.0962
32 bars Case 17	.0250	.2240	.1773	.2897	.3880	.1661	.2785	.3768	.1549	.2673	.3656
	.0500	.2183	.1622	.2735	.3588	.1511	.2623	.3477	.1399	.2512	.3365
	.0750	.2125	.1464	.2500	.3249	.1353	.2389	.3138	.1242	.2278	.3027
	.1000	.2068	.1296	.2159	.2843	.1186	.2049	.2733	.1076	.1939	.2623
	.1250	.2011	.1083	.1769	.2362	.0974	.1660	.2253	.0865	.1551	.2144
	.1500	.1953	.0749	.1289	.1852	.0641	.1181	.1744	.0533	.1073	.1636
	.1750	.1896	.0355	.0781	.1155	.0249	.0674	.1048	.0142	.0567	.0942

(*Contd.*)

Table 5.3C (*Contd.*)

Reinforcement detailing	$\dfrac{d'_b}{B}$	Values of q_{cy}	Values of q_{sy}							
			M15			M20			M25		
			Fe 415	Fe 500	Fe 550	Fe 415	Fe 500	Fe 550	Fe 415	Fe 500	Fe 550
38 bars Case 18	.0250	.2240	.1654	.2790	.3783	.1542	.2678	.3671	.1430	.2566	.3559
	.0500	.2183	.1517	.2641	.3503	.1405	.2530	.3392	.1294	.2418	.3280
	.0750	.2125	.1372	.2420	.3177	.1261	.2309	.3066	.1150	.2198	.2955
	.1000	.2068	.1220	.2092	.2783	.1110	.1982	.2673	.0999	.1872	.2563
	.1250	.2011	.1022	.1716	.2315	.0913	.1606	.2205	.0803	.1497	.2096
	.1500	.1953	.0703	.1250	.1818	.0595	.1141	.1710	.0486	.1033	.1601
	.1750	.1896	.0325	.0755	.1134	.0218	.0648	.1026	.0110	.0540	.0919
42 bars Case 19	.0250	.2240	.1594	.2736	.3734	.1482	.2623	.3622	.1369	.2511	.3510
	.0500	.2183	.1463	.2594	.3460	.1352	.2482	.3349	.1240	.2370	.3237
	.0750	.2125	.1326	.2379	.3140	.1215	.2268	.3029	.1104	.2157	.2918
	.1000	.2068	.1181	.2058	.2753	.1071	.1948	.2643	.0960	.1837	.2532
	.1250	.2011	.0991	.1689	.2291	.0882	.1579	.2181	.0772	.1469	.2071
	.1500	.1953	.0680	.1229	.1801	.0571	.1121	.1592	.0463	.1012	.1584
	.1750	.1896	.0310	.0742	.1123	.0202	.0635	.1015	.0094	.0527	.0907

(Contd.)

The interaction curves for different reinforcement detailing has been presented for axial force and uniaxial moment for design of column section under axial force and biaxial moments. The method involves iteration where area of steel is assumed and adequacy of section is checked by using interaction equation for biaxial moments. It gives conservative estimate of area of steel. Therefore, alternate interaction curves for axial force and biaxial moments have been proposed which facilitate to determine the area of steel more accurately without any iteration. The interaction curves for (i) axial force and uniaxial moment and (ii) axial force and biaxial moments for design of column section are described below.

5.4.1 Interaction Curves for Axial Force and Uniaxial Moment

It has been obtained by plotting axial force P_u, uniaxial moment M_{uxo} or M_{uyo}, effective cover to reinforcement d'_d or d'_b and area of steel A_s expressed in non-dimensional form as $P_u / f_{ck} A_g$, $M_{uxo} / f_{ck} Z_x$ or $M_{uyo} / f_{ck} Z_y$, d'_d / D or d'_b / B and $p f_y / f_{ck}$ where p is defined as $100 A_s / A_g$. Charts R1.1 to R1.19.2 show interaction curves for different reinforcement detailing where axial force $P_u / f_{ck} A_g$ and uniaxial moment $M_{uxo} / f_{ck} Z_x$ or $M_{uyo} / f_{ck} Z_y$ are plotted on the left-hand and right-hand sides of the horizontal axis respectively and curves for different values of $p f_y / f_{ck}$ and d'_d / D or d'_b / B are plotted on the left-hand and right-hand sides of the Chart. Charts R1.1 to R1.13 are valid for any value of D/B ratio as the reinforcement detailing are geometrically similar about the axis of bending for any value of D/B ratio. Charts R1.14 to R1.19 are valid for D/B ratio considered because of geometrical dissimilarity of reinforcement detailing about the axis of bending for different values of D/B ratio. However, it gives reasonably accurate value of uniaxial moment for D/B values in their close vicinity. Therefore, a range of values of D/B ratio has been recommended for these Charts. The index of Charts R1.1 to R1.19 is given on Table 5.4.

Table 5.4 Index for Charts R1.1 to R1.19 for axial force and uniaxial moment

Reinforcement detailing	Chart No.	Chart used for	Page No.
Case 1	R1.1	M_{uxo}, M_{uyo}	157
Case 2	R1.2	M_{uxo}, M_{uyo}	158

(Contd.)

Table 5.4 (Contd.)

Reinforcement detailing	Chart No.	Chart used for	Page No.
Case 3	R1.3	M_{uxo}, M_{uyo}	159
Case 4	R1.4	M_{uxo}, M_{uyo}	160
Case 5	R1.5.1	M_{uxo}	161
Case 5A	R1.5.2	M_{uyo}	162
Case 5B	R1.5.3	M_{uyo}	163
Case 5C	R1.5.4	M_{uyo}	164

(Contd.)

Table 5.4 (*Contd.*)

Reinforcement detailing	Chart No.	Chart used for	Page No.
Case 5D	R1.5.5	M_{uyo}	165
 Case 6	R1.6.1	M_{uyo}	166
	R1.6.2	M_{uyo}	167
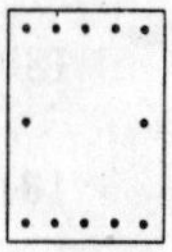 Case 7	R1.7.1	M_{uxo}	168
	R1.7.2	M_{uyo}	169
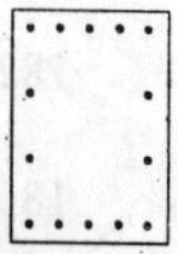 Case 8	R1.8.1	M_{uxo}	170
	R.1.8.2	M_{uyo}	171
 Case 9	R1.9.1	M_{uxo}	172
	R1.9.2	M_{uyo}	173
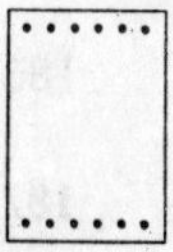 Case 10	R1.10.1	M_{uxo}	174
	R1.10.2	M_{uyo}	175

(*Contd.*)

Table 5.4 (*Contd.*)

Reinforcement detailing	Chart No.	Chart used for	Page No.
 20 bars Case 11	R1.11.1	M_{uxo}	176
	R1.11.2	M_{uyo}	177
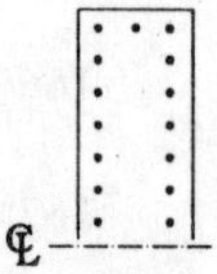 30 bars Case 12	R1.12.1	M_{uxo}	178
	R1.12.2	M_{uyo}	179
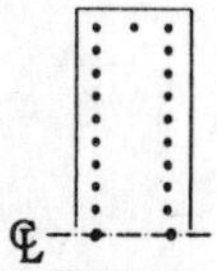 40 bars Case 13	R1.13.1	M_{uxo}	180
	R1.13.2	M_{uyo}	181
14 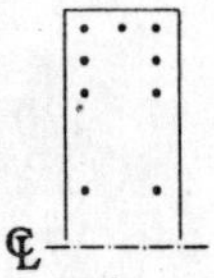 18 bars Case 14	R1.14.1	M_{uxo}	182
	R1.14.2	M_{uyo}	183
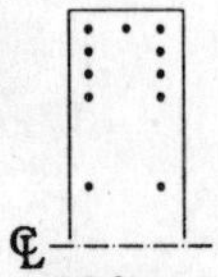 22 bars Case 15	R1.15.1	M_{uxo}	184
	R1.15.2	M_{uyo}	185
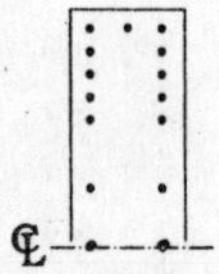 28 bars Case 16	R1.16.1	M_{uxo}	186
	R1.16.2	M_{uyo}	187

(*Contd.*)

Table 5.4 (*Contd.*)

Reinforcement detailing	Chart No.	Chart used for	Page No.
32 bars Case 17	R1.17.1	M_{uxo}	188
	R1.17.2	M_{uyo}	189
38 bars Case 18	R1.18.1	M_{uxo}	190
	R1.18.2	M_{uyo}	191
42 bars Case 19	R1.19.1	M_{uxo}	192
	R1.19.2	M_{uyo}	193

The use of the chart is simple as shown by directed arrows p_1, p_2, p_3 and and p_4 (Chart R1.1) where point p_1 represents $P_u / f_{ck} A_g$, point p_2 represents $p f_y / f_{ck}$, point p_3 represents d'_d / D or d'_b / B and point p_4 represents $M_{uxo} / f_{ck} Z_x$ or $M_{uyo} / f_{ck} Z_y$.

The use of design charts is explained with the help of the following examples.

Example 5.3: Design a biaxially eccentrically loaded rectangular column section for the following data:

Column section, $B \times D$ = 350 mm × 700 mm

Ultimate axial load, P_u = 3675 kN.m

Ultimate moments, M_{ux} = 500 kN.m

M_{uy} = 250 kN.m

Grade of concrete : M25

Grade of steel : Fe 415

Solution

Consider 25 mm φ bars uniformly distributed on four faces at clear cover of 40 mm.

∴ Effective cover, $d' = 40 + 12.5 = 52.5$ mm

∴ $$\frac{d'_d}{D} = \frac{52.5}{700} = 0.075$$

and $$\frac{d'_b}{B} = \frac{52.5}{350} = 0.15$$

Area of the section,
$$A_g = BD = 350 \times 700 = 245000 \text{ mm}^2$$

Section Modulus,
$$Z_x = \frac{BD^2}{6} = \frac{350 \times 700^2}{6} = 28583333 \text{ mm}^3$$

$$Z_y = \frac{B^2 D}{6} = \frac{350^2 \times 700}{6} = 14291667 \text{ mm}^3$$

Compute
$$\frac{P_u}{f_{ck} A_g} = \frac{3675 \times 10^3}{25 \times 245000} = 0.6$$

$$\frac{M_{ux}}{f_{ck} Z_x} = \frac{500 \times 10^6}{25 \times 28583333} = 0.7$$

$$\frac{M_{uy}}{f_{ck} Z_y} = \frac{250 \times 10^6}{25 \times 14291667} = 0.7$$

Design for reinforcement is made by assuming the area of steel and then checking the adequacy of the section. Assume percentage area of steel, $p = 5.0$

Compute
$$\frac{p f_y}{f_{ck}} = \frac{5.0 \times 415}{25} = 83.0$$

Uniaxial moment capacities M_{uxo} and M_{uyo} are determined with the use of Chart R1.4 for $d'_d / D = 0.075$ and $d'_b / B = 0.15$ respectively as,

$$\frac{M_{uxo}}{f_{ck} Z_x} = 1.08$$

$$\frac{M_{uyo}}{f_{ck} Z_y} = 0.9$$

The assumed area of steel is adequate if the following interaction condition is satisfied.

$$\left(\frac{M_{ux}}{M_{uxo}}\right)^\alpha + \left(\frac{M_{uy}}{M_{uyo}}\right)^\alpha \leq 1$$

where
$$\alpha = 0.667 + 1.667 \frac{P_u}{P_{uz}}$$

$$\frac{P_{uz}}{f_{ck} A_g} = 0.446 + \left(0.75 \frac{f_y}{f_{ck}} - 0.446\right)\frac{p}{100}$$

$$= 0.446 + \left(0.75 \times \frac{415}{25} - 0.446\right) \times \frac{5.0}{100} = 1.0462$$

∴
$$\alpha = 0.667 + \frac{1.667 \times 0.6}{1.0462} = 1.623$$

and
$$\left(\frac{M_{ux}}{M_{uxo}}\right)^\alpha + \left(\frac{M_{uy}}{M_{uyo}}\right)^\alpha = \left(\frac{0.7}{1.08}\right)^{1.623} + \left(\frac{0.7}{0.9}\right)^{1.623}$$

$$= 1.16 > 1.0 \text{ (Hence it is unsafe)}.$$

As the strength of the section is less than its required value, the assumed percentage of reinforcement is inadequate. Therefore, consider increased value of $p = 5.75$.

Compute
$$\frac{p f_y}{f_{ck}} = \frac{5.75 \times 415}{25} = 95.45$$

Uniaxial movement capacities M_{uxo} and M_{uyo} are determined with the use of Chart R1.4 for $d'_d / D = 0.075$ and $d'_b / B = 0.15$ respectively as,

$$\frac{M_{uxo}}{f_{ck} Z_x} = 1.24$$

$$\frac{M_{uyo}}{f_{ck} Z_y} = 1.02$$

Compute
$$\frac{P_{uz}}{f_{ck} A_g} = 0.446 + \left(0.75 \frac{f_y}{f_{ck}} - 0.446\right)\frac{p}{100}$$

$$= 0.446 + \left(0.75 \times \frac{415}{25} - 0.446\right) \times \frac{5.75}{100} = 1.136$$

and
$$\alpha = 0.667 + \frac{1.667 \times 0.6}{1.136} = 1.547$$

Compute
$$\left(\frac{M_{ux}}{M_{uxo}}\right)^\alpha + \left(\frac{M_{uy}}{M_{uyo}}\right)^\alpha = \left(\frac{0.7}{1.24}\right)^{1.547} + \left(\frac{0.7}{1.02}\right)^{1.547}$$

$$= 0.972 < 1.0 \text{ (Hence it is safe)}$$

As the strength of the section is nearly equal to its required value, the assumed percentage of reinforcement is adopted.

$$\therefore \quad A_s = \frac{5.75 \times 350 \times 700}{100} = 14090 \text{ mm}^2$$

Provide $32 \times 25\ \phi$ bars (15708 mm²). This may cause congestion of reinforcement which may be avoided by providing $24 \times 28\ \phi$ bar (14778 mm²) as shown in Fig. Ex. 5.3.

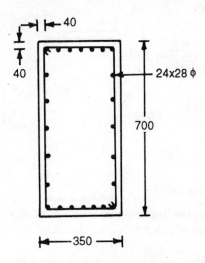

Fig. Ex. 5.3 Reinforcement detailing

Example 5.4: Design a biaxially eccentrically loaded braced square column deforming in single curvature for the following data.

Ultimate axial load, P_u	= 1500 kN
Ultimate moment about x-axis:	
at bottom, M_{ux1}	= 90 kN.m
at top, M_{ux2}	= 50 kN.m
Ultimate moment about y-axis:	
at bottom, M_{uy1}	= 60 kN.m
at top, M_{uy2}	= 40 kN.m
Unsupported length of column, l	= 8 m
Effective length about x-axis, l_{ex}	= 5.2 m
Effective length about y-axis, l_{ey}	= 6.35 m
Column section, $B \times D$	= 350 mm × 350 mm

Grade of concrete : M25
Grade of steel : Fe 415

Solution

Moment due to minimum eccentricities:

$$M_{uxe} = P_u e_{x,\,min}$$

and

$$M_{uye} = P_u e_{x,\,min}$$

where $e_{y,\,min} = \dfrac{1}{500} + \dfrac{B}{30}$ or 20 mm whichever is more

$= \dfrac{800}{500} + \dfrac{350}{30}$ or 20 mm whichever is more = 27.67 mm

and $e_{x,\,min} = \dfrac{1}{500} + \dfrac{D}{30}$, or 20 mm whichever is less

$= \dfrac{8000}{500} + \dfrac{350}{30}$, or 20 mm whichever is less = 27.67 mm

∴ $M_{uxe} = 1500 \times 27.67$ kN.mm $= 41.505$ kN.m $< M_{ux}$

and $M_{uye} = 1500 \times 27.67$ kN.mm $= 41.505$ kN.m $< M_{uy}$

Check for short/long column:

$$\dfrac{l_{ex}}{D} = \dfrac{5200}{350} = 14.86 > 12$$

$$\dfrac{l_{ey}}{B} = \dfrac{6350}{350} = 18.143 > 12$$

Hence it is a long column about both axes of bending.
Additional moments due to slenderness effects,

$$M_{ax} = P_u e_y = P_u \dfrac{B}{2000} \left(\dfrac{l_{ey}}{D}\right)^2$$

$$= 1500 \times \dfrac{350}{2000} \times 18.145^2 \text{ kN.mm} = 86.53 \text{ kN.m}$$

and

$$M_{ay} = P_u e_x = P_u \dfrac{B}{2000} \left(\dfrac{l_{ex}}{D}\right)^2$$

$$= 1500 \times \dfrac{350}{2000} \times 14.86^2 \text{ kN.mm} = 57.97 \text{ kN.m}$$

The additional moments M_{ax} and M_{ay} are modified by multiplying with a factor K given by,

$$K = \frac{P_{uz} - P_u}{P_{uz} - P_{ub}}$$

where P_{uz} and P_{ub} are determined for the assumed area of longitudinal reinforcement and its distribution. Consider percentage of reinforcement $p = 4.5$. It is distributed uniformly on four faces with clear cover of 40 mm. Consider 25 mm ϕ bars.

$\therefore$ Effective cover, $d' = 40 + \dfrac{25}{2} = 52.5$ mm

Compute $P_{uz} = 0.446 f_{ck} A_g + (0.75 f_y - 0.446 f_{ck}) \dfrac{pA_g}{100}$

$= 0.446 \times 25 \times 350 \times 350 + (0.75 \times 415 - 0.446 \times 25) \times 4.5 \times 350 \times \dfrac{350}{100}$ N

$= 3020.18$ kN

and $P_{ub} = q_c f_{ck} A_g + q_s p A_g$

where for $\dfrac{d'_d}{D} = \dfrac{d'_b}{B} = 0.15$

$q_c = q_{cx} = q_{cy} = 0.1953$

and $q_s = q_{sx} = q_{sy} = 0.1898$

$\therefore P_{ub} = P_{ubx} = P_{uby} = 0.1953 \times 25 \times 350 \times 350 + 0.1898 \times 4.5 \times 350 \times 350$ N $= 702.734$ kN

Compute $K = K_x = K_y = \dfrac{P_{uz} - P_u}{P_{uz} - P_{ub}} = \dfrac{3020.18 - 15000}{3020.18 - 702.734} = 0.656$

Modified additional moments are, $M_{ax} = k_x M_{ax} = 0.656 \times 86.43 = 56.698$ kN.m

and $M_{ay} = k_y M_{ay} = 0.656 \times 57.97 = 38.028$ kN.m

The ultimate design moments are, $M_{ux} = M'_{ux} + M'_{ax}$

and $M_{uy} = M'_{uy} + M'_{ay}$

where $M'_{ux} = 0.6 M_{ux,max} + 0.4 M_{ux,min} = 0.6 \times 90 + 0.4 \times 50$

$= 74$ kN.m $> 0.4 M_{ux,max}$

$\therefore M_{ux} = 74 + 56.698 = 130.7$ kN.m

and $M'_{uy} = 0.6 M_{uy,max} + 0.4 M_{uy,min} = 0.6 \times 60 + 0.4 \times 40$

$$= 52 \text{ kN.m} > 0.4 \, M_{uy,\,max}$$

$$\therefore \quad M_{uy} = 52 + 38.028 = 90.028 \text{ kN.m}$$

The area of steel is determined with the use of appropriate design charts for axial load and biaxial moments as follows.

Compute
$$\frac{P_u}{f_{ck} A_g} = \frac{1500 \times 10^3}{25 \times 350 \times 350} = 0.49$$

$$\frac{M_{ux}}{f_{ck} Z_x} = \frac{130.7 \times 10^6}{25 \times 350 \times 350^2 / 6} = 0.732$$

$$\frac{M_{uy}}{f_{ck} Z_y} = \frac{90.028 \times 10^6}{25 \times 350 \times 350^2 / 6} = 0.504$$

Design for reinforcement is made by assuming the area of steel and then checking the adequacy of the section. Assume percentage area of steel, $p = 4.5$.

Compute
$$\frac{p f_y}{f_{ck}} = \frac{4.5 \times 415}{25} = 74.7$$

Uniaxial moment capacities M_{uxo} and M_{uyo} are determined with the use of Chart R1.4 for $d'_d / D = d'_b / B = 0.15$ as,

$$\frac{M_{uxo}}{f_{ck} Z_x} = \frac{M_{uyo}}{f_{ck} Z_y} = 0.925$$

The assumed area of steel is adequate if the following condition is satisfied.

$$\left(\frac{M_{ux}}{M_{uxo}}\right)^\alpha + \left(\frac{M_{uy}}{M_{uyo}}\right)^\alpha \leq 1$$

where
$$\alpha = 0.667 + 1.667 \frac{P_u}{P_{uz}}$$

$$\frac{P_{uz}}{f_{ck} A_g} = 0.446 + \left(0.75 \frac{f_y}{f_{ck}} - 0.446\right) \frac{p}{100}$$

$$= 0.446 + \left(0.75 \times \frac{415}{25} - 0.446\right) \times \frac{4.5}{100} = 0.9862$$

$$\therefore \quad \alpha = 0.667 + \frac{1.667 \times 0.49}{0.9862} = 1.4953$$

and
$$\left(\frac{M_{ux}}{M_{uxo}}\right)^\alpha + \left(\frac{M_{uy}}{M_{uyo}}\right)^\alpha = \left(\frac{0.7156}{0.925}\right)^{1.4953} + \left(\frac{0.4932}{0.925}\right)^{1.4953} = 1.072 > 1.0 \text{ (Hence it is unsafe)}$$

As the strength of section is less than its required value, the assumed percentage of reinforcement is inadequate. Therefore consider increased value of $p = 5.0$.

Compute

$$\frac{pf_y}{f_{ck}} = \frac{5 \times 415}{25} = 83.0$$

Uniaxial moment capacities M_{uxo} and M_{uyo} are determined with the use of Chart R1.4 for $d'_d/D = d'_b/B = 0.15$ as,

$$\frac{M_{uxo}}{f_{ck} Z_x} = \frac{M_{uyo}}{f_{ck} Z_y} = 1.04$$

The assumed area of steel is adequate if the following condition is satisfied.

$$\left(\frac{M_{ux}}{M_{uxo}}\right)^\alpha + \left(\frac{M_{uy}}{M_{uyo}}\right)^\alpha \leq 1$$

where

$$\alpha = 0.667 + 1.667 \frac{P_u}{P_{uz}} < 1.0 > 2.0$$

$$\frac{P_{uz}}{f_{ck} A_g} = 0.446 + \left(0.75 \frac{f_y}{f_{ck}} - 0.446\right)\frac{p}{100}$$

$$= 0.446 + \left(0.75 \times \frac{415}{25} - 0.446\right) \times \frac{5}{100} = 1.0462$$

$\therefore$

$$\alpha = 0.667 + \frac{1.667 \times 0.49}{1.0462} = 1.448$$

and

$$\left(\frac{M_{ux}}{M_{uxo}}\right)^\alpha + \left(\frac{M_{uy}}{M_{uyo}}\right)^\alpha = \left(\frac{0.7156}{1.04}\right)^{1.448} + \left(\frac{0.4932}{1.04}\right)^{1.448} = 0.92 < 1 \text{ (Hence it is safe)}$$

As the strength of section is nearly equal to its required value, the assumed percentage of reinforcement is adopted.

$\therefore$

$$A_s = 5 \times 350 \times \frac{350}{100} = 6125 \text{ mm}^2$$

Provide 4×28 mm ϕ + 8×25 mm ϕ bars ($A_s = 6390$ mm^2) as shown in Fig. Ex. 5.4.

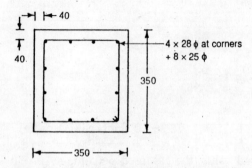

Fig. Ex. 5.4 Reinforcement detailing

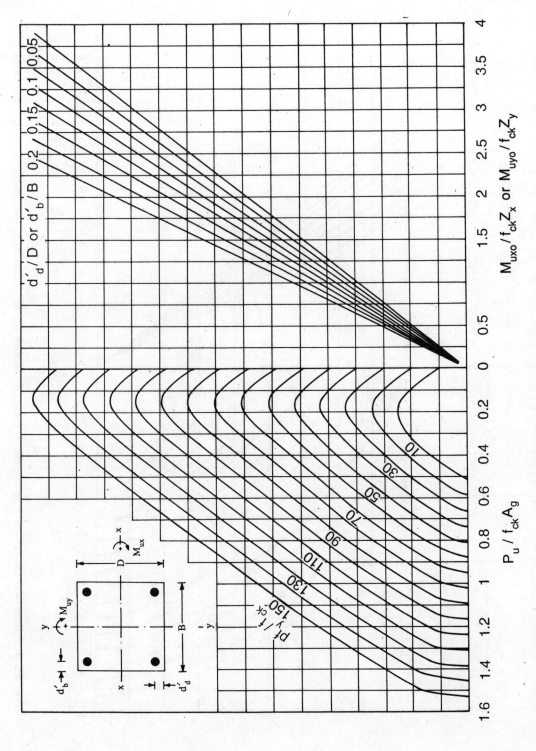

Chart R1.1 Interaction curves for rectangular section of column for M_{uxo} and M_{uyo} (Reinforcement detailing type 1)

158 *Handbook of Reinforced Concrete Design*

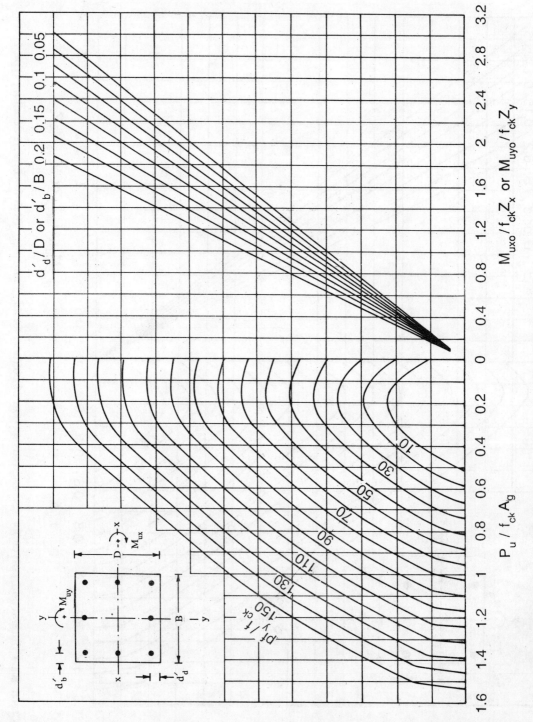

Chart R1.2 Interaction curves for rectangular section of column for M_{uxo} and M_{uyo} (Reinforcement detailing type 2)

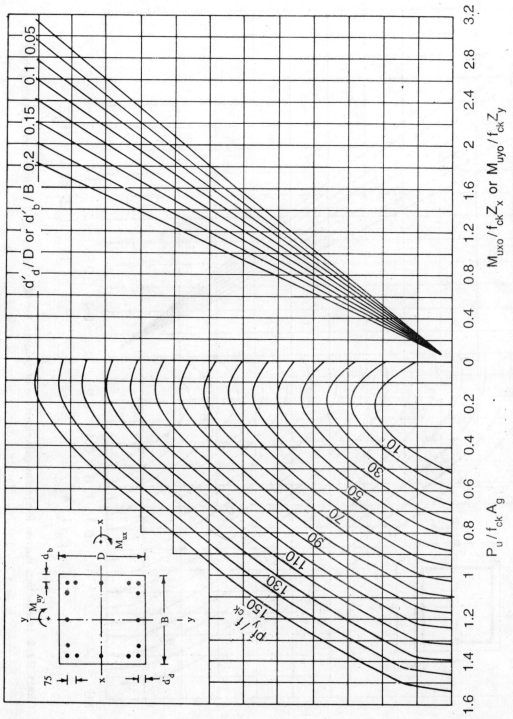

Chart R1.3 Interaction curves for rectangular section of column for M_{uxo} and M_{uyo} (Reinforcement detailing type 3)

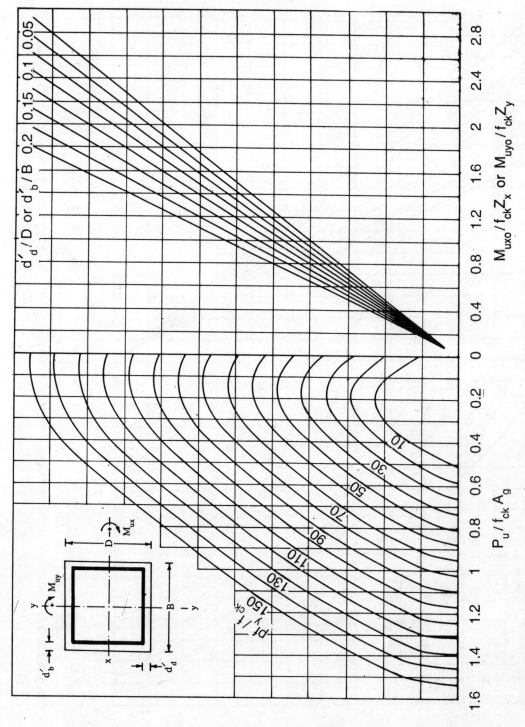

Chart RI.4 Interaction curves for rectangular section of column for M_{uxo} and M_{uyo} (Reinforcement detailing type 4)

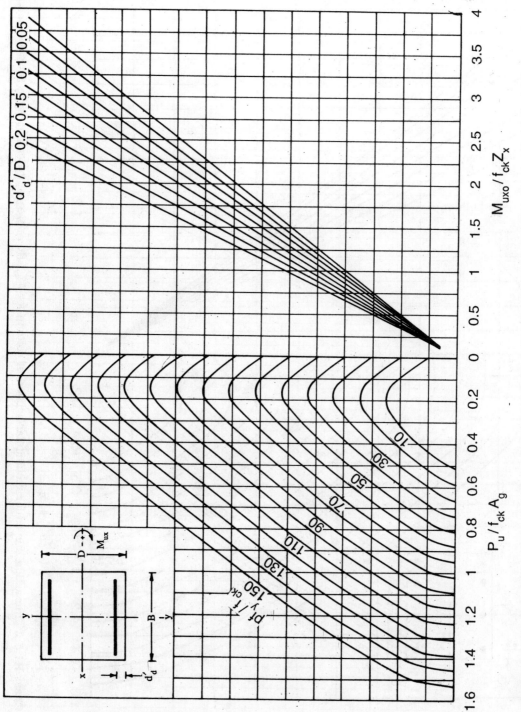

Chart R1.5.1 Interaction curves for rectangular section of column for M_{uxo} (Reinforcement detailing type 5)

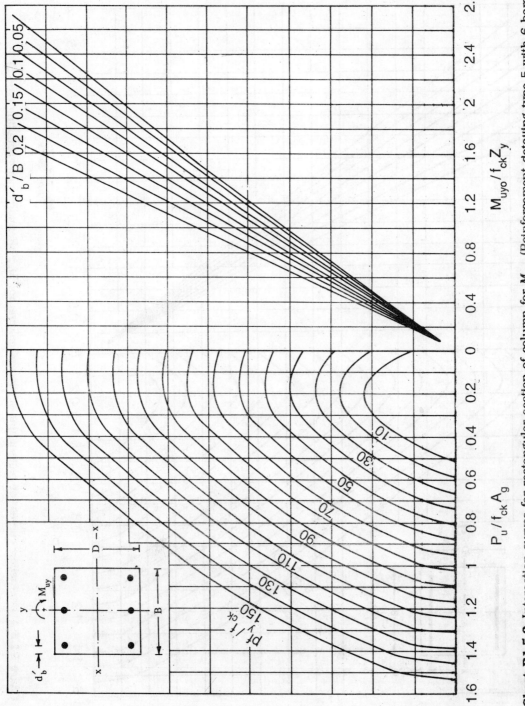

Chart R1.5.2 Interaction curves for rectangular section of column for M_{uyo} (Reinforcement detailing type 5 with 6 bars)

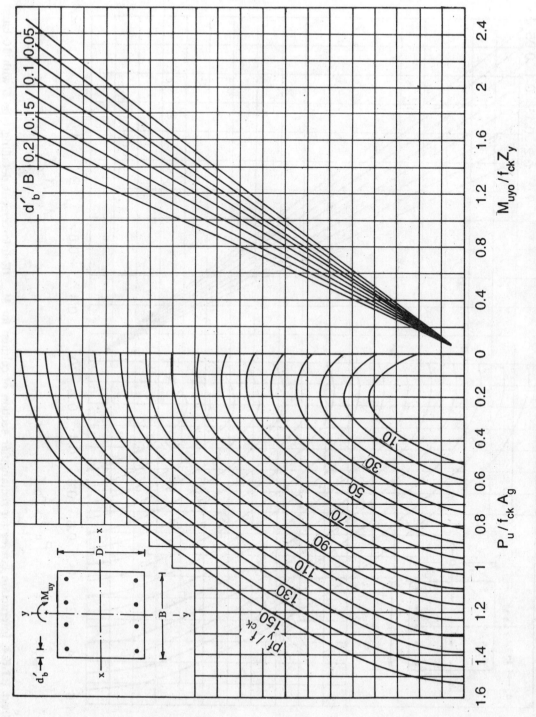

Chart R1.5.3 Interaction curves for rectangular section of column for M_{uyo} (Reinforcement detailing type 5 with 8 bars)

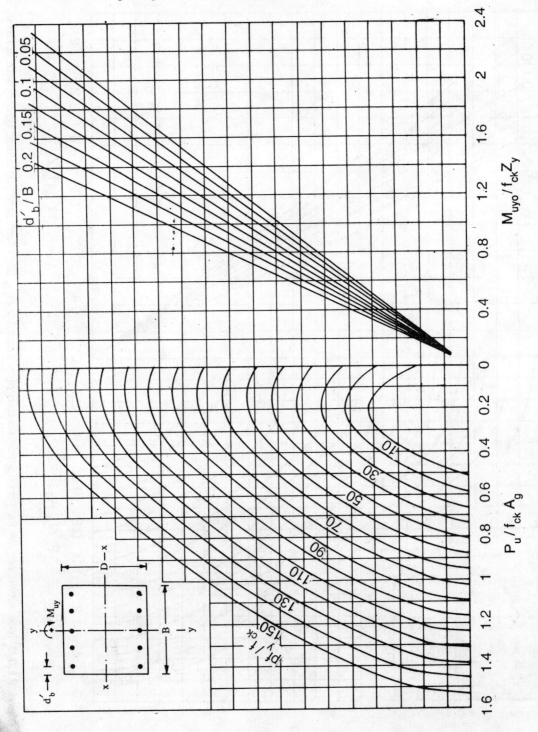

Chart R1.5.4 Interaction curves for rectangular section of column for M_{uyo} (Reinforcement detailing type 5 with 10 bars)

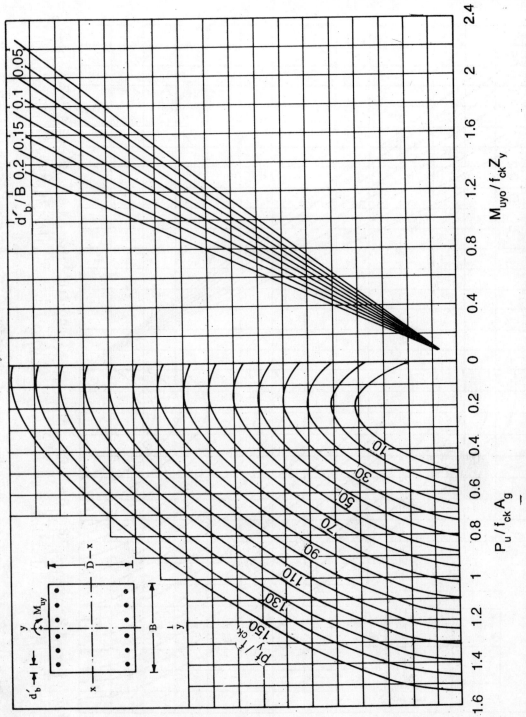

Chart R1.5.5 Interaction curves for rectangular section of column for M_{uyo} (Reinforcement detailing type 5 with 12 bars)

166   Handbook of Reinforced Concrete Design

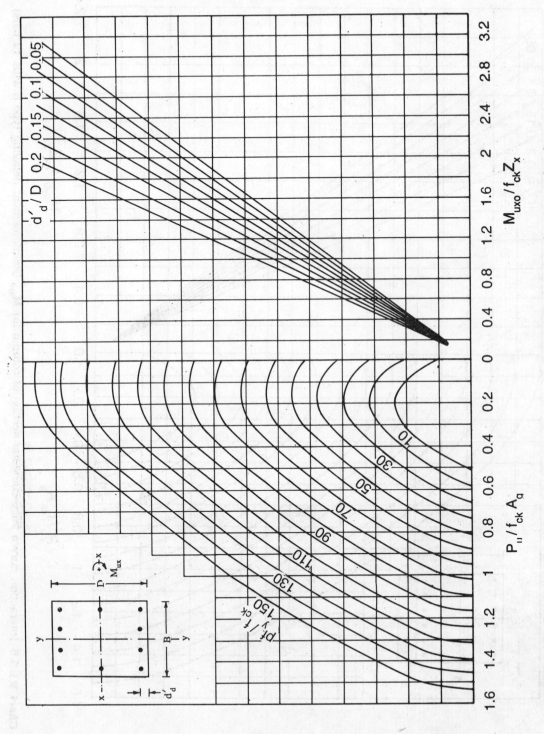

Chart **R1.6.1** Interaction curves for rectangular section of column for M_{uxo} (Reinforcement detailing type 6)

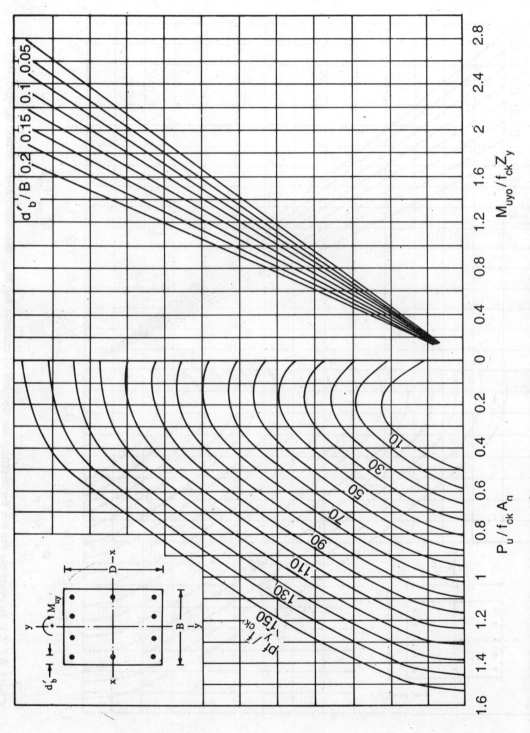

Chart R1.6.2 Interaction curves for rectangular section of column for M_{uyo} (Reinforcement detailing type 6)

168 Handbook of Reinforced Concrete Design

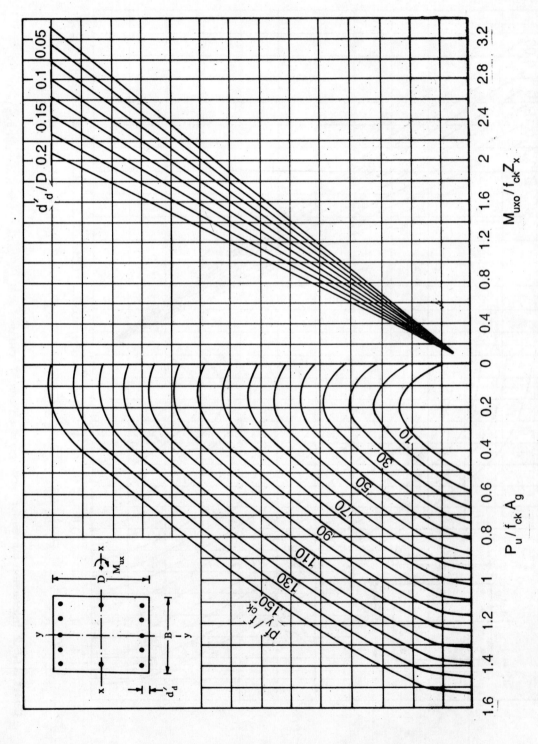

Chart R1.7.1 Interaction curves for rectangular section of column for M_{uxo} (Reinforcement detailing type 1)

Columns 169

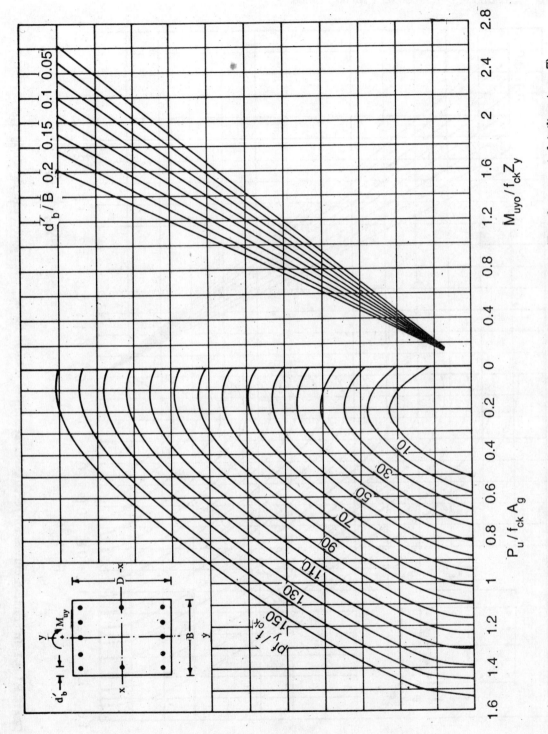

Chart R1.7.2 Interaction curves for rectangular section of column for M_{uyo} (Reinforcement detailing type 7)

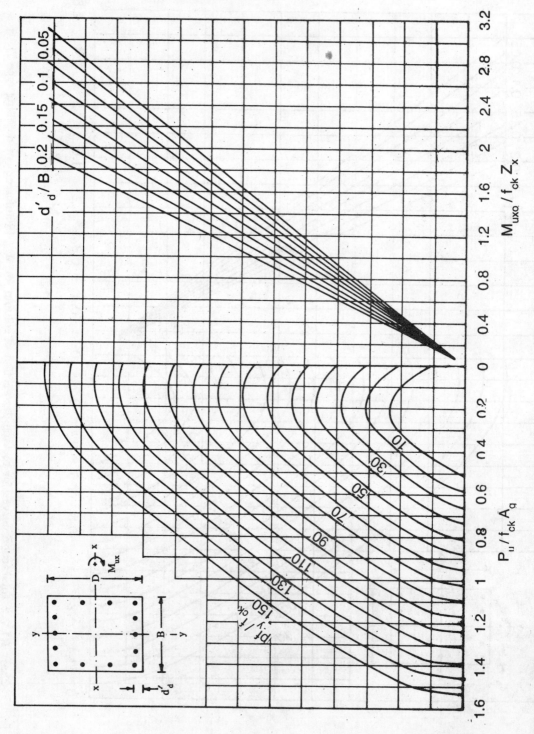

Chart R1.8.1 Interaction curves for rectangular section of column for M_{uxo} (Reinforcement detailing type 8)

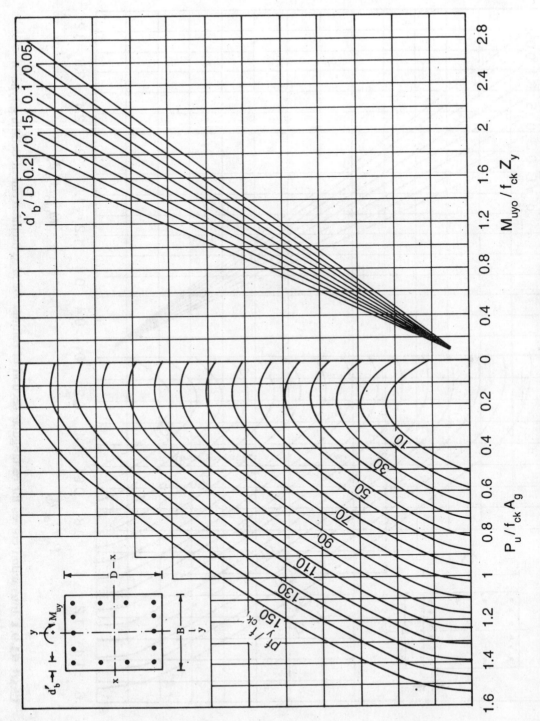

Chart R1.8.2 Interaction curves for rectangular section of column for M_{uyo} (Reinforcement detailing type 8)

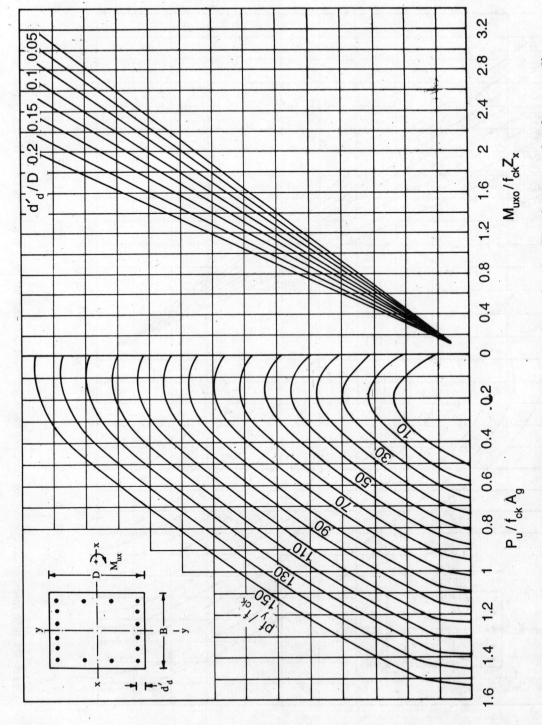

Chart R1.9.1 Interaction curves for rectangular section of column for M_{uxo} (Reinforcement detailing type 9)

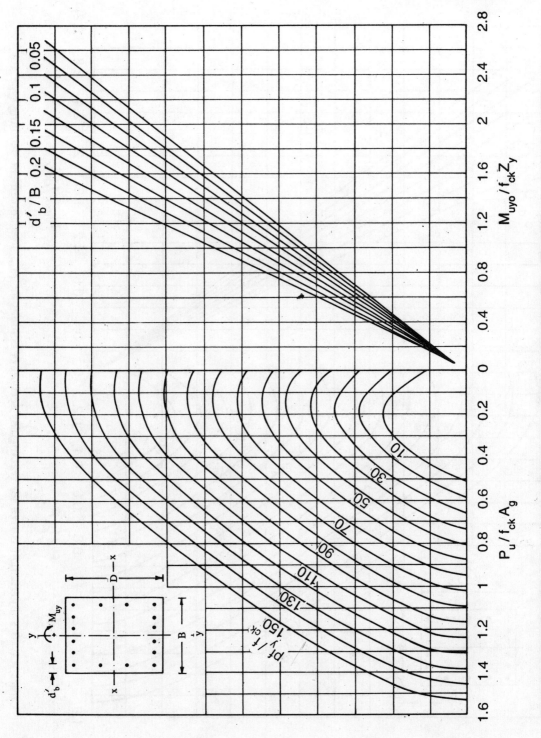

Chart R1.9.2 Interaction curves for rectangular section of column for M_{uyo} (Reinforcement detailing type 9)

174 Handbook of Reinforced Concrete Design

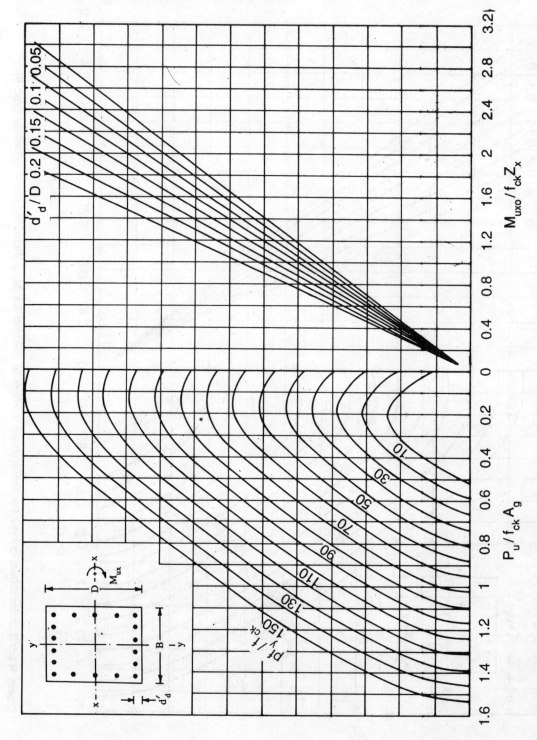

Chart R1.10.1 Interaction curves for rectangular section of column for M_{uxo} (Reinforcement detailing type 10)

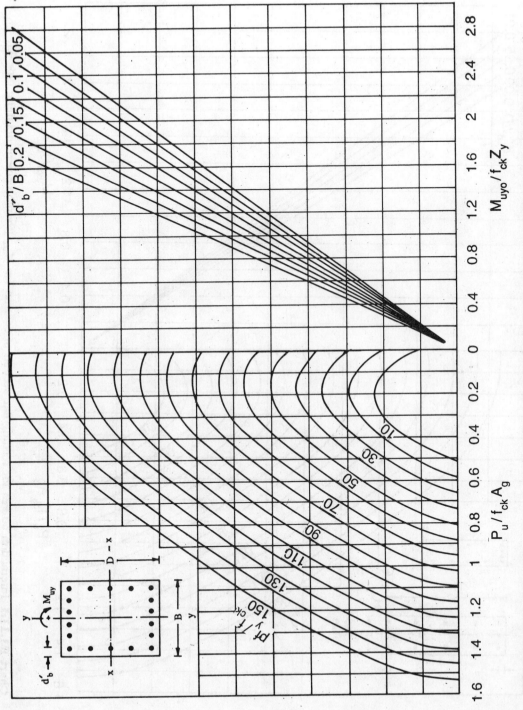

Chart R1.10.2 Interaction curves for rectangular section of column for M_{uyo} (Reinforcement detailing type 10)

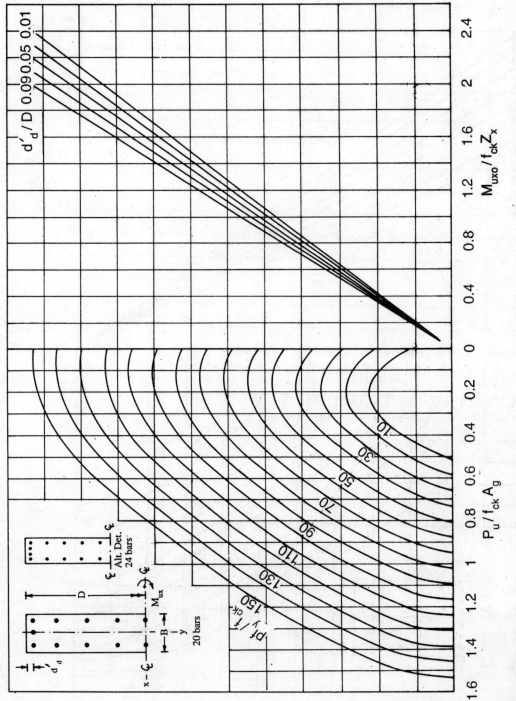

Chart R1.11.1 Interaction curves for rectangular section of column for M_{uxo} (Reinforcement detailling type 1)

Columns 177

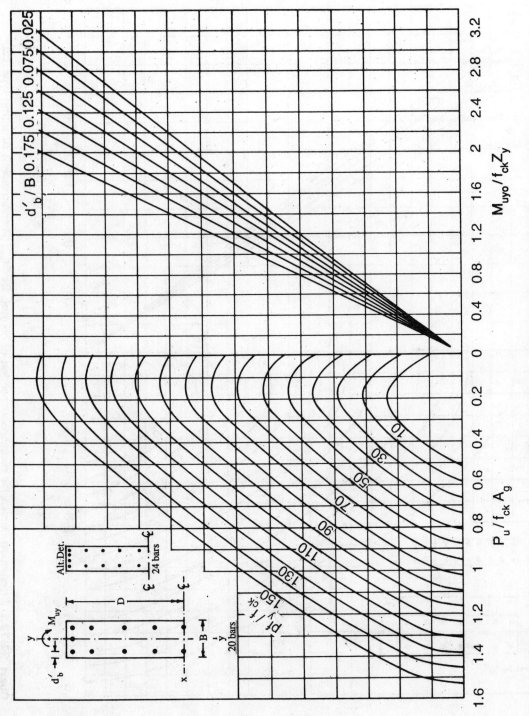

Chart R1.11.2 Interaction curves for rectangular section of column for M_{uyo} (Reinforcement detailing type 11)

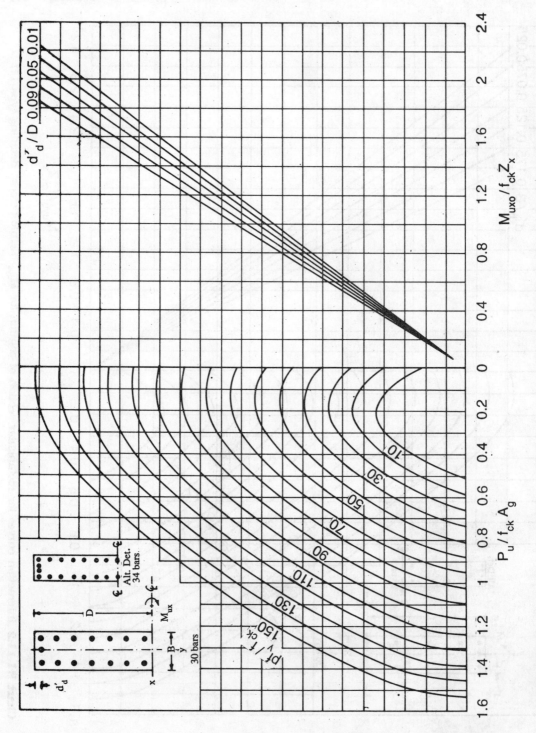

Chart R1.12.1 Interaction curves for rectangular section of column for M_{uxo} (Reinforcement detailing type 12)

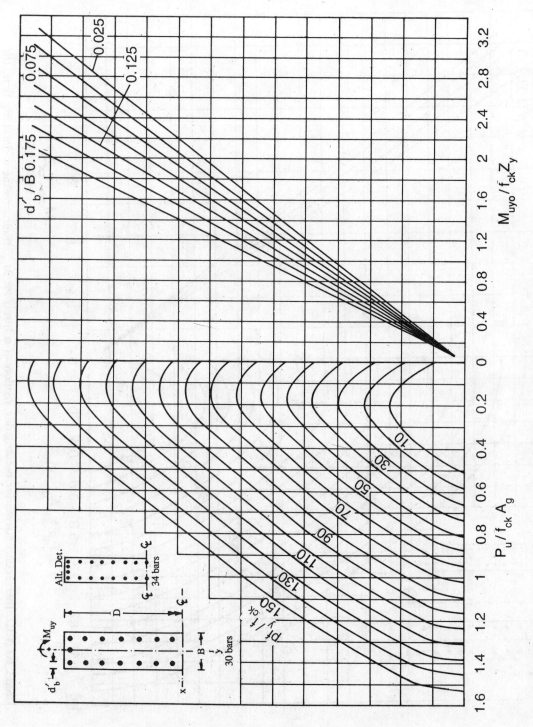

Chart R1.12.2 Interaction curves for rectangular section of column for M_{uyo} (Reinforcement detailing type 12)

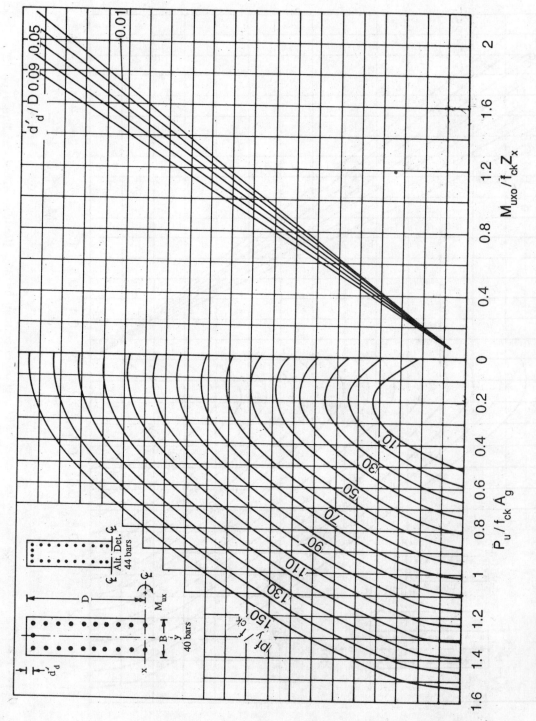

Chart K1.13.1 Interaction curves for rectangular section of column for M_{uxo} (Reinforcement detailing type 13)

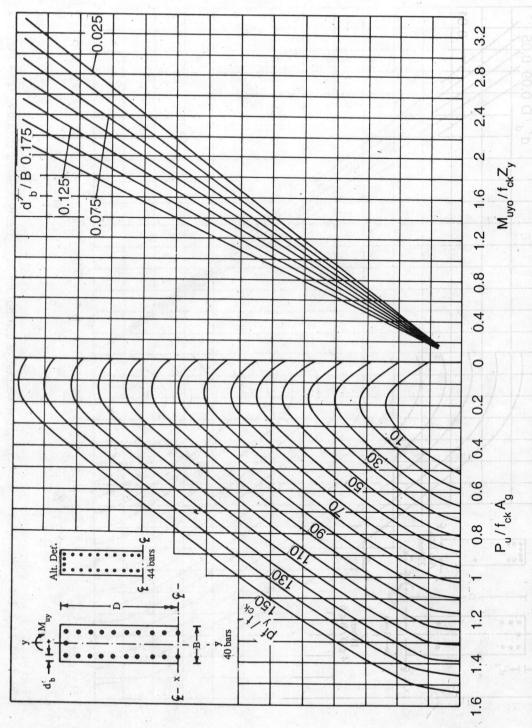

Chart R1.13.2 Interaction curves for rectangular section of column for M_{uyo} (Reinforcement detailing type 13)

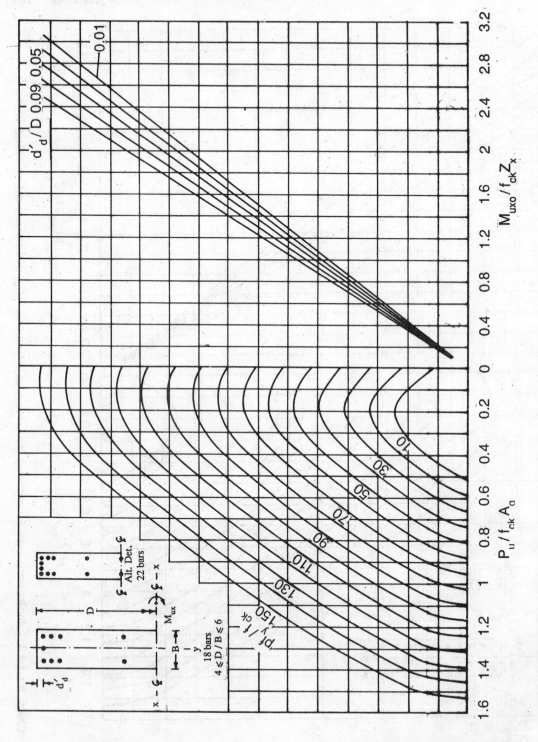

Chart R1.14.1 Interaction curves for rectangular section of column for M_{uxo} (Reinforcement detailing type 14)

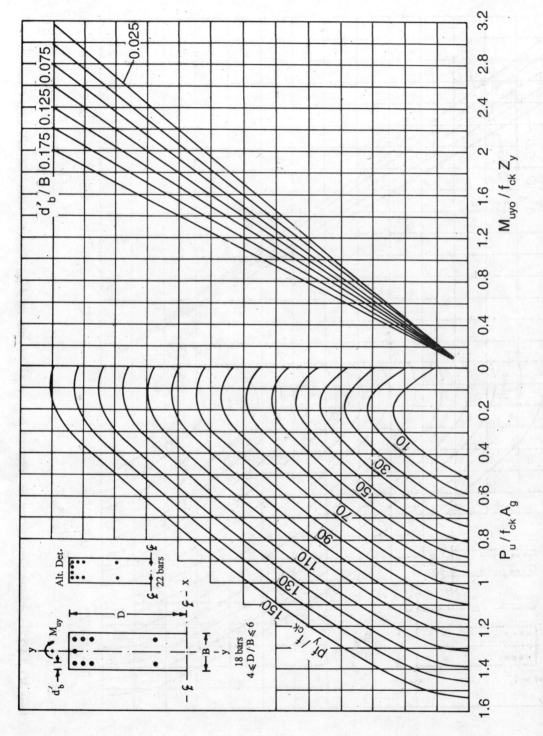

Chart R1.14.2 Interaction curves for rectangular section of column for M_{uyo} (Reinforcement detailing type 14)

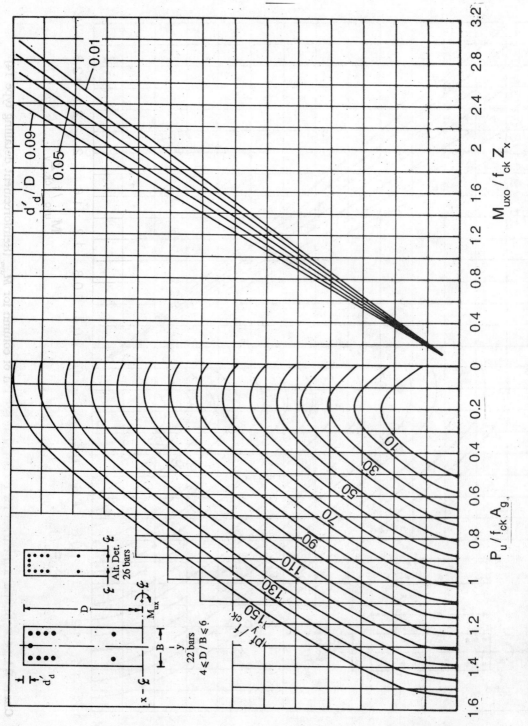

Chart R1.15.1 Interaction curves for rectangular section of column for M_{uxo} (Reinforcement detailing type 15)

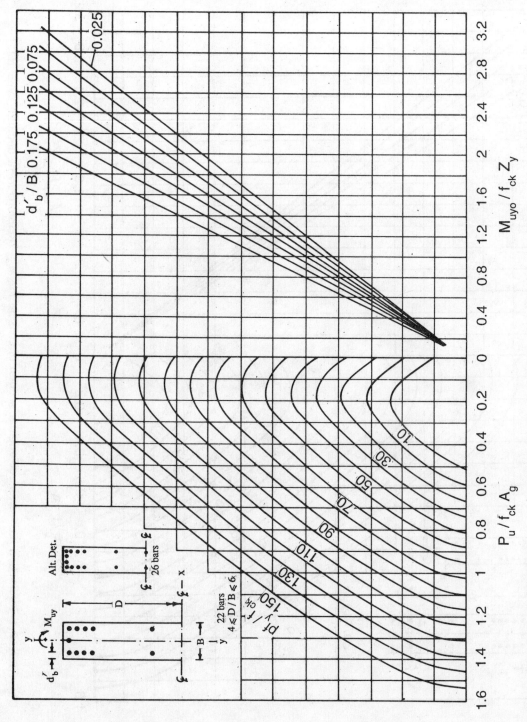

Chart R1.15.2 Interaction curves for rectangular section of column for M_{uyo} (Reinforcement detailing type 15)

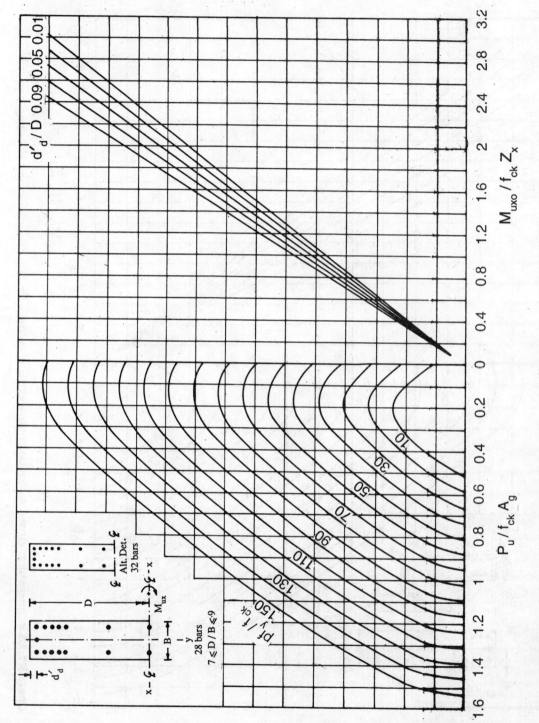

Chart R1.16.1 Interaction curves for rectangular section of column for M_{uxo} (Reinforcement detailing type 16)

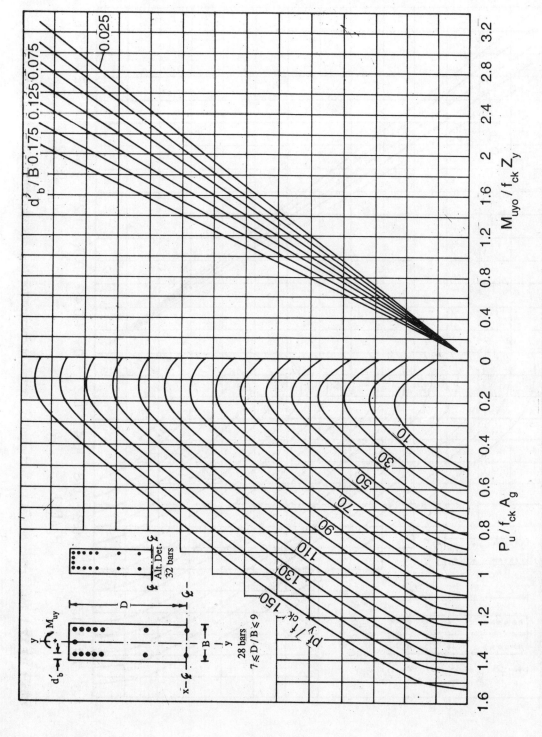

Chart R1.16.2 Interaction curves for rectangular section of column for M_{uyo} (Reinforcement detailing type 16)

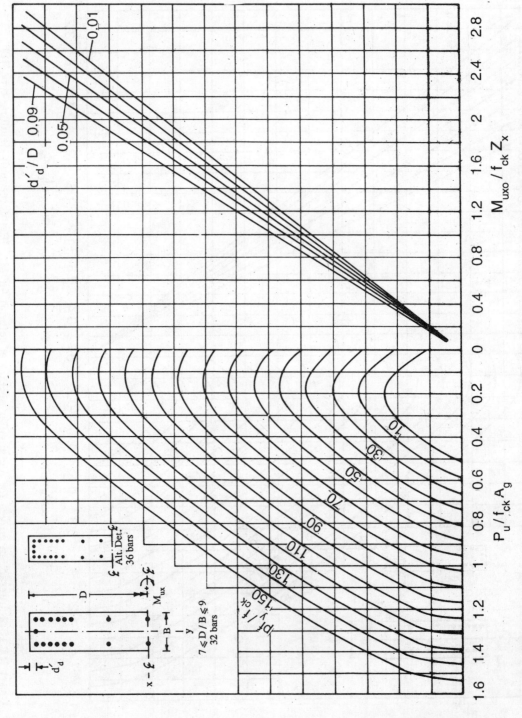

Chart R1.17.1 Interaction curves for rectangular section of column for M_{uxo} (Reinforcement detailing type 17)

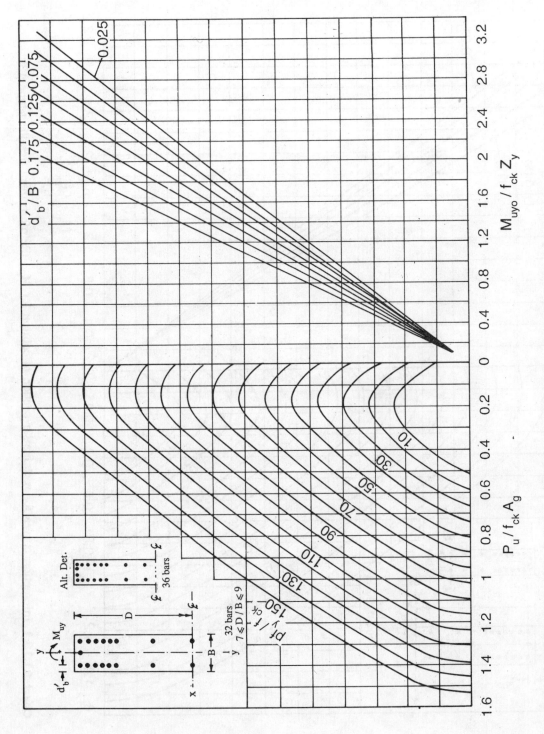

Chart R1.17.2 Interaction curves for rectangular section of column for M_{uyo} (Reinforcement detailing type 17)

190 Handbook of Reinforced Concrete Design

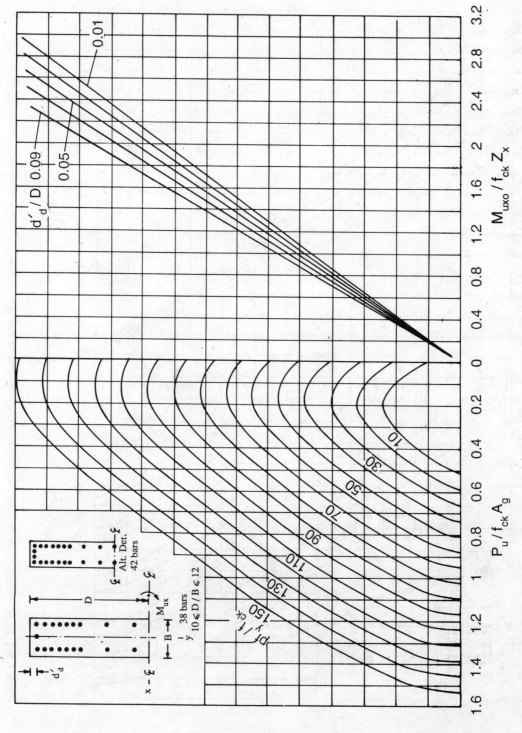

Chart R1.18.1 Interaction curves for rectangular section of column for M_{uxo} (Reinforcement detailing type 18)

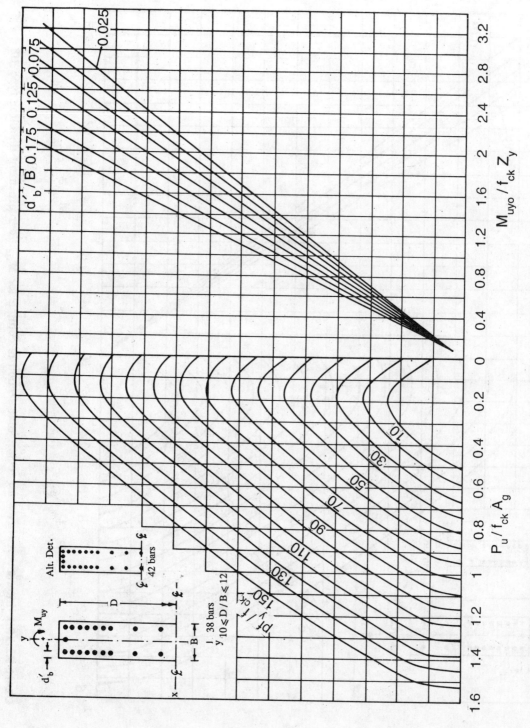

Chart R1.18.2 Interaction curves for rectangular section of column for M_{uyo} (Reinforcement detailing type 18)

192 Handbook of Reinforced Concrete Design

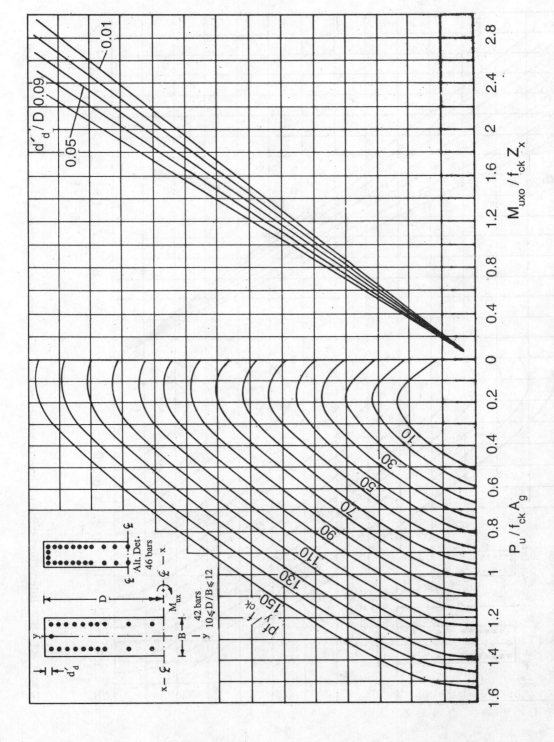

Chart R1.19.1 Interaction curves for rectangular section of column for M_{uxo} (Reinforcement detailing type 19)

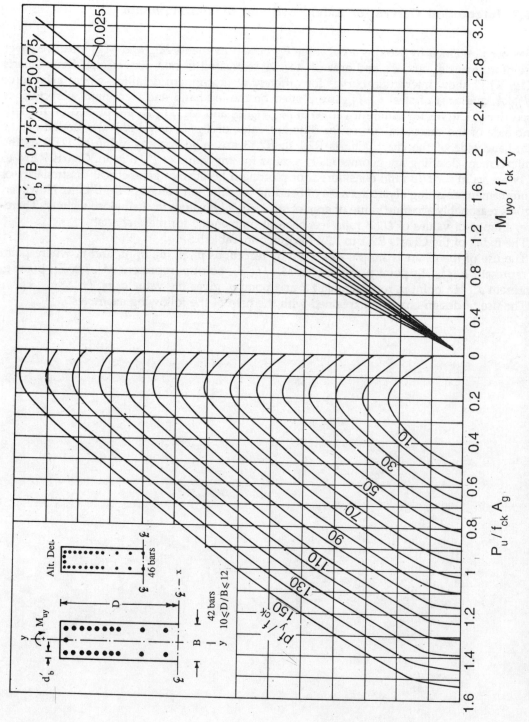

Chart R1.19.2 Interaction curves for rectangular section of column for M_{uyo} (Reinforcement detailing type 19)

5.4.2 Interaction Curves for Axial Load and Biaxial Moments

It has been obtained by plotting axial force P_u, biaxial moments M_{ux} and M_{uy}, effective cover to reinforcement d'_d and d'_b, and area of steel A_s expressed in non-dimensional form. Charts R2.1 to R2.19 show interaction curves for different reinforcement detailing where axial force $P_u/f_{ck}A_g$ and area of steel pf_y/f_{ck} are plotted on the left-hand and right-hand vertical axes respectively. Curves for different values of $M_{ux}/f_{ck}Z_x$ and $M_{uy}/f_{ck}Z_y$ are plotted on the left-hand side of the chart and curves for different value of d'_d/D and d'_b/B are plotted on the right-hand side of the chart. Charts R2.1 to R2.13 are valid for any value of D/B as the reinforcement detailing are geometrically similar for any value of D/B ratio. Charts R2.14 to R2.19 are valid for D/B ratio considered for plotting it because of geometrical dissimilarity of reinforcement detailing about the axes of bending for different values of D/B ratio. However, it gives reasonably accurate value of area of steel for D/B values in their close vicinity. Therefore, a range of values of D/B ratio has been recommended for these charts.

The index of the Charts R2.1 to R2.19 is given in Table 5.5.

The use of the Chart is simple as shown by directions p_1, p_2, p_3, p_4, p_5 and p_6 where point p_1 represent $P_u/f_{ck}A_g$, point p_2 represents $M_{ux}/f_{ck}Z_x$, point p_3 represent $M_{uy}/f_{ck}Z_y$, point p_4 represent d'_d/D, point p_5 represent d'_b/B and point p_6 gives the value of pf_y/f_{ck}.

The use of design charts is explained with the help of the following examples.

Table 5.5 Index for Charts R2.1 to R2.19 for axial force and biaxial moments

Reinforcement detailing	Chart	d'_d/D	d'_b/B	Page No.
Case 1	R2.1.1	0.075, 0.1, 0.125	0.075, 0.1, 0.125	203
	R2.1.2	0.075, 0.1, 0.125	0.15, 0.175, 0.2	204
	R2.1.3	0.15, 0.175, 0.2	0.075, 0.1, 0.125	205
	R2.1.4	0.15, 0.175, 0.2	0.15, 0.175, 0.2	206
Case 2	R2.2.1	0.05, 0.075, 0.1	0.05, 0.075, 0.1	207
	R2.2.2	0.05, 0.075, 0.1	0.125, 0.15, 0.175	208
	R2.2.3	0.125, 0.15, 0.175	0.05, 0.075, 0.1	209
	R2.2.4	0.125, 0.15, 0.175	0.125, 0.15, 0.175	210
Case 3	R2.3.1	0.05, 0.075, 0.1	0.05, 0.075, 0.1	211
	R2.3.2	0.05, 0.075, 0.1	0.125, 0.15, 0.175	212
	R2.3.3	0.125, 0.15, 0.175	0.05, 0.075, 0.1	213
	R2.3.4	0.125, 0.15, 0.175	0.125, 0.15, 0.175	214
Case 4	R2.4.1	0.05, 0.075, 0.1	0.05, 0.075, 0.1	215
	R2.4.2	0.05, 0.075, 0.1	0.125, 0.15, 0.175	216
	R2.4.3	0.125, 0.15, 0.175	0.05, 0.075, 0.1	217
	R2.4.4	0.125, 0.15, 0.175	0.125, 0.15, 0.175	218
Case 5A	R2.5.1	0.1, 0.125, 0.15	0.1, 0.125, 0.15	219
Case 5B	R2.5.2	0.1, 0.125, 0.15	0.1, 0.125, 0.15	220
Case 5C	R2.5.3	0.1, 0.125, 0.15	0.1, 0.125, 0.15	221

(Contd.)

Table 5.5 (*Contd.*)

Reinforcement detailing	Chart	d'_d/D	d'_b/B	Page No.
Case 5D 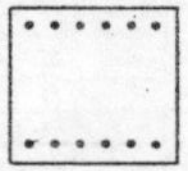	R2.5.4	0.1, 0.125, 0.15	0.1, 0.125, 0.15	222
Case 6 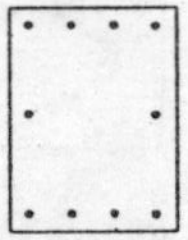	R2.6	0.1, 0.125, 0.15	0.1, 0.125, 0.15	223
Case 7 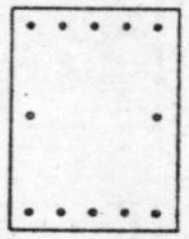	R2.7	0.1, 0.125, 0.15	0.1, 0.125, 0.15	224
Case 8 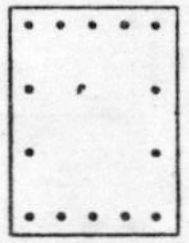	R2.8	0.1, 0.125, 0.15	0.1, 0.125, 0.15	225
Case 9 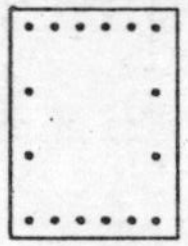	R2.9	0.1, 0.125, 0.15	0.1, 0.125, 0.15	226
Case 10 	R2.10	0.1, 0.125, 0.15	0.1, 0.125, 0.15	227
Case 11 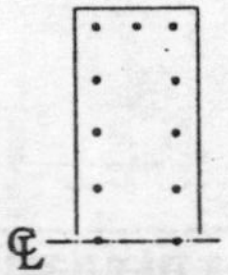 (20 bars)	R2.11	0.015, 0.035, 0.055	0.125, 0.15, 0.175	228

(*Contd.*)

Table 5.5 (*Contd.*)

Reinforcement detailing	Chart	d'_d/D	d'_b/B	Page No.
Case 12, 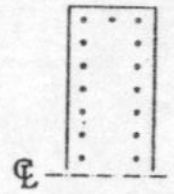 30 bars	R2.12	0.015, 0.035, 0.055	0.125, 0.15, 0.175	229
Case 13, 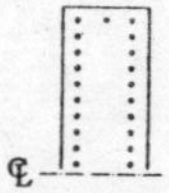 40 bars	R2.13	0.015, 0.035, 0.055	0.125, 0.15, 0.175	230
Case 14, 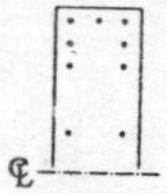 18 bars	R2.14	0.015, 0.035, 0.055	0.125, 0.15, 0.175	231
Case 15, 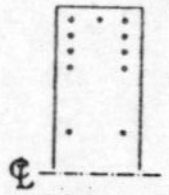 22 bars	R2.15	0.015, 0.035, 0.055	0.125, 0.15, 0.175	232
Case 16, 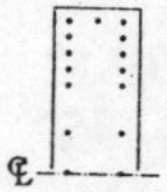 28 bars	R2.16	0.015, 0.035, 0.055	0.125, 0.15, 0.175	233
Case 17, 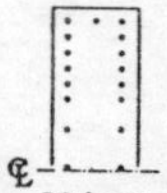 32 bars	R2.17	0.015, 0.035, 0.055	0.125, 0.15, 0.175	234
Case 18, 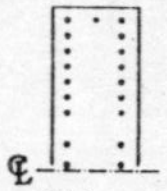 38 bars	R2.18	0.015, 0.035, 0.055	0.125, 0.15, 0.175	235
Case 19, 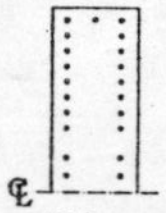 42 bars	R2.19	0.015, 0.035, 0.055	0.125, 0.15, 0.175	236

Example 5.5: Design a biaxially eccentrically loaded column section of Example 5.3 for the following data:

$$\text{Column section, } B \times D = 350 \text{ mm} \times 700 \text{ mm}$$
$$\text{Ultimate axial load, } P_u = 3675 \text{ kN}$$
$$\text{Ultimate moments, } M_{ux} = 500 \text{ kN.m}$$
$$M_{uy} = 250 \text{ kN.m}$$
$$\text{Grade of concrete} : \text{M25}$$
$$\text{Grade of steel} : \text{Fe 415}$$

Solution

Consider 25 mm ϕ bars uniformly distributed on four faces at clear cover of 40 mm.

$\therefore$ Effective cover, $d' = 40 + 25/2 = 52.5$ mm

and $d'_d/D = 52.5/700 = 0.075$

and $d'_b/B = 52.5/350 = 0.15$

Area of the section, $A_g = BD = 350 \times 700 = 245000 \text{ mm}^2$

Section Modulus, $Z_x = BD^2/6 = 350 \times 700^2/6 = 28583333 \text{ mm}^3$

$Z_y = B^2D/6 = 350^2 \times 700/6 = 14291667 \text{ mm}^3$

Compute
$$\frac{P_u}{f_{ck} A_g} = \frac{3675 \times 10^3}{25 \times 245000} = 0.6$$

$$\frac{M_{ux}}{f_{ck} Z_x} = \frac{500 \times 10^6}{25 \times 28583333} = 0.7$$

$$\frac{M_{uy}}{f_{ck} Z_y} = \frac{250 \times 10^6}{25 \times 14291667} = 0.7$$

Area of steel is determined with the use of Chart R2.4.2 for $d'_d/D = 0.075$ and $d'_b/B = 0.15$ as,

$$p f_y / f_{ck} = 92.75$$

$\therefore$ $p = 92.75 \times 25 / 415 = 5.587$

and $A_{st} = 5.587 \times 350 \times 700 / 100 \text{ mm}^2 = 13690 \text{ mm}^2$

Provide 24×28 ϕ bars (14778 mm²) as shown in Fig. Ex. 5.5

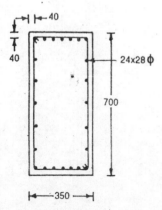

Fig. Ex. 5.5 Reinforcement detailing

Area of steel has also been determined with the use of computer program as 13650 mm². Therefore the use of interaction curve for axial load and uniaxial moment results in conservative estimate of area of steel where as the use of interaction curves for axial load and biaxial moments results in more accurate design.

Example 5.6: Design a biaxially eccentrically loaded braced square column deforming in single curvature of Example 5.4 for the following data.

Ultimate axial load, P_u	= 1500 kN
Ultimate moment about x-axis:	
at bottom, M_{ux1}	= 90 kN.m
at top, M_{ux2}	= 50 kN.m
Ultimate moment about y-axis:	
at bottom, M_{uy1}	= 60 kN.m
at top, M_{uy2}	= 40 kN.m
Unsupported length of column, l	= 8 m
Effective length about x axis, l_{ex}	= 5.2 m
Effective length about y-axis, l_{ey}	= 6.35 m
Column section, $B \times D$	= 350 mm × 350 mm
Grade of concrete	: M25
Grade of steel	: Fe 415

Solution

Moment due to minimum eccentricities:

$$M_{uxe} = P_u e_{y,\,min}$$

and

$$M_{uye} = P_u e_{x,\,min}$$

where
$$e_{y,\,min} = \frac{l}{500} + \frac{B}{30} \text{ or } 20 \text{ mm whichever is more}$$

$$= \frac{8000}{500} + \frac{350}{30} \text{ or } 20 \text{ mm whichever is more}$$

$$= 27.67 \text{ mm}$$

$$e_{x,min} = \frac{l}{500} + \frac{D}{30}, \text{ or } 20 \text{ mm whichever is less}$$

$$= \frac{8000}{500} + \frac{350}{30}, \text{ or } 20 \text{ mm whichever is less}$$

$$= 27.67 \text{ mm}$$

$\therefore \qquad M_{uxe} = 1500 \times 27.67 \text{ kN.mm} = 41.505 \text{ kN.m} < M_{ux}$

and $\qquad M_{uye} = 1500 \times 27.67 \text{ kN.mm} = 41.505 \text{ kN.m} < M_{uy}$

Check for short/long column:

$$l_{ex}/D = 5200/350 = 14.86 > 12$$

$$l_{ey}/B = 6350/350 = 18.143 > 12$$

Hence it is a long column about both axes of bending.

Additional moments due to slenderness effects,

$$M_{ax} = P_u e_y = P_u \frac{B}{2000}\left(\frac{l_{ey}}{D}\right)^2$$

$$= 1500 \times \frac{350}{2000} \times 18.143^2 \text{ kN.mm} = 86.406 \text{ kN.m}$$

and $\qquad M_{ay} = P_u e_x = P_u \frac{B}{2000}\left(\frac{l_{ex}}{D}\right)^2$

$$= 1500 \times \frac{350}{2000} \times 14.86^2 \text{ kN.mm} = 57.97 \text{ kN.m}$$

The additional moments M_{ax} and M_{ay} are modified by multiplying with a factor K given by,

$$K = \frac{P_{uz} - P_u}{P_{uz} - P_{ub}}$$

where P_{uz} and P_{ub} are determined for the assumed area of longitudinal reinforcement and its distribution. Consider percentage of reinforcement $p = 4.5$. It is distributed uniformly on four faces with clear cover of 40 mm. Consider 25 mm ϕ bars.

∴ Effective cover, $d' = 40 + 25/2 = 52.5$ mm

Compute $P_{uz} = 0.446 f_{ck} A_g + (0.75 f_y - 0.446 f_{ck}) pA_g / 100$

$= 0.446 \times 25 \times 350 \times 350 + (10.75 \times 415 - 0.446 \times 25) \times 4.5 \times 350 \times 350 / 100$ N

$= 3020.18$ kN

and $P_{ub} = q_c f_{ck} A_g + q_s pA_g$

where for $d'_d / D = d'_b / B = 52.5/3500 = 0.15$

$q_c = q_{cx} = q_{cy} = 0.1953$

and $q_s = q_{sx} = q_{sy} = 0.1898$

∴ $P_{ub} = P_{ubx} = P_{uby} = 0.1953 \times 25 \times 350 \times 350 + 0.1898 \times 4.5 \times 350 \times 350$ N $= 702.734$ kN

Compute $K = K_x = K_y = \dfrac{P_{uz} - P_u}{P_{uz} - P_{ub}} = \dfrac{3020.18 - 1500.0}{3020.18 - 702.734} = 0.656$

Modified additional moments are,

$M_{ax} = k_x M_{ax} = 0.656 \times 86.43 = 56.698$ kN.m

and $M_{ay} = k_y M_{ay} = 0.656 \times 57.97 = 38.028$ kN.m

The ultimate design moments are,

$M_{ux} = M'_{ux} + M'_{ax}$

and $M_{uy} = M'_{uy} + M'_{ay}$

where $M'_{ux} = 0.6 M_{ux, max} + 0.4 M_{ux, min}$

$= 0.6 \times 90 + 0.4 \times 50$

$= 74$ kN.m $> 0.4 M_{ux, max}$

∴ $M_{ux} = 74 + 56.698 = 130.7$ kN.m

and
$$M'_{uy} = 0.6 M_{uy, max} + 0.4 M_{uy, min}$$
$$= 0.6 \times 60 + 0.4 \times 40 = 52 \text{ kN.m} > 0.4 M_{uy, max}$$

$\therefore$
$$M_{uy} = 52 + 38.028 = 90.028 \text{ kN.m}$$

The area of steel is determined with the use of appropriate design charts for axial load and biaxial moments as follows.

Compute,
$$\frac{P_u}{f_{ck} A_g} = \frac{1500 \times 10^3}{25 \times 350 \times 350} = 0.49$$

$$\frac{M_{ux}}{f_{ck} Z_x} = \frac{130.7 \times 10^6}{25 \times 350 \times 350^2 / 6} = 0.732$$

$$\frac{M_{uy}}{f_{ck} Z_y} = \frac{90.028 \times 10^6}{25 \times 350 \times 350^2 / 6} = 0.504$$

For the above computed values, area of steel is determined with the help of Chart R2.4.4 for $d'_d / B = d'_b / B = 0.15$ as,

$$pf_y / f_{ck} = 79.5$$

$\therefore$
$$p = 79.5 f_{ck} / f_y$$
$$= 79.5 \times 25 / 415 = 4.79$$

As the assumed value of p is approximately equal to its computed value, the value of K as computed above is acceptable.

$\therefore$
$$A_s = p A_g / 100 = 4.79 \times 350 \times 350 / 100 = 5867.8 \text{ mm}^2$$

Provide 12 bars of 25 mm ϕ ($A_s = 5890 \text{ mm}^2$) as shown in Fig. Ex. 5.6.

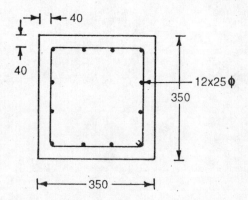

Fig. Ex. 5.6 Reinforcement detailing

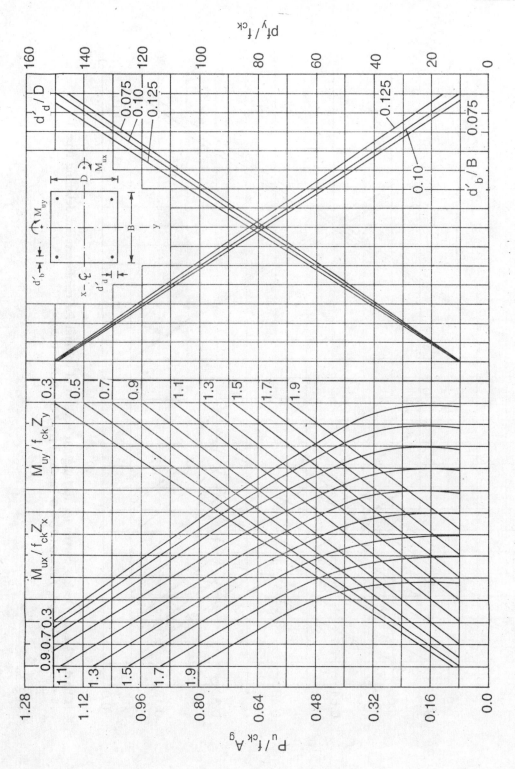

Chart R2.1.1 Design curves for rectangular section of column for reinforcement detailing type 1 ($d'_d/D = 0.075 - 0.125$ and $d'_b/B = 0.075 - 0.125$)

204 Handbook of Reinforced Concrete Design

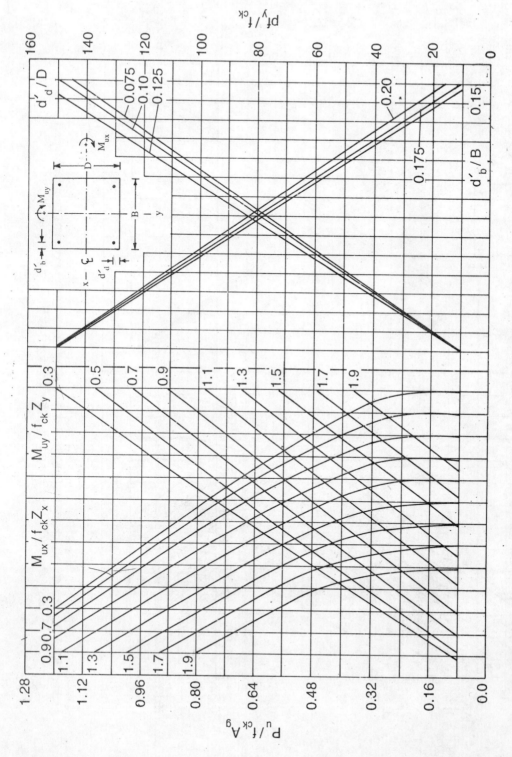

Chart R2.1.2 Design curves for rectangular section of column for reinforcement detailing type 1 ($d'_d/D = 0.075 - 0.125$ and $d'_b/B = 0.15 - 0.20$)

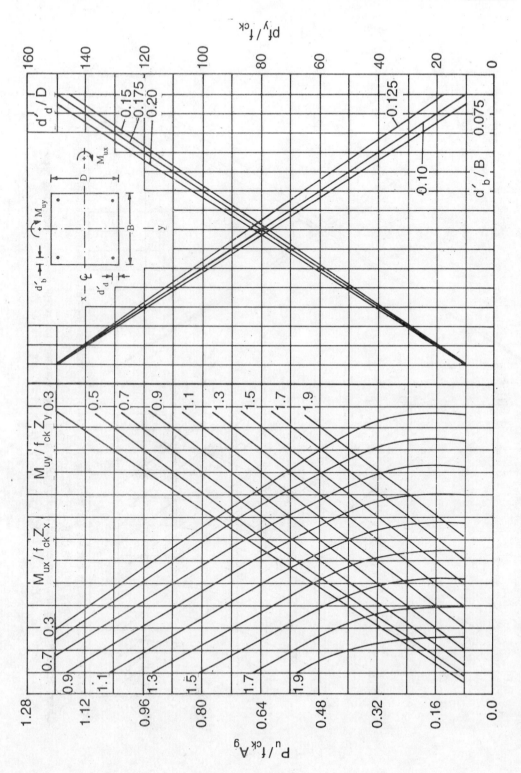

Chart R2.1.3 Design curves for rectangular section of column for reinforcement detailing type 1 ($d'_d/D = 0.15 - 0.20$ and $d'_b/B = 0.075 - 0.125$)

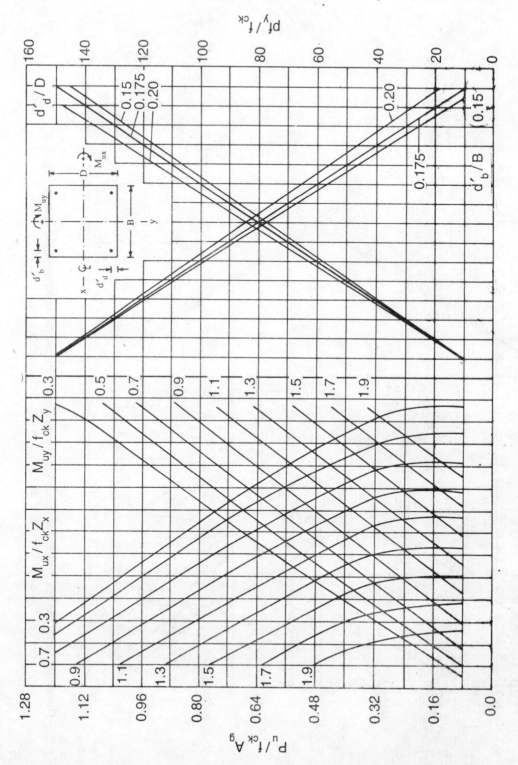

Chart R2.1.4 Design curves for rectangular section of column for reinforcement detailing type 1 ($d'_d/D = 0.15 - 0.20$ and $d'_b/B = 0.15 - 0.20$)

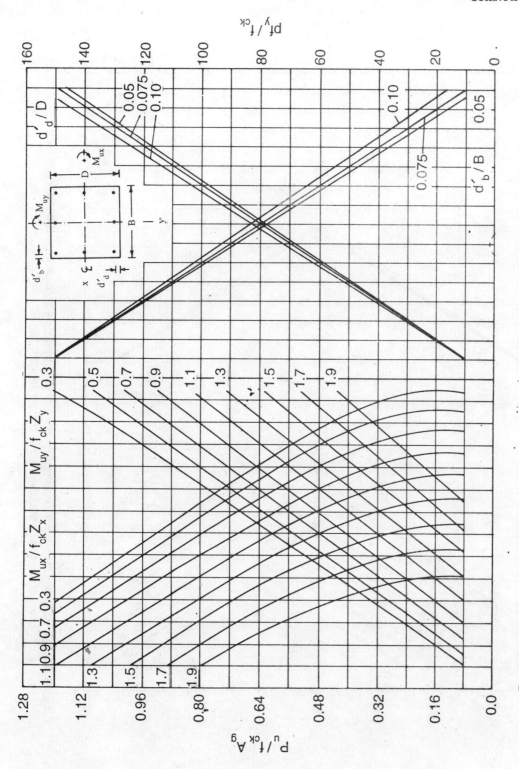

Chart R2.2.1 Design curves for rectangular section of column for reinforcement detailing type 2 ($d'_d / D = 0.05 - 0.10$ and $d'_b / B = 0.05 - 0.10$)

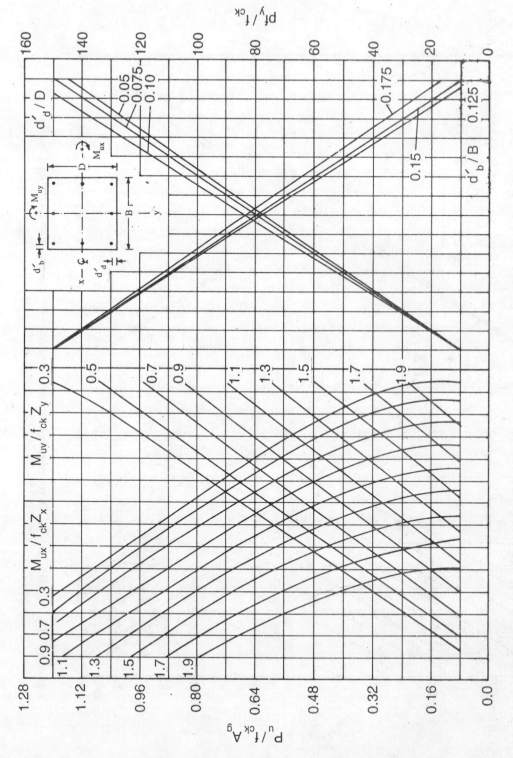

Chart K2.2.2 Design curves for rectangular section of column for reinforcement detailing type 2 ($d'_d / D = 0.05 - 0.10$ and $d'_b / B = 0.125 - 0.175$)

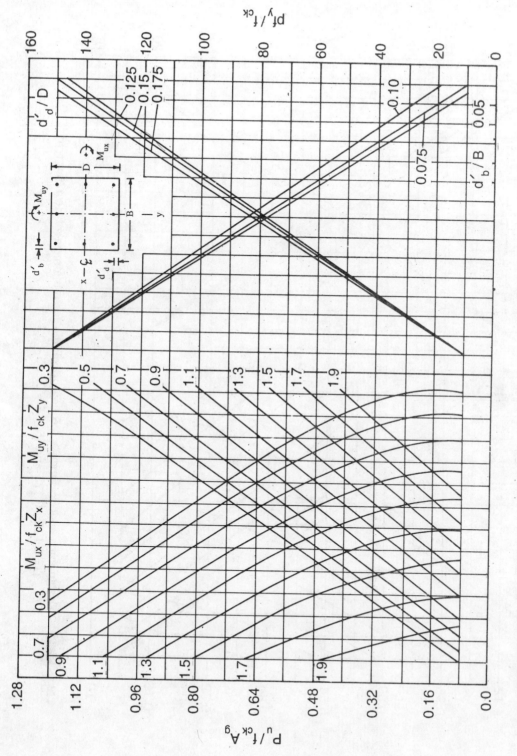

Chart R2.2.3 Design curves for rectangular section of column for reinforcement detailing type 2 ($d'_d/D = 0.125 - 0.175$ and $d'_b/B = 0.05 - 0.10$)

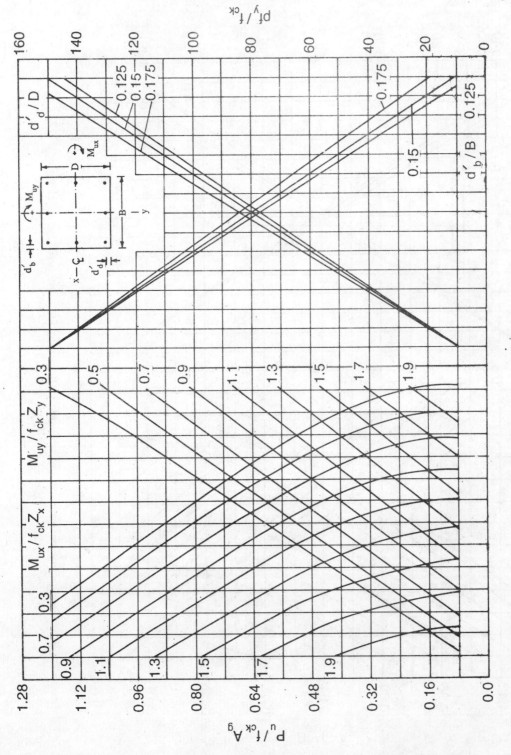

Chart R2.2.4 Design curves for rectangular section of column for reinforcement detailing type 2 ($d'_d / D = 0.125 - 0.175$ and $d'_b / B = 0.125 - 0.175$)

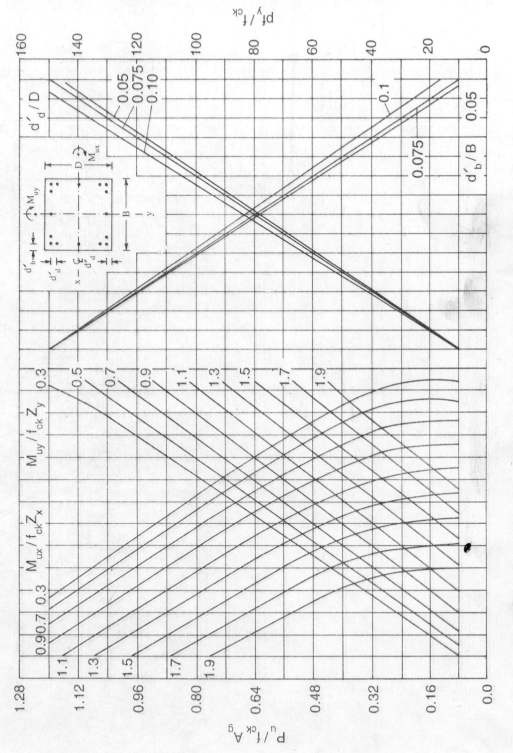

Chart R2.3.1 Design curves for rectangular section of column for reinforcement detailing type 3 ($d'_d/D = 0.05 - 0.10$ and $d'_b/B = 0.05 - 0.10$)

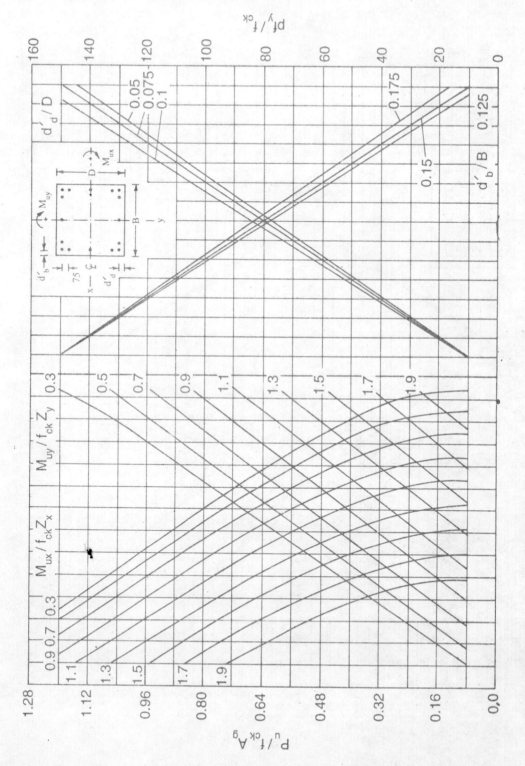

Chart R2.3.2 Design curves for rectangular section of column for reinforcement detailing type 3. ($d'_d / D = 0.05 - 0.10$ and $d'_b / B = 0.125 - 0.175$)

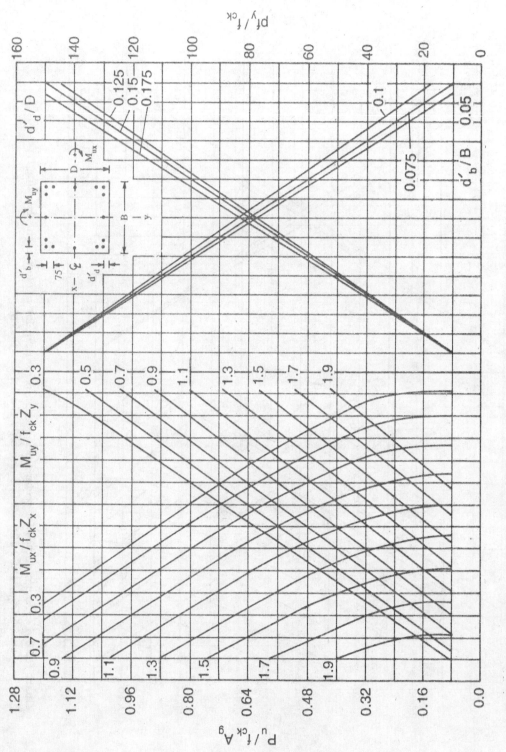

Chart R2.3.3 Design curves for rectangular section of column for reinforcement detailing type 3 ($d'_d / D = 0.125 - 0.175$ and $d'_b / B = 0.05 - 0.10$)

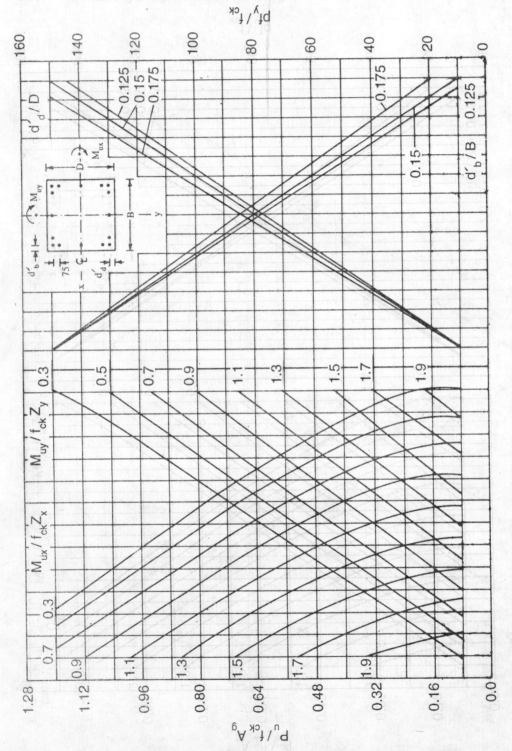

Chart R2.3.4 Design curves for rectangular section of column for reinforcement detailing type 3 ($d'_d / D = 0.125 - 0.175$ and $d'_b / B = 0.125 - 0.175$)

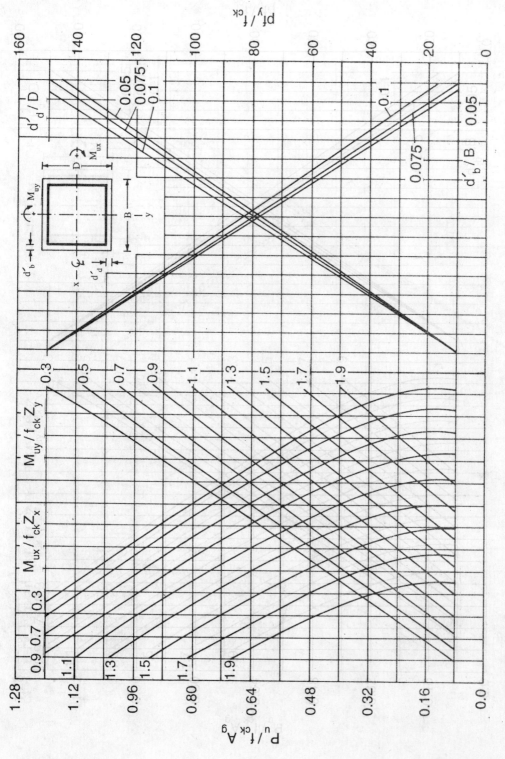

Chart R2.4.1 Design curves for rectangular section of column for reinforcement detailing type 4 ($d'_d / D = 0.05 - 0.10$ and $d'_b / B = 0.05 - 0.10$).

216 Handbook of Reinforced Concrete Design

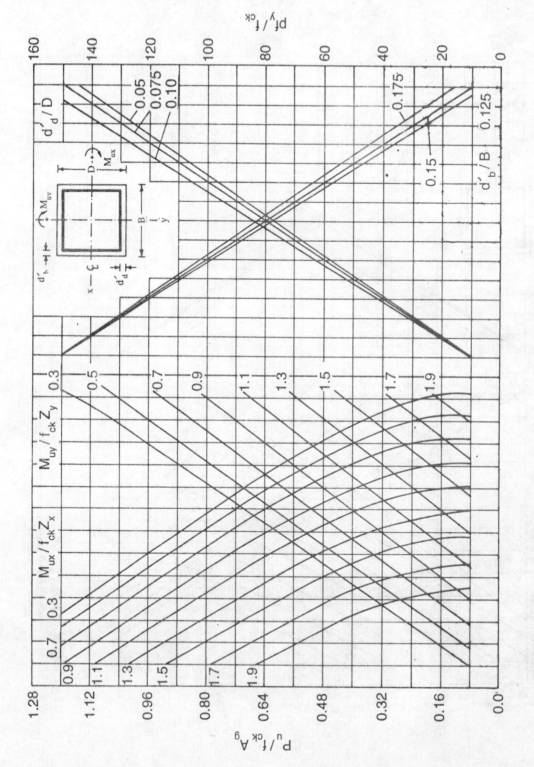

Chart R2.4.2 Design curves for rectangular section of column for reinforcement detailing type 4 ($d'_d / D = 0.05 – 0.10$ and $d'_b / B = 0.125 – 0.175$)

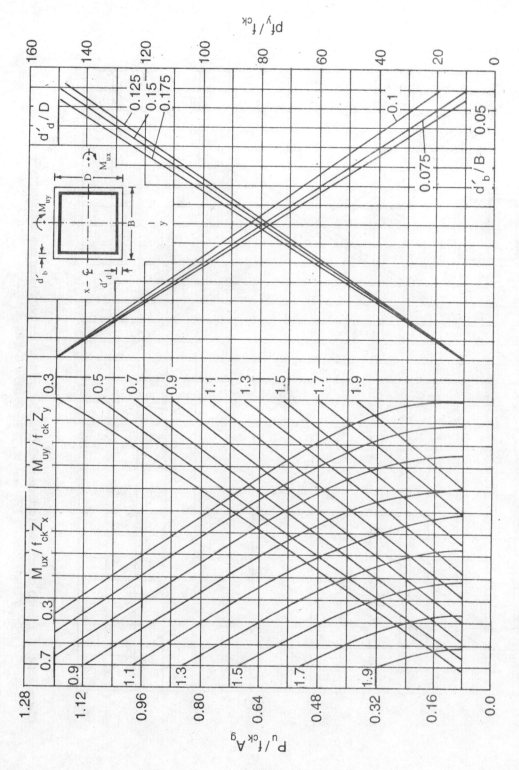

Chart R2.4.3 Design curves for rectangular section of column for reinforcement detailing type 4· ($d'_d / D = 0.125 - 0.175$ and $d'_b / B = 0.05 - 0.10$)

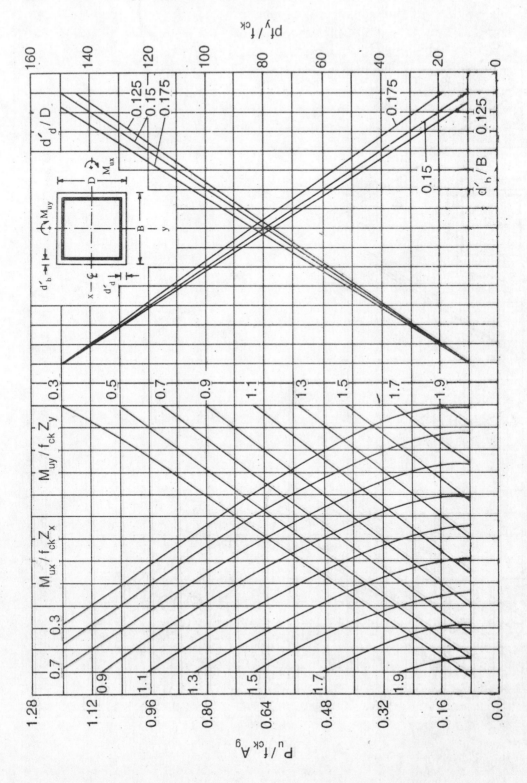

Chart R2.4.4 Design curves for rectangular section of column for reinforcement detailing type 4 ($d'_d / D = 0.125 - 0.175$ and $d'_b / B = 0.125 - 0.175$)

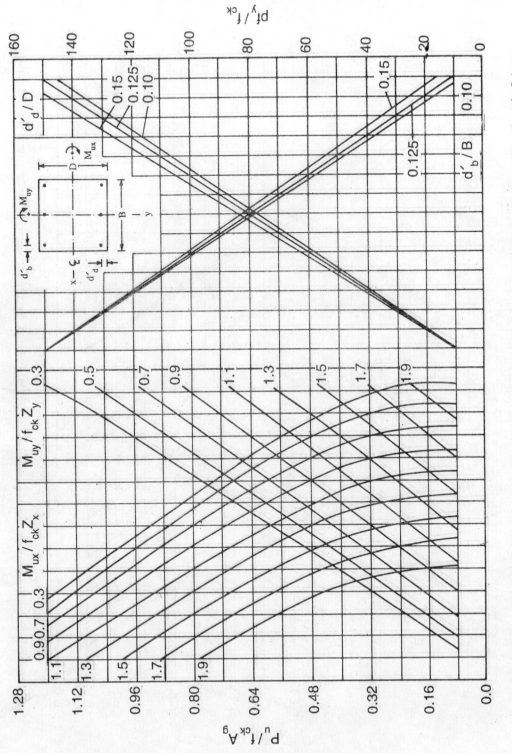

Chart R2.5.1 Design curves for rectangular section of column for reinforcement detailing type 5 with 6 bars ($d'_d / D = 0.10 - 0.15$ and $d'_b / B = 0.10 - 0.15$)

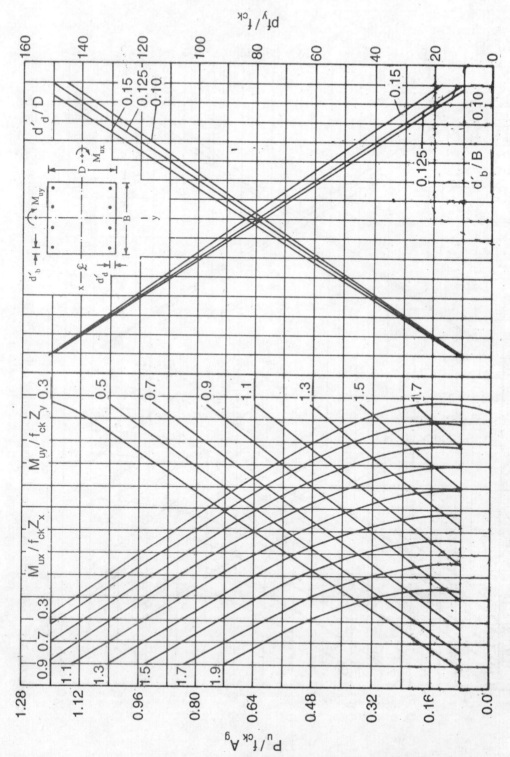

Chart R2.5.2 Design curves for rectangular section of column for reinforcement detailing type 5 with 8 bars ($d'_d / D = 0.10 - 0.15$ and $d'_b / B = 0.1 - 0.15$)

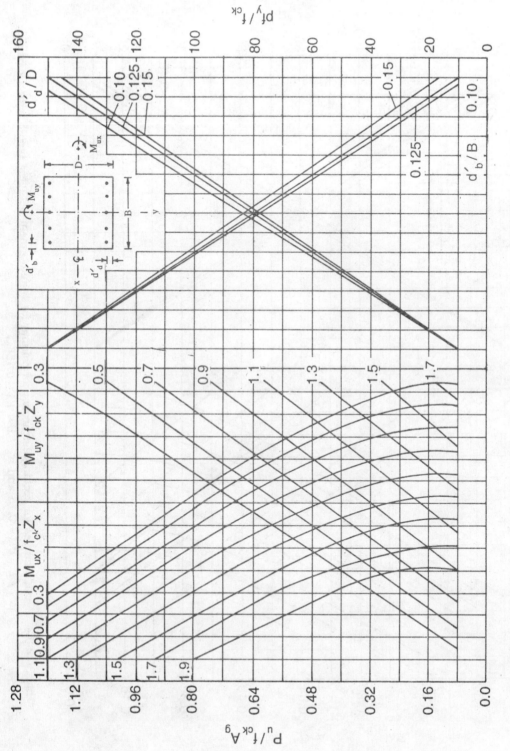

Chart R2.5.3 Design curves for rectangular section of column for reinforcement detailing type 5 with 10 bars ($d'_d / D = 0.10 - 0.15$ and $d'_b / B = 0.1 - 0.15$)

222 Handbook of Reinforced Concrete Design

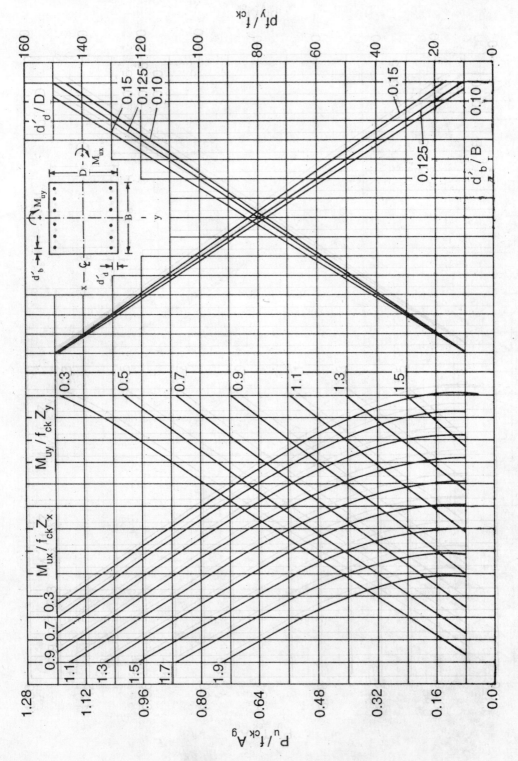

Chart R2.5.4 Design curves for rectangular section of column for reinforcement detailing type 5 with 12 bars ($d'_d/D = 0.1 - 0.15$ and $d'_b/B = 0.1 - 0.15$)

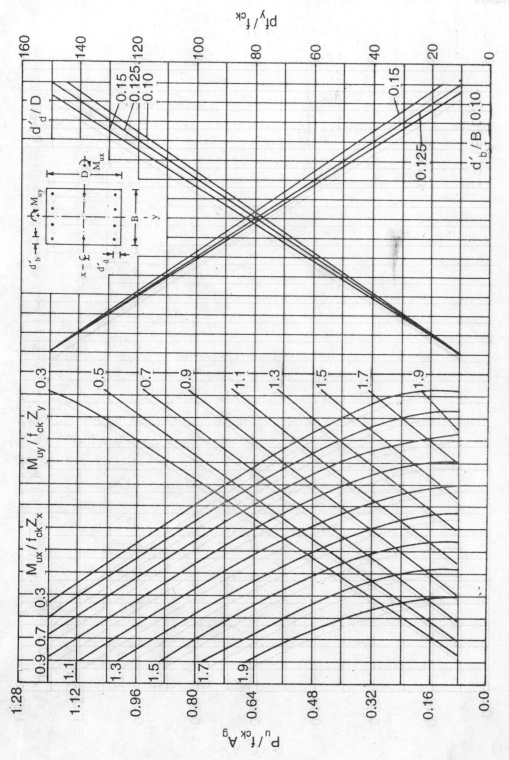

Chart R2.6 Design curves for rectangular section of column for reinforcement detailing type 6 ($d'_d/D = 0.10 - 0.15$ and $d'_b/B = 0.10 - 0.15$)

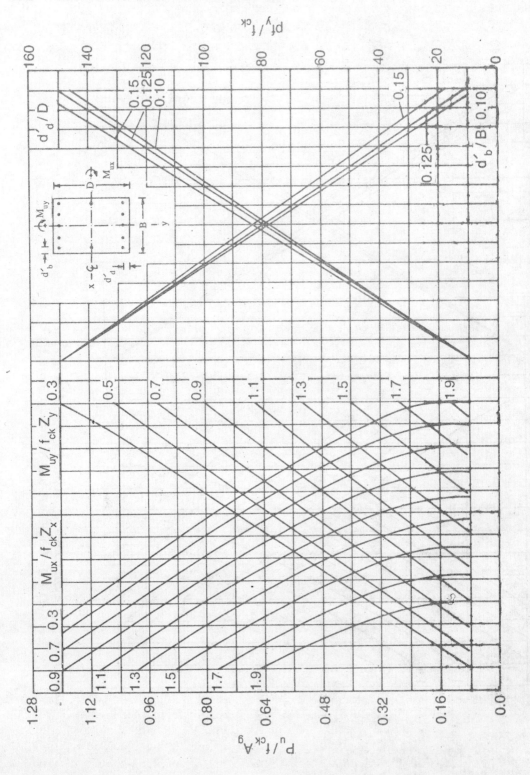

Chart R2.7 Design curves for rectangular section of column for reinforcement detailing type 7 ($d'_d/D = 0.10 - 0.15$ and $d'_b/B = 0.10 - 0.15$)

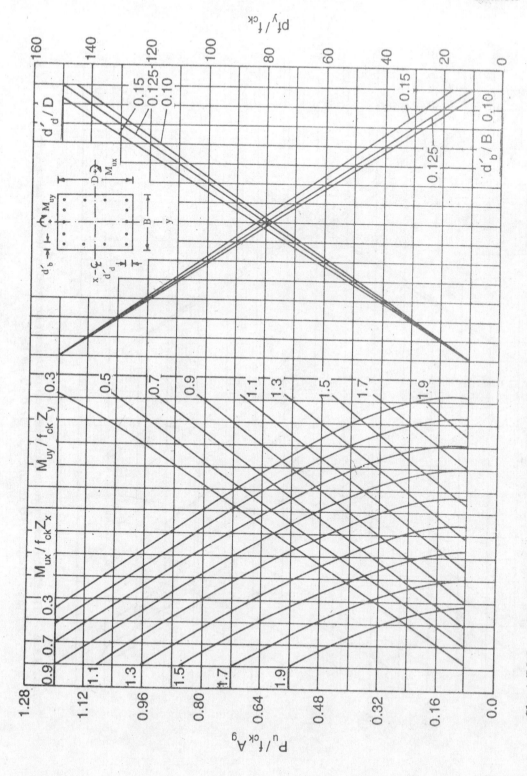

Chart R2.8 Design curves for rectangular section of column for reinforcement detailing type 8 ($d'_d/D = 0.10 - 0.15$ and $d'_b/B = 0.10 - 0.15$).

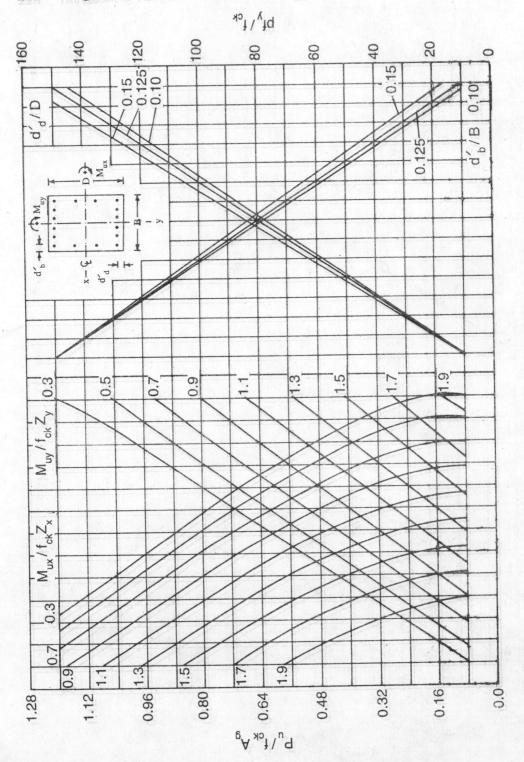

Chart R2.9 Design curves for rectangular section of column for reinforcement detailing type 9 ($d'_d/D = 0.10 - 0.15$ and $d'_b/B = 0.10 - 0.15$)

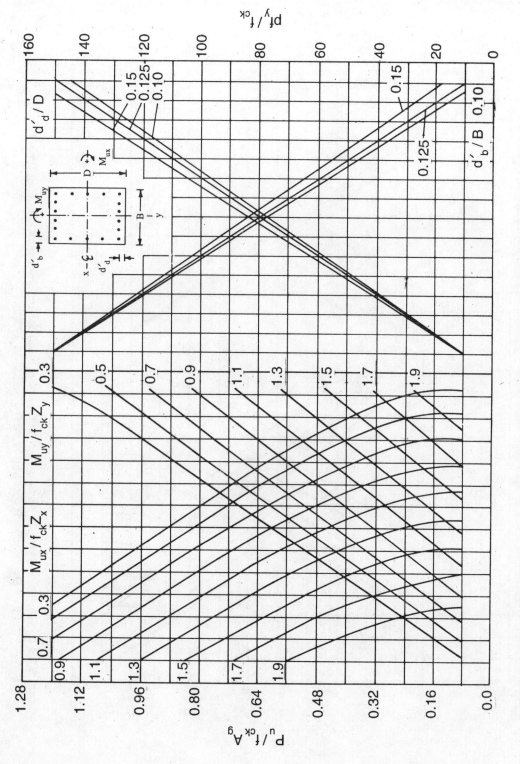

Chart R2.10 Design curves for rectangular section of column for reinforcement detailing type 10 ($d'_d / D = 0.10 - 0.15$ and $d'_b / B = 0.10 - 0.15$)

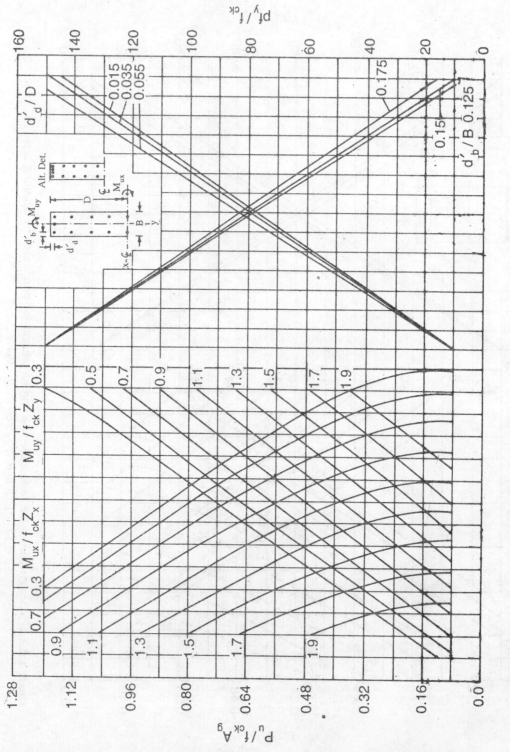

Chart R2.11 Design curves for rectangular section of column for reinforcement detailing type 11 ($d'_d/D = 0.015 - 0.055$ and $d'_b/B = 0.125 - 0.175$)

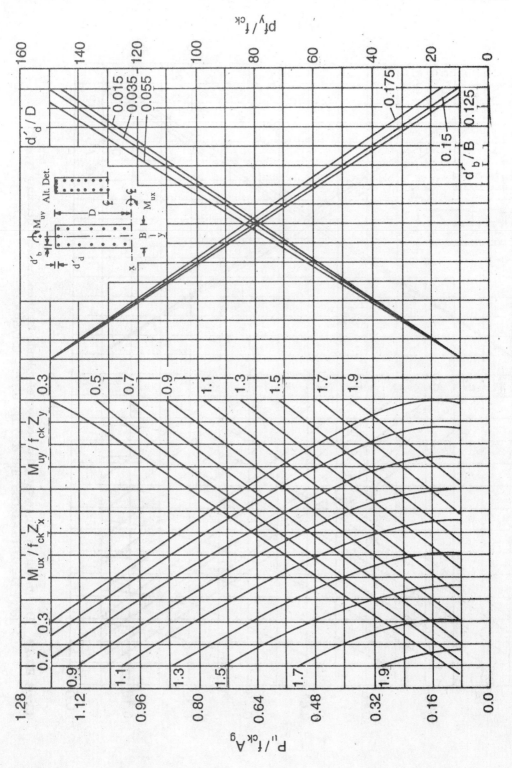

Chart R2.12 Design curves for rectangular section of column for reinforcement detailing type 12 ($d'_d/D = 0.015 - 0.055$ and $d'_b/B = 0.125 - 0.175$)

230 Handbook of Reinforced Concrete Design

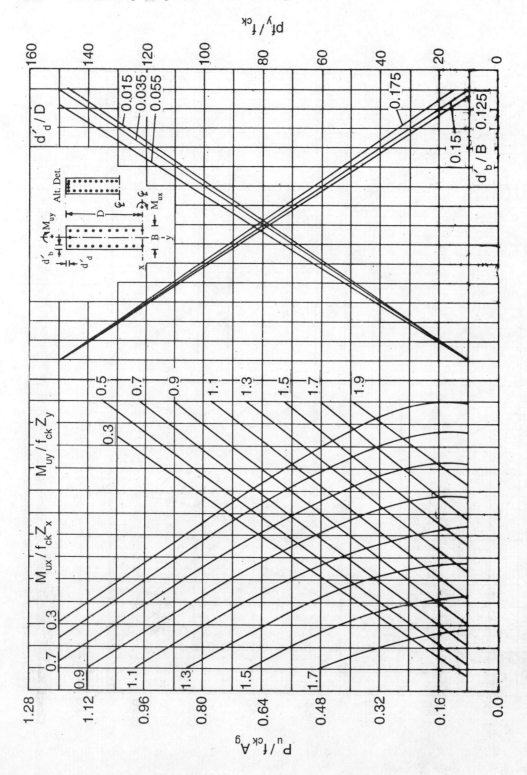

Chart R2.13 Design curves for rectangular section of column for reinforcement detailing type 13 ($d'_d/D = 0.015 - 0.055$ and $d'_b/B = 0.125 - 0.175$)

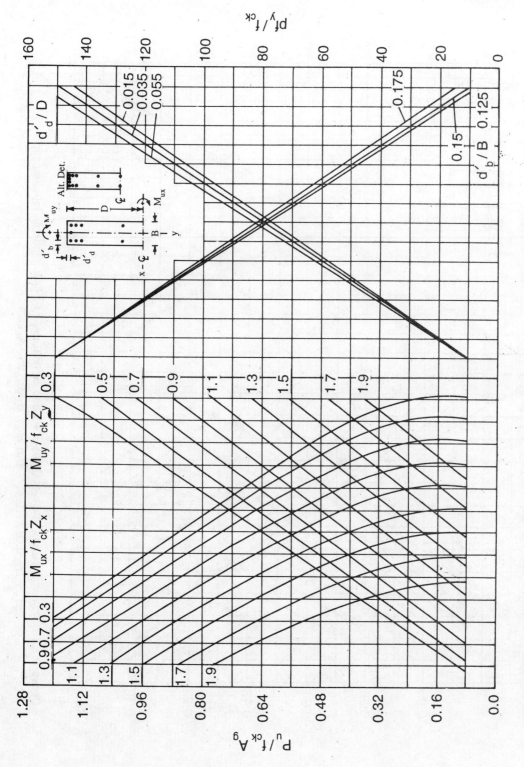

Chart R2.14 Design curves for rectangular section of column for reinforcement detailing type 14 ($d'_d / D = 0.015 - 0.055$ and $d'_b / B = 0.125 - 0.175$)

232 Handbook of Reinforced Concrete Design

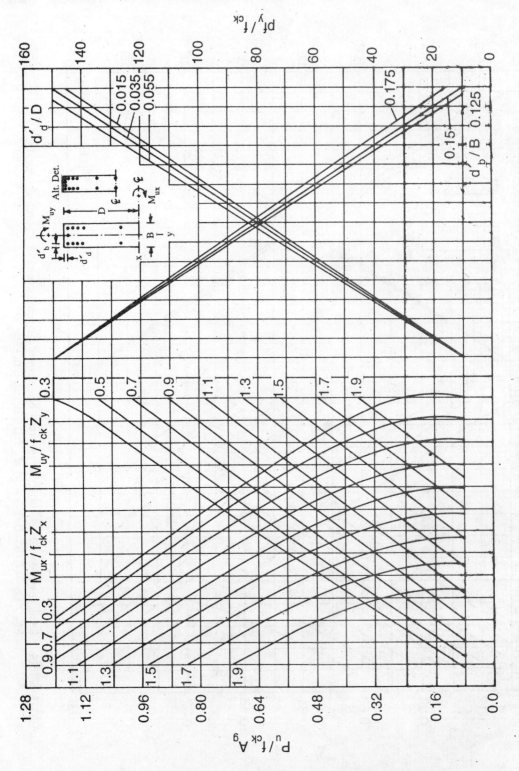

Chart R2.15 Design curves for rectangular section of column for reinforcement detailing type 15 ($d'_d / D = 0.015 - 0.055$ and $d'_b / B = 0.125 - 0.175$)

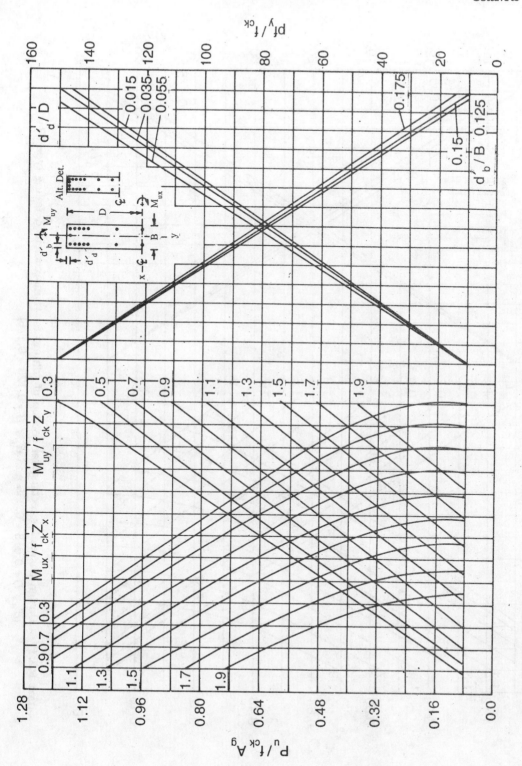

Chart R2.16 Design curves for rectangular section of column for reinforcement detailing type 16 ($d'_d/D = 0.015 - 0.055$ and $d'_b/B = 0.125 - 0.175$)

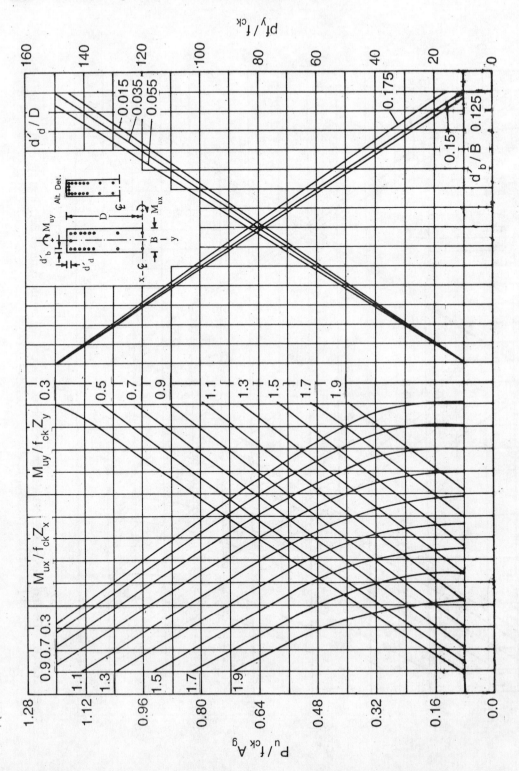

Chart R2.17 Design curves for rectangular section of column for reinforcement detailing type 17 ($d'_d/D = 0.015 - 0.055$ and $d'_b/B = 0.125 - 0.175$)

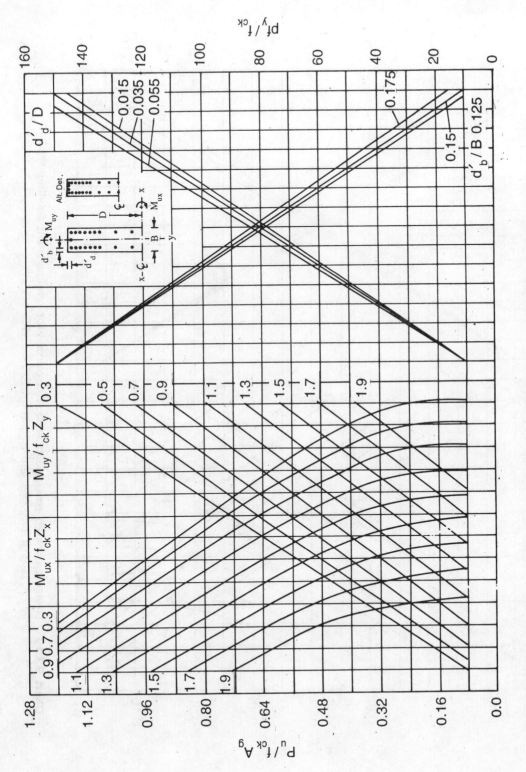

Chart R2.18 Design curves for rectangular section of column for reinforcement detailing type 18 ($d'_d / D = 0.015 - 0.055$ and $d'_b / B = 0.125 - 0.175$)

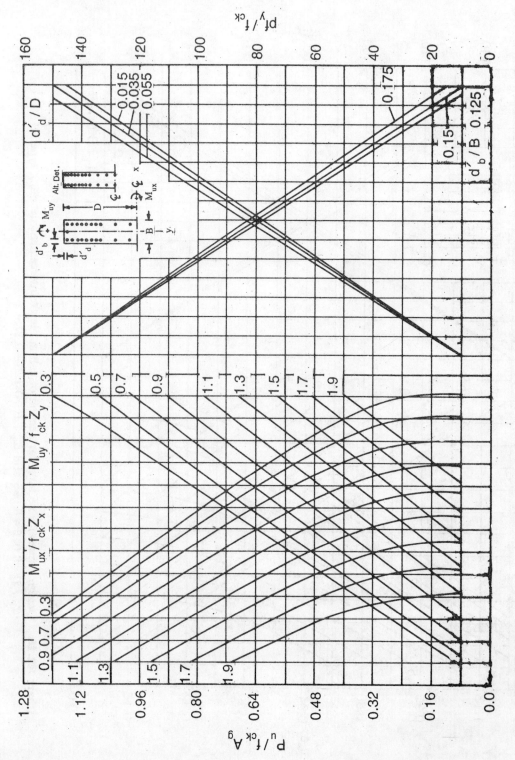

Chart R2.19 Design curves for rectangular section of column for reinforcement detailing type 19 ($d'_d / D = 0.015 - 0.055$ and $d'_b / B = 0.125 - 0.175$)

5.5 L-SECTION OF COLUMNS

The L-section of a column is geometrically antisymmetrical about x-y axis as shown in Fig. 5.24. Interaction curves for biaxial moments for such a typical section are shown in the figure. It is antisymmetrical about x-y axis. Therefore, only half of the interaction curves either above or below the x-y axis may be considered. The use of interaction curves above the x-y axis can be made for moments falling below the x-y axis by reversing their sign. Similarly the use of interaction curves below the x-y axis can be made for moments falling above the x-y axis by reversing their sign.

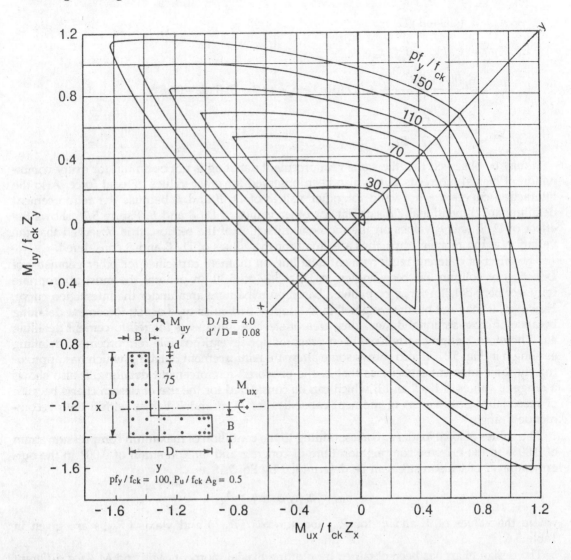

Fig. 5.24 Interaction curves for biaxial moments for a typical L-section of column

The L-section of a column may have large variation in the geometry defined by the ratio of depth (D) and width (B) of the section. It may have many possible ways of reinforcement detailing. Different reinforcement detailing for a typical geometry of L-section is shown in Fig. 5.25. For studying the effect of these reinforcement detailing on moment capacities, interaction curves for biaxial moments for a particular value of axial force have been shown in the figure. For interaction curves, moments M_{ux} and M_{uy}, axial force P_u and area of steel A_s have been expressed in non-dimensional form as $P_u/f_{ck}A_g$, $M_{ux}/f_{ck}Z_x$, $M_{uy}/f_{ck}Z_y$ and pf_y/f_{ck} respectively where,

$$p = 100\,A_s/A_g$$

$$A_g = (2D-B)D$$

$$Z_x, Z_y = I_{xx}/y_o = I_{yy}/x_o$$

$$I_{xx}, I_{yy} = \frac{BD^3 + DB^3 - B^4}{12} + BD(0.5D - y_o)^2 + (D-B)B(y_o - 0.5B)^2$$

$$x_o, y_o = \frac{D^2 + (D-B)B}{2(2D-B)}$$

It may be observed that the same reinforcement detailing is not optimum for every combination of biaxial moments. It has also been observed for other values of axial force. Also the interaction curves are not valid for other values of D/B and B because the reinforcement detailing are dimensionally dissimilar for other values of D/B and B. Figure 5.26 shows the effect of D/B and B values on the moment capacities of the section. It is observed that the variation in the moment capacities are small for the values of D/B and B considered.

The effect of different reinforcement detailing on moment capacities for other geometry of L-section of column has been shown in Figs 5.27 to 5.30. In general the most appropriate reinforcement detailing may be evolved based on maximum area under the interaction curve. However, it may not be always possible to adopt a particular type of reinforcement detailing because of cross-sectional dimensions, area of steel, requirements for reinforcement detailing etc. Therefore, design charts have been prepared for several possible reinforcement detailing as shown in Fig. 5.31. It also shows some alternate reinforcement detailing which has approximately same moment capacity as that of the proposed reinforcement detailing. It also shows a range of values of D/B and B which can be considered for the use of design charts because of reasonably close values of moment capacities for D/B and B values within their recommended range.

The values of axial load, P_{ub} corresponding to the condition of maximum compression strain of 0.0035 in the extreme compression fibre of concrete and tension strain of 0.002 in the outermost layer of tenison steel can be determined by Eq. 5.18 as,

$$P_{ub} = q_c f_{ck} A_g + q_s\, p\, A_g$$

where the values of q_c and q_s for P_{ub} about x-axis (P_{ubx}) and y-axis (P_{uby}) are given in Table 5.6.

The design chart has been obtained by plotting biaxial moments M_{ux} and M_{uy} for different values of area of steel A_s and a particular value of axial force P_u expressed in non-dimensional

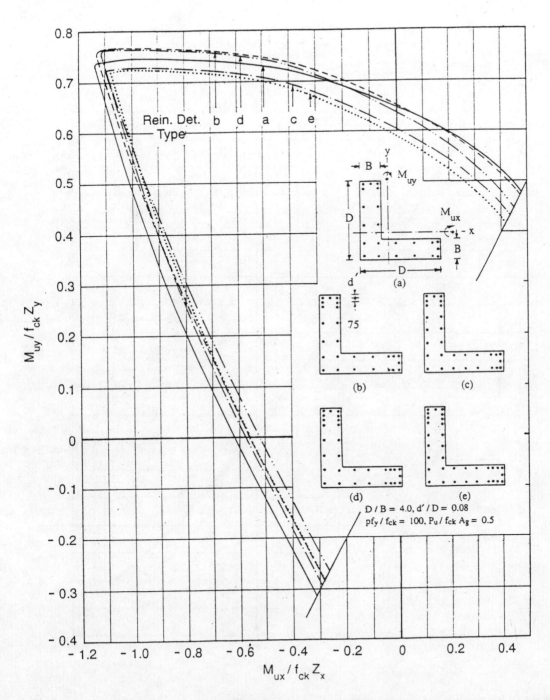

Fig. 5.25 Interaction curves for different types of reinforcement detailing for a typical L-section of column ($D/B = 4.0$)

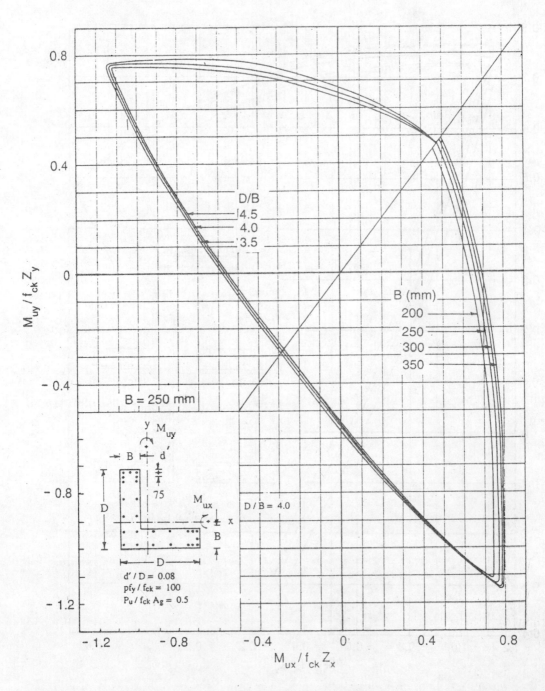

Fig. 5.26 Interaction curves for different values of D/B and B for a typical L section of column

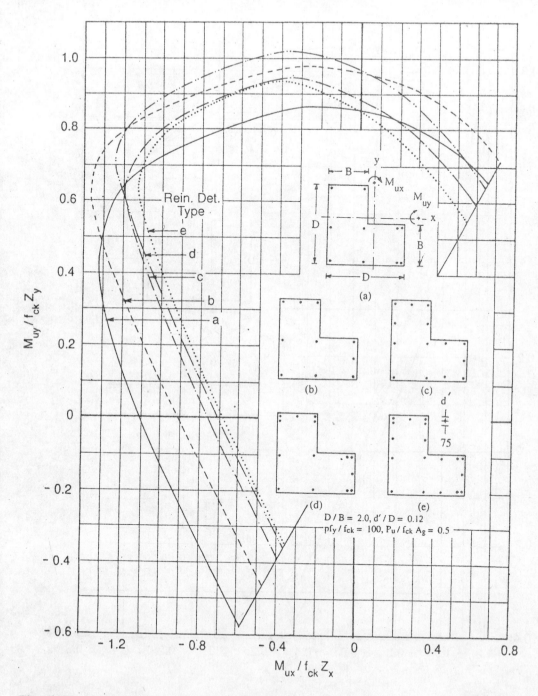

Fig. 5.27 Interaction curves for different types of reinforcement detailing for a typical L-section of column ($D/B = 2.0$)

242 *Handbook of Reinforced Concrete Design*

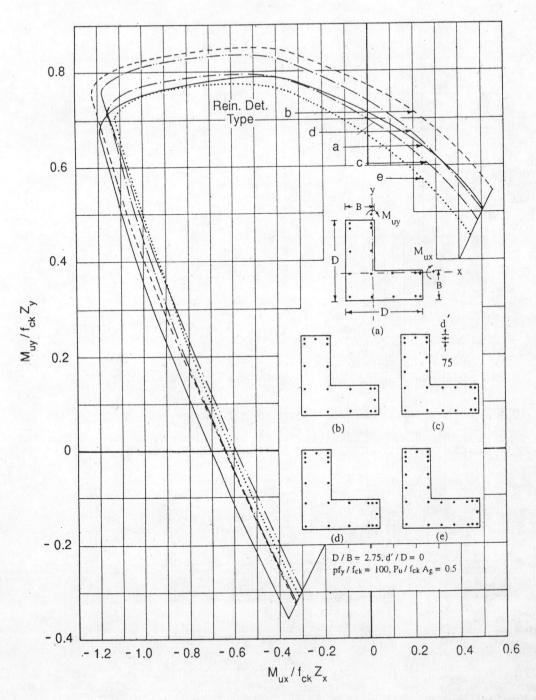

Fig. 5.28 Interaction curves for different types of reinforcement detailing for a typical L-section of column ($D/B = 2.75$)

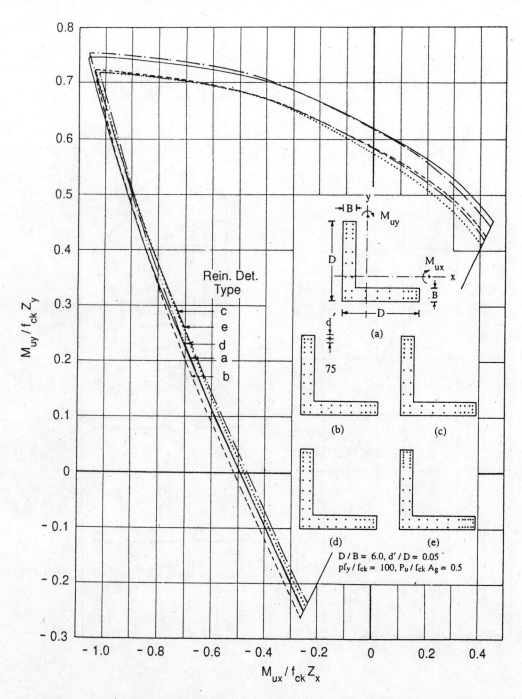

Fig. 5.29 Interaction curves for different types of reinforcement detailing for a typical L-section of column ($D/B = 6.0$)

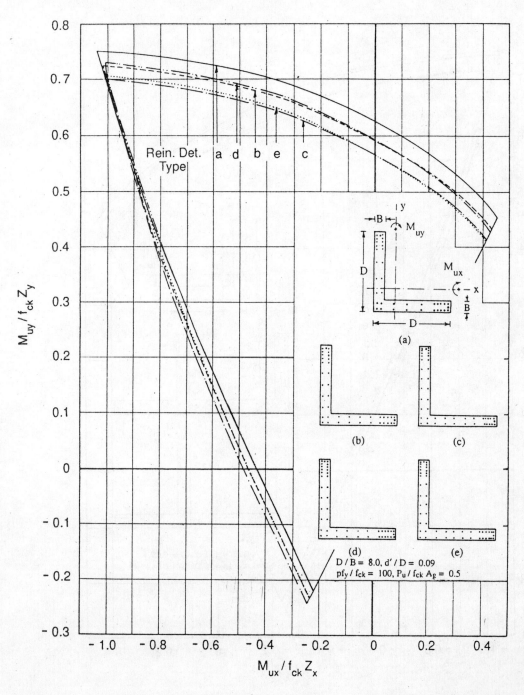

Fig. 5.30 Interaction curves for different types of reinforcement detailing for a typical L-section of column ($D/B = 8.0$)

Columns 245

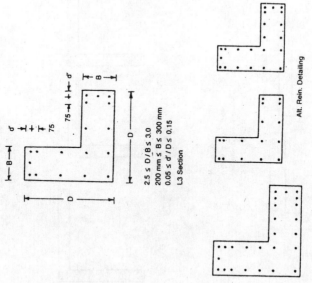

2.5 ≤ D / B ≤ 3.0
200 mm ≤ B ≤ 300 mm
0.05 ≤ d' / D ≤ 0.15
L3 Section

2.5 ≤ D / B ≤ 3.0
200 mm ≤ B ≤ 300 mm
0.05 ≤ d' / D ≤ 0.15
L4 Section Alt. Rein. Detailing

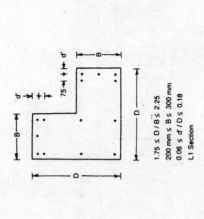
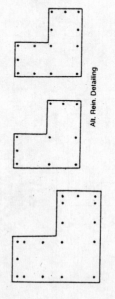

1.75 ≤ D / B ≤ 2.25
200 mm ≤ B ≤ 300 mm
0.06 ≤ d' / D ≤ 0.18
L1 Section

1.75 ≤ D / B ≤ 2.25
200 mm ≤ B ≤ 300 mm
0.06 ≤ d' / D ≤ 0.18
L2 Section Alt. Rein. Detailing

(contd.)

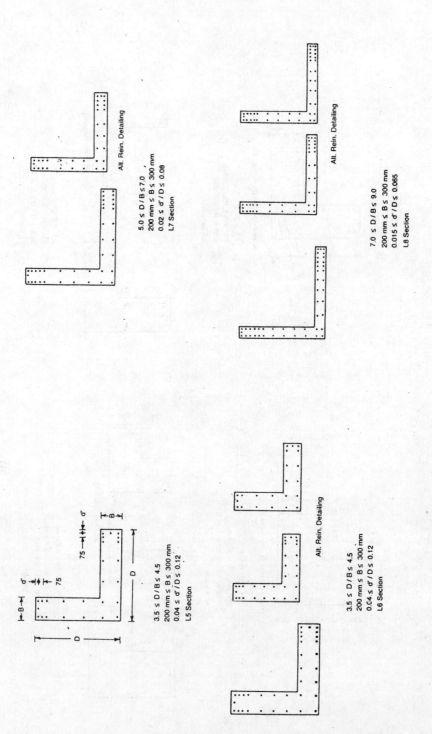

Fig. 5.31 Reinforcement detailing for different geometry of L section of column

Table 5.6 Values of q_c and q_s for L-shape of column

Cross-sections	Values of $\frac{d'}{D}$	Values of q_c	Values of q_c and q_s for P_{ubx} and P_{uby} for positive moments M_{ux} and M_{uy}								
			Values of q_s for concrete and steel of grade								
			M20			M25			M30		
			Fe 415	Fe 500	Fe 550	Fe 415	Fe 500	Fe 550	Fe 415	Fe 500	Fe 550
L1 Section	0.0600	0.1512	0.2707	0.3662	0.4245	0.2614	0.3569	0.4152	0.2521	0.3476	0.4059
	0.0900	0.1448	0.0748	0.1523	0.2022	0.0669	0.1443	0.1942	0.0589	0.1363	0.1862
	0.1200	0.1375	-0.1443	-0.0909	-0.0671	-0.1523	-0.0988	-0.0750	-0.1602	-0.1068	-0.0829
	0.1500	0.1319	-0.3946	-0.3854	-0.3735	-0.4024	-0.3932	-0.3813	-0.4102	-0.4011	-0.3891
	0.1800	0.1260	-0.6923	-0.7121	-0.7115	-0.6999	-0.7196	-0.7191	-0.7074	-0.7272	-0.7266
L2 Section	0.0600	0.1512	0.3209	0.3700	0.4046	0.3113	0.3604	0.3950	0.3017	0.3508	0.3854
	0.0900	0.1448	0.0818	0.1168	0.1449	0.0738	0.1088	0.1369	0.0657	0.1007	0.1288
	0.1200	0.1375	-0.1821	-0.1658	-0.1579	-0.1898	-0.1734	-0.1656	-0.1974	-0.1810	-0.1732
	0.1500	0.1319	-0.4764	-0.4945	-0.4959	-0.4835	-0.5015	-0.5029	-0.4905	-0.5085	-0.5099
	0.1800	0.1260	-0.8144	-0.8550	-0.8653	-0.8206	-0.8612	-0.8715	-0.8267	-0.8673	-0.8776
L3 Section	0.0500	0.1334	-0.1309	-0.0776	-0.0337	-0.1390	-0.0857	-0.0418	-0.1471	-0.0938	-0.0500
	0.0750	0.1299	-0.3413	-0.2939	-0.2558	-0.3491	-0.3017	-0.2636	-0.3569	-0.3095	-0.2715
	0.1000	0.1264	-0.5622	-0.5264	-0.4960	-0.5696	-0.5339	-0.5035	-0.5771	-0.5413	-0.5109
	0.1250	0.1229	-0.7914	-0.7776	-0.7553	-0.7984	-0.7847	-0.7624	-0.8054	-0.7917	-0.7694
	0.1500	0.1194	-0.9965	-1.0474	-1.0427	-1.0029	-1.0539	-1.0491	-1.0094	-1.0603	-1.0556

(Contd.)

Table 5.6 (Contd.)

Cross-sections	Values of $\frac{d'}{D}$	Values of q_c	Values of q_c and q_s for P_{ubx} and P_{uby} for positive moments M_{ubx} and M_{uy}								
			Values of q_s for concrete and steel of grade								
			M20			M25			M30		
			Fe 415	Fe 500	Fe 550	Fe 415	Fe 500	Fe 550	Fe 415	Fe 500	Fe 550
L4 Section	0.0500	0.1334	−0.1094	−0.0739	−0.0451	−0.1174	−0.0819	−0.0530	−0.1253	−0.0898	−0.0609
	0.0750	0.1299	−0.3373	−0.3125	−0.2996	−0.3447	−0.3199	−0.3069	−0.3520	−0.3272	−0.3143
	0.1000	0.1264	−0.5801	−0.5814	−0.5752	−0.5871	−0.5884	−0.5821	−0.5940	−0.5953	−0.5891
	0.1250	0.1229	−0.8462	−0.8716	−0.8710	−0.8527	−0.8781	−0.8775	−0.8592	−0.8846	−0.8840
	0.1500	0.1194	−1.0995	−1.1812	−1.1927	−1.1055	−1.1872	−1.1987	−1.1115	−1.1932	−1.2047
L5 Section	0.0400	0.1260	−0.4202	−0.3747	−0.3333	−0.4279	−0.3824	−0.3410	−0.4356	−0.3901	−0.3487
	0.0600	0.1234	−0.5988	−0.5606	−0.5269	−0.6062	−0.5680	−0.5343	−0.6135	−0.5754	−0.5417
	0.0800	0.1208	−0.7814	−0.7612	−0.7298	−0.7886	−0.7684	−0.7370	−0.7959	−0.7756	−0.7442
	0.1000	0.1182	−0.9049	−0.9675	−0.9470	−0.9120	−0.9746	−0.9540	−0.9190	−0.9816	−0.9611
	0.1200	0.1155	−1.0203	−1.1347	−1.1779	−1.0271	−1.1415	−1.1847	−1.0340	−1.1483	−1.1915
L6 Section	0.0400	0.1260	−0.4435	−0.4091	−0.3779	−0.4511	−0.4167	−0.3855	−0.4587	−0.4242	−0.3930
	0.0600	0.1234	−0.6320	−0.6070	−0.5873	−0.6392	−0.6143	−0.5946	−0.6465	−0.6216	−0.6018
	0.0800	0.1208	−0.8249	−0.8220	−0.8093	−0.8318	−0.8289	−0.8162	−0.8388	−0.8358	−0.8231
	0.1000	0.1182	−0.9630	−1.0499	−1.0456	−0.9696	−0.0565	−1.0522	−0.9762	−1.0631	−1.0588
	0.1200	0.1155	−1.1010	−1.2403	−1.2949	−1.1073	−1.2467	−1.3012	−1.1137	−1.2530	−1.3076

(Contd.)

Table 5.6 (Contd.)

Cross-sections	Values of $\frac{d'}{D}$	Values of q_c	Values of q_c and q_s for P_{ubx} and P_{uby} for positive moments M_{ux} and M_{uy}									
			Values of q_s for concrete and steel of grade									
			M20				M25				M30			
			Fe 415	Fe 500	Fe 550	Fe 415	Fe 500	Fe 550	Fe 415	Fe 500	Fe 550
L7 Section	0.2000	0.1228	−0.4857	−0.4284	−0.3816	−0.4937	−0.4364	−0.3897	−0.5017	−0.4444	−0.3977
	0.0350	0.1209	−0.6167	−0.5679	−0.5230	−0.6245	−0.5757	−0.5308	−0.6324	−0.5836	−0.5387
	0.0500	0.1191	−0.7078	−0.7104	−0.6729	−0.7154	−0.7181	−0.6805	−0.7231	−0.7258	−0.6882
	0.0650	0.1172	−0.7888	−0.8459	−0.8284	−0.7962	−0.8534	−0.8359	−0.8037	−0.8608	−0.8433
	0.0800	0.1153	−0.8697	−0.9574	−0.9821	−0.8770	−0.9647	−0.9893	−0.8842	−0.9719	−0.9966
L8 Section	0.0150	0.1198	−0.5860	−0.5462	−0.4972	−0.5940	−0.5541	−0.5052	−0.6020	−0.5621	−0.5132
	0.0275	0.1183	−0.6518	−0.6620	−0.6167	−0.6596	−0.6698	−0.6245	−0.6674	−0.6777	−0.6324
	0.0400	0.1168	−0.7138	−0.7607	−0.7413	−0.7215	−0.7685	−0.7491	−0.7293	−0.7762	−0.7568
	0.0525	0.1153	−0.7769	−0.8470	−0.8568	−0.7845	−0.8547	−0.8645	−0.7921	−0.8623	−0.8721
	0.0650	0.1137	−0.8373	−0.9246	−0.9531	−0.8449	−0.9322	−0.9607	−0.8524	−0.9397	−0.9682

(Contd.)

Table 5.6 (Contd.)

Cross-sections	Values of $\frac{d'}{D}$	Values of q_c	Values of q_c and q_s for P_{ubx} and P_{uby} for negative moments M_{ux} and M_{uy}									
			Values of q_s for concrete and steel of grade									
			M20			M25			M30			
			Fe 415	Fe 500	Fe 550	Fe 415	Fe 500	Fe 550	Fe 415	Fe 500	Fe 550	
L1 Section	0.0600	0.2808	0.7059	0.8391	0.9156	0.6934	0.8266	0.9031	0.6809	0.8141	0.8906	
	0.0900	0.2734	0.7619	0.8833	0.9551	0.7491	0.8705	0.9423	0.7363	0.8577	0.9295	
	0.1200	0.2669	0.8214	0.9280	0.9905	0.8083	0.9149	0.9774	0.7952	0.9017	0.9643	
	0.1500	0.2587	0.8744	0.9684	1.0264	0.8610	0.9550	1.0130	0.8476	0.9416	0.9996	
	0.1800	0.2508	0.9274	1.0120	1.0507	0.9138	0.9983	1.0370	0.9001	0.9847	1.0234	
L2 Section	0.0600	0.2808	0.8188	0.9608	1.0403	0.8070	0.9489	1.0285	0.7951	0.9371	1.0167	
	0.0900	0.2734	0.8859	1.0166	1.0918	0.8737	1.0045	1.0796	0.8616	0.9923	1.0675	
	0.1200	0.2669	0.9568	1.0734	1.1399	0.9443	1.0609	1.1274	0.9318	1.0484	1.1148	
	0.1500	0.2587	1.0222	1.1269	1.1894	1.0092	1.1139	1.1764	0.9963	1.1010	1.1635	
	0.1800	0.2508	1.0887	1.1845	1.2291	1.0753	1.1711	1.2157	1.0619	1.1578	1.2023	
L3 Section	0.0500	0.3049	1.1699	1.3914	1.4805	1.1560	1.3775	1.4667	1.1422	1.3637	1.4528	
	0.0750	0.3020	1.1988	1.4413	1.5361	1.1848	1.4273	1.5220	1.1707	1.4132	1.5080	
	0.1000	0.2986	1.2294	1.4736	1.5920	1.2152	1.4593	1.5778	1.2009	1.4451	1.5635	
	0.1250	0.2948	1.2540	1.5014	1.6270	1.2396	1.4870	1.6126	1.2252	1.4726	1.5982	
	0.1500	0.2906	1.2713	1.5209	1.6568	1.2568	1.5063	1.6422	1.2423	1.4918	1.6277	

(Contd.)

Table 5.6 (Contd.)

Values of q_c and q_s for P_{ubx} and P_{uby} for negative moments M_{ux} and M_{uy}

Values of q_s for concrete and steel of grade

Cross-sections	Values of $\dfrac{d'}{D}$	Values of q_c	M20			M25			M30		
			Fe 415	Fe 500	Fe 550	Fe 415	Fe 500	Fe 550	Fe 415	Fe 500	Fe 550
L4 Section	0.0500	0.3049	1.3125	1.5396	1.6301	1.2985	1.5256	1.6161	1.2845	1.5116	1.6021
	0.0750	0.3020	1.3527	1.6003	1.6964	1.3385	1.5860	1.6821	1.3242	1.5718	1.6679
	0.1000	0.2986	1.3953	1.6445	1.7637	1.3808	1.6300	1.7492	1.3663	1.6155	1.7347
	0.1250	0.2948	1.4328	1.6851	1.8115	1.4181	1.6705	1.7968	1.4035	1.6558	1.7822
	0.1500	0.2906	1.4641	1.7185	1.8550	1.4493	1.7037	1.8402	1.4345	1.6889	1.8254
L5 Section	0.0400	0.3129	1.3419	1.6354	1.8005	1.3276	1.6211	1.7862	1.3133	1.6069	1.7719
	0.0600	0.3155	1.3679	1.6629	1.8283	1.3535	1.6485	1.8139	1.3391	1.6341	1.7995
	0.0800	0.3088	1.3928	1.6861	1.8547	1.3783	1.6716	1.8402	1.3637	1.6570	1.8257
	0.1000	0.3108	1.4181	1.7086	1.8746	1.4035	1.6940	1.8599	1.3888	1.6793	1.8453
	0.1200	0.3039	1.4441	1.7316	1.8936	1.4294	1.7168	1.8788	1.4146	1.7021	1.8640
L6 Section	0.0400	0.3129	1.4743	1.7775	1.9419	1.4597	1.7629	1.9273	1.4451	1.7483	1.9128
	0.0600	0.3155	1.5021	1.8112	1.9760	1.4873	1.7965	1.9613	1.4726	1.7818	1.9466
	0.0800	0.3088	1.5275	1.8410	2.0090	1.5127	1.8262	1.9942	1.4979	1.8114	1.9793
	0.1000	0.3108	1.5533	1.8681	2.0359	1.5383	1.8531	2.0209	1.5234	1.8382	2.0060
	0.1200	0.3039	1.5798	1.8935	2.0622	1.5647	1.8784	2.0472	1.5496	1.8633	2.0321

(Contd.)

Table 5.6 (*Contd.*)

Cross-sections	Values of $\frac{d'}{D}$	Values of q_c	Values of q_c and q_s for P_{ubx} and P_{uby} for negative moments M_{ux} and M_{uy}									
			Values of q_s for concrete and steel of grade										
			M20			M25			M30		
			Fe 415	Fe 500	Fe 550	Fe 415	Fe 500	Fe 550	Fe 415	Fe 500	Fe 550
L7 Section	0.2000	0.3201	1.3700	1.6931	1.8793	1.3557	1.6788	1.8650	1.3414	1.6645	1.8507
	0.0350	0.3301	1.3869	1.7144	1.9023	1.3725	1.7000	1.8879	1.3581	1.6856	1.8735
	0.0500	0.3250	1.4014	1.7347	1.9226	1.3869	1.7202	1.9081	1.3725	1.7058	1.8936
	0.0650	0.3198	1.4152	1.7545	1.9409	1.4007	1.7400	1.9264	1.3862	1.7255	1.9118
	0.0800	0.3147	1.4294	1.7691	1.9590	1.4148	1.7546	1.9444	1.4002	1.7400	1.9298
L8 Section	0.0150	0.3329	1.3485	1.6835	1.8828	1.3341	1.6691	1.8684	1.3197	1.6547	1.8540
	0.0275	0.3287	1.3605	1.6998	1.8996	1.3461	1.6853	1.8852	1.3316	1.6709	1.8707
	0.0400	0.3245	1.3727	1.7164	1.9133	1.3582	1.7019	1.8988	1.3437	1.6874	1.8843
	0.0525	0.3203	1.3834	1.7307	1.9265	1.3689	1.7162	1.9120	1.3544	1.7017	1.8975
	0.0650	0.3160	1.3941	1.7437	1.9401	1.3796	1.7292	1.9255	1.3650	1.7146	1.9110

form as $M_{ux}/f_{ck}Z_x$, $M_{uy}/f_{ck}Z_y$, pf_y/f_{ck} and $P_u/f_{ck}A_g$ respectively (Fig. 5.32), where p is defined as $100A_s/A_g$. Only half of the antisymmetrical interaction curves about the axis of antisymmetry have been shown in the figure. The values of $P_u/f_{ck}A_g$ considered for design charts are 0.1, 0.3, 0.4, 0.5, 0.6, 0.7, 0.8 and 1.0. Design curves for two consequtive values of $P_u/f_{ck}A_g$ are shown on one chart. It also shows curves for three values of ratio of effective cover to reinforcement (d') to depth of section (D) appropriate for each geometry of the column section as given in Fig. 5.31. The interaction curves have been plotted for intermediate value of d'/D. For other values of d'/D, necessary corrections are made with the help of d'/D curves as explained below.

The design charts for different geometry and reinforcement detailing are shown in Charts L1.1 to L8.4. The index of the charts is given in Table 5.7.

The use of the charts for design is simple as explained with the help of Fig. 5.32. Consider that the area of steel is required to be determined for negative value of moment $M_{ux}/f_{ck}Z_x$ and positive value of moment $M_{uy}/f_{ck}Z_y$ and for $P_u/f_{ck}A_g$ = 0.1 and 0.3. As the point corresponding to the given set of moments does not lie on the interaction curves for $P_u/f_{ck}A_g$ = 0.3, area of steel shall be determined by reversing the sign of moments due to antisymmetry of the interaction curves. On design charts for $P_u/f_{ck}A_g$ = 0.1 and 0.3, point a is plotted for the given value of moment $M_{ux}/f_{ck}Z_x$. On the vertical line drawn from point a, point b is plotted corresponding to the given value of d'/D from the interpolation of d'/D curves. A horizontal line is drawn from point b to intersect the intermediate d'/D curve at point c. Pcint c indicates the modified value of $M_{ux}/f_{ck}Z_x$ for the given value of d'/D for determining the area of steel with the use of curves plotted for the intermediate value of d'/D. Similarly, point a' is plotted for the given value of moment $M_{uy}/f_{ck}Z_y$. On the horizontal line drawn from point a', point b' is plotted corresponding to the given value of d'/D from the interpolation of d'/D curves. A vertical line is drawn from point b', to intersect the intermediate d'/D curve at point c'. Point c' indicates the modified value of $M_{uy}/f_{ck}Z_y$ for the given value of d'/D for determining the area of steel with the use of curves plotted for intermediate value of d'/D. The area of steel is given by point d which is obtained by the intersection of vertical line drawn from point c and horizontal line drawn from point c'.

The use of design charts is explained with the help of following examples.

Example 5.7: Design a biaxial eccecentrically loaded L - section of column for the following data:

Column section: D/B = 4

B = 250 mm

D = 1000 mm

Ultimate axial load, P_u = 4000 kN

Ultimate moments, M_{ux} = 750 kN.m

M_{uy} = 750 kN.m

Grade of concrete : M25

Grade of steel : Fe 415

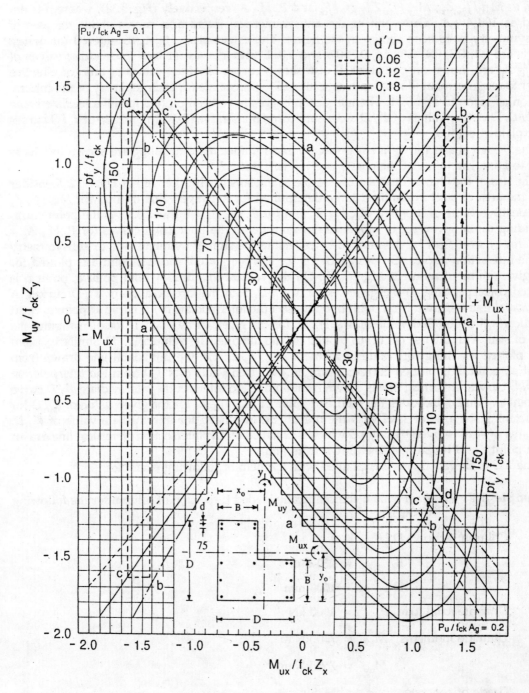

Fig. 5.32 Method for design of L section of column with the use of a typical design chart L1.1

Columns

Table 5.7 Index for Charts L1.1 to L8.4

Reinforcement Detailing	Chart No.	$P_u / f_{ck} A_g$	Page No.
L 1 Section	L1.1	0.1 and 0.3	262
	L1.2	0.4 and 0.5	263
	L1.3	0.6 and 0.7	264
	L1.4	0.8 and 1.0	265
L 2 Section	L2.1	0.1 and 0.3	266
	L2.2	0.4 and 0.5	267
	L2.3	0.6 and 0.7	268
	L2.4	0.8 and 1.0	269
L 3 Section	L3.1	0.1 and 0.3	270
	L3.2	0.4 and 0.5	271
	L3.3	0.6 and 0.7	272
	L3.4	0.8 and 1.0	273
L 4 Section	L4.1	0.1 and 0.3	274
	L4.2	0.4 and 0.5	275
	L4.3	0.6 and 0.7	276
	L4.4	0.8 and 1.0	277
L 5 Section	L5.1	0.1 and 0.3	278
	L5.2	0.4 and 0.5	279
	L5.3	0.6 and 0.7	280
	L5.4	0.8 and 1.0	281
L 6 Section	L6.1	0.1 and 0.3	282
	L6.2	0.4 and 0.5	283
	L6.3	0.6 and 0.7	284
	L6.4	0.8 and 1.0	285
L 7 Section	L7.1	0.1 and 0.3	286
	L7.2	0.4 and 0.5	287
	L7.3	0.6 and 0.7	288
	L7.4	0.8 and 1.0	289
L 8 Section	L8.1	0.1 and 0.3	290
	L8.2	0.4 and 0.5	291
	L8.3	0.6 and 0.7	292
	L8.4	0.8 and 1.0	293

Solution

Consider 28 mm ϕ bars with clear cover of 40 mm.

$\therefore$ Effective cover, $d' = 40 + 14 = 54$ mm

and $d'/D = 54/1000 = 0.054$

Area of the section,

$$A_g = (2D - B)B = (2 \times 1000 - 250) \times 250 = 437500 \text{ mm}^2$$

Centre of gravity of the section,

$$x_o = y_o = \frac{D^2 + (D-B)/B}{2(2D-B)}$$

$$= \frac{1000^2 + (1000 - 250) \times 250}{2 \times (2 \times 1000 - 250)} = 339.2857 \text{ mm}$$

Moment of inertia about x and y axes of the section,

$$I_{xx}, I_{yy} = \frac{BD^3 + DB^3 - B^4}{12} + BD(0.5D - y_o)^2 + (D-B)B(y_o - 0.5B)^2$$

$$= \frac{250 \times 1000^3 + 1000 \times 250^3 - 250^4}{12} + 250 \times 1000 \times (0.5 \times 1000 - 339.2857)^2$$

$$+ (1000 - 250) \times 250 \times (339.2857 - 0.5 \times 250)^2$$

$$= 21809.894 \times 10^6 \text{ mm}^4$$

Section modulus about x and y axes,

$$Z_x, Z_y = I_{xx}/y_o = I_{yy}/x_o$$

$$= \frac{21809.894 \times 10^6}{339.2857} = 6428.18 \times 10^4 \text{ mm}^3$$

Compute

$$\frac{P_u}{f_{ck} A_g} = \frac{4000 \times 10^3}{20 \times 437500} = 0.4571$$

$$\frac{M_{ux}}{f_{ck} Z_x} = \frac{750 \times 10^6}{20 \times 6428.18 \times 10^4} = 0.5834$$

$$\frac{M_{uy}}{f_{ck} Z_y} = \frac{750 \times 10^6}{20 \times 6428.18 \times 10^4} = 0.5834$$

Area of steel is determined from the linear interpolation of area of steel determined for $P_u/f_{ck} A_g = 0.4$ and 0.5 from Chart L5.2. The design moments $M_{ux}/f_{ck} Z_x$ and $M_{uy}/f_{ck} Z_y$ are considered negative and positive respectively according to the sign convention for moments for the chart. The use of chart for determining the area of steel has been explained with the help of Fig. 5.32. Area of steel for $P_u/f_{ck}A_g = 0.4$ and 0.5 are determined as,

For $\dfrac{P_u}{f_{ck}A_g} = 0.4$, $\dfrac{pf_y}{f_{ck}} = 66.0$

For $\dfrac{P_u}{f_{ck}A_g} = 0.5$, $\dfrac{pf_y}{f_{ck}} = 74.0$

∴ For $\dfrac{P_u}{f_{ck}A_g} = 0.4571$, $\dfrac{pf_y}{f_{ck}} = 66.0 + \dfrac{74.0 - 66.0}{0.5 - 0.4} \times (0.4571 - 0.4) = 70.568$

∴ $\qquad p = 70.563 \times 20 / 415 = 3.4$

and $\qquad A_s = 3.4 \times 437500 / 100 = 14875 \text{ mm}^2$

The diameter of reinforcement bars is chosen to provide 26 bars in accordance with recommended reinforcement detailing. Accordingly 28 mm ϕ bars ($A_s = 16008 \text{ mm}^2$) are provided as shown in Fig. Ex. 5.7.

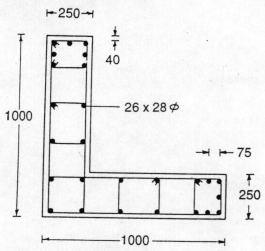

Fig. Ex. 5.7 Reinforcement detailing

Example 5.8: Design a biaxial eccentrically loaded braced L-shaped column deforming in single curvature for the following data.

Cross-section: D/B	= 2
B	= 250 mm
D	= 500 mm
Ultimate axial load, P_u	= 2000 kN

Ultimate moment about x-axis,

    at bottom, M_{ux1} = 325.0 kN.m

    at top, M_{ux2} = 200.0 kN.m

Ultimate moment about y-axis,

    at bottom, M_{uy1} = 250.0 kN

    at top, $|M_{ux2}$ = 200.0 kN.m

Unsupported length of column, l = 9 m

Effective length about x-axis, l_{ex} = 6.9 m

Effective length about y-axis, l_{ey} = 7.95 m

Grade of concrete   : M25

Grade of steel   : Fe 415

Solution

Moment due to minimum of eccentricities:

$$e_{min} = e_{x,min} = e_{y,min}$$

$$= \frac{1}{500} + \frac{D}{30} \text{ or } 20 \text{ mm whichever is more}$$

$$= \frac{9000}{500} + \frac{500}{30} \text{ or } 20 \text{ mm whichever is more}$$

$$= 34.67 \text{ mm}$$

$\therefore$ $M_{uxe} = M_{uye} = P_u e_{min} = 2000 \times 34.67$ kN.mm

$$= 69.34 \text{ kN.m} < M_{ux} \text{ and } M_{uy}$$

Check for short/long column:

$$l_{ex}/D = 6900/500 = 13.8 > 12$$

$$l_{ey}/D = 7950/500 = 15.91 > 12$$

Hence it is a slender column about both axes of bending.

Additional moments due to slenderness effects,

$$M_{ax} = P_u e_y = P_u \frac{D}{2000}\left(\frac{l_{ey}}{D}\right)^2$$

$$= 2000 \times \frac{500}{2000} \times (15.91)^2 \text{ kN.mm} = 126.564 \text{ kN.m}$$

and $\quad M_{ay} = P_u e_x = P_u \dfrac{D}{2000} \left(\dfrac{l_{ex}}{D}\right)^2$

$\quad\quad\quad\quad\quad = 2000 \times \dfrac{500}{2000} \times 13.8^2 \text{ kN.m} = 95.22 \text{ kN.m}$

Additional moments M_{ax} and M_{ay} are reduced by multiplying with a factor K given by,

$$K = \dfrac{P_{uz} - P_u}{P_{uz} - P_{ub}}$$

where P_{uz} and P_{ub} are determined for the assumed area of longitudinal reinforcement and its distribution. Consider percentage of reinforcement, $p = 5.0$ provided as reinforcement detailing type 1 (Fig. 5.31) with clear cover of 40 mm. Consider 28 mm bars.

$\therefore\quad\quad\quad\quad$ Effective cover, $d' = 40 + 28/2 = 54$ mm

and $\quad\quad\quad\quad d'/D = 54/500 = 0.108$

Area of the section,

$$A_g = (2D - B) = (2 \times 500 - 250) \times 250 = 187500 \text{ mm}^2$$

Centre of gravity of the section,

$$x_o = y_o = \dfrac{D^2 + (D-B)B}{2(2D-B)} = \dfrac{500^2 + (500-250) \times 250}{2 \times (2 \times 500 - 250)} = 208.333 \text{ mm}$$

Moment of intertia about x and y-axes of the section,

$$I_{xx} = I_{yy} = \dfrac{BD^3 + DB^3 - B^4}{12} + BD(0.5D - y_o)^2 + (D-B)B(y_o - 0.5B)^2$$

$$= \dfrac{250 \times 500^3 + 500 \times 250^3 - 250^4}{12} + 250 \times 500 \times (0.5 \times 500 - 208.333)^2$$

$$\quad + (500 - 250) \times 250 \times (208.333 - 0.5 \times 250)^2$$

$$= 358.072 \times 10^8 \text{ mm}^4$$

Section modulus about x and y axes of the section,

$$Z_x, Z_y = \dfrac{I_{xx}}{y_o} = \dfrac{I_{yy}}{x_o} = \dfrac{358.072 \times 10^6}{208.33} = 17187500 \text{ mm}^3$$

Compute $\quad P_{uz} = 0.466 f_{ck} A_g + (0.75 f_y - 0.446 f_{ck}) p A_g / 100$

$\quad\quad\quad\quad = 0.446 \times 25 \times 187500 + (0.75 \times 415 - 0.446 \times 25) \times 0.05 \times 187500$ N

$\quad\quad\quad\quad = 4904.063$ kN

$\quad\quad P_{ub} = q_c f_{ck} A_g + q_s p A_g$

where q_c and q_s are determined from Table 5.6 as,

$$q_{cx} = 0.2734 + \frac{0.2669 - 0.2734}{0.03} \times 0.018 = 0.2695$$

$$q_{sx} = 0.7491 + \frac{0.8083 - 0.7491}{0.03} \times 0.018 = 0.7846$$

$$q_{cy} = 0.1448 + \frac{0.1375 - 0.1448}{0.03} \times 0.018 = 0.1404$$

$$q_{sy} = 0.0669 + \frac{-0.1523 - 0.0669}{0.03} \times 0.018 = -0.046$$

$\therefore \quad P_{ubx} = 0.2695 \times 25 \times 187500 \times 0.784 \times 5 \times 187500 \text{ N} = 1998.28 \text{ kN}$

$P_{uby} = 0.1404 \times 25 \times 187500 - 0.0646 \times 5 \times 187500 \text{ N} = 597.56 \text{ kN}$

and
$$K_x = \frac{4904.063 - 2000}{4904.063 - 1998.28} = 0.9994$$

$$K_y = \frac{4904.063 - 2000}{4904.063 - 597.56} = 0.6743$$

Reduced additional moments are,

$$M'_{ax} = K_x M_{ax} = 0.9994 \times 126.564 = 126.49 \text{ kN.m}$$

and
$$M'_{ay} = K_y M_{ay} = 0.6743 \times 95.22 = 64.2 \text{ kN.m}$$

The ultimate design moments are,

$$M_{ux} = M'_{ux} + M'_{ux}$$

and
$$M_{uy} = M'_{uy} + M'_{uy}$$

where $M'_{ux} = 0.6 M_{ux,\,max} + 0.4 M_{uy,\,min}$

$\qquad\qquad\quad = 0.6 \times 325 + 0.4 \times 200$

$\qquad\qquad\quad = 275 \text{ kN.m} > 0.4 M_{ux,\,max}$

$\therefore \qquad\qquad M_{ux} = 275 + 126.49 = 401.49 \text{ kN.m}$

and $\quad M'_{uy} = 0.6 M_{uy,\,max} + 0.4 M_{uy,\,min}$

$\qquad\qquad\quad = 0.6 \times 250 + 0.4 \times 200$

$\qquad\qquad\quad = 230 \text{ kN.m} > 0.4 M_{uy,\,max}$

and
$$M_{uy} = 230 + 64.2 = 294.2 \text{ kN.m}$$

Compute
$$\frac{P_u}{f_{ck} A_g} = \frac{2000 \times 10^3}{25 \times 187500} = 0.4267$$

$$\frac{M_{ux}}{f_{ck} Z_x} = \frac{401.49 \times 10^6}{25 \times 17187500} \approx 0.934$$

$$\frac{M_{uy}}{f_{ck} Z_y} = \frac{294.2 \times 10^6}{25 \times 17187500} = 0.685$$

Area of steel is determined from the linear interpolation of area of steel determined for $P_u/f_{ck} A_g = 0.4$ and $P_u/f_{ck} A_g = 0.5$ from Chart L1.2. The design moments $M_{ux}/f_{ck} Z_x$ and $M_{uy}/f_{ck} Z_y$ are considered negative and positive respectively, according to the sign convention for moments for the chart. The use of chart for determining the area of steel has been explained with the help of Fig. 5.32. Area of steel for $P_u/f_{ck} A_g = 0.4$ and 0.5 are determined as

For $\dfrac{P_u}{f_{ck} A_g} = 0.4$, $\dfrac{p f_y}{f_{ck}} = 72.0$

For $\dfrac{P_u}{f_{ck} A_g} = 0.5$, $\dfrac{p f_y}{f_{ck}} = 92.0$

∴ For $\dfrac{P_u}{f_{ck} A_g} = 0.4267$, $\dfrac{p f_y}{f_{ck}} = 72.0 + \dfrac{92.0 - 72.0}{0.5 - 0.4} \times (0.4267 - 0.4) = 77.34$

∴
$$p = \frac{77.34 \times 25}{415} = 4.659$$

and
$$A_s = \frac{4.659 \times 187500}{100} = 8735.6 \text{ mm}^2$$

The diameter of reinforcement bars is chosen to provide 14 bars in accordance with recommended reinforcement detailing. Accordingly 4×30 mm ϕ + 10×28 mm ϕ bars (A_s = 8985 mm^2) are provided as shown in Fig. Ex. 5.8. Reinforcement bars of 30 mm ϕ are placed at the edges which shall result in conservative design.

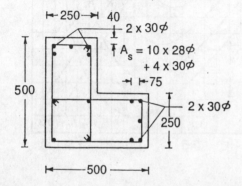

Fig. Ex. 5.8 Reinforcement detailing

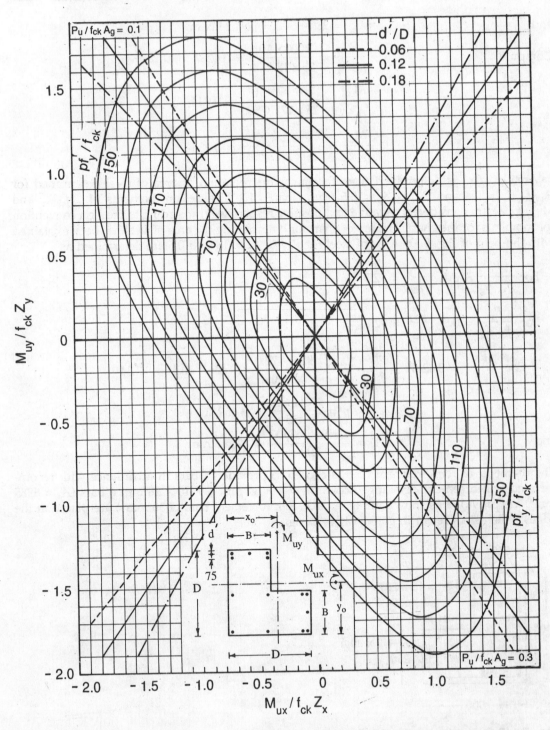

Chart L1.1 Design curves for L1 section of column for $P_u/f_{ck}A_g = 0.1$ and 0.3 $(1.75 \leq D/B \leq 2.25)$

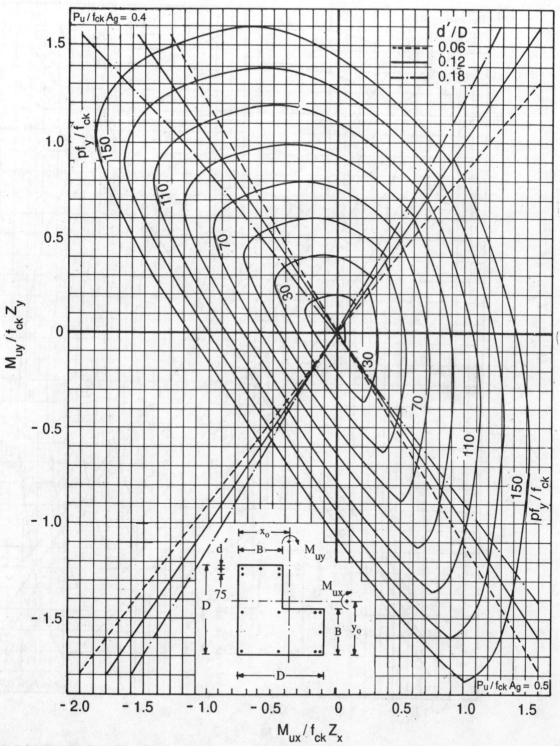

Chart L1.2 Design curves for L1 section of column for $P_u / f_{ck} A_g = 0.4$ and 0.5 ($1.75 \leq D / B \leq 2.25$)

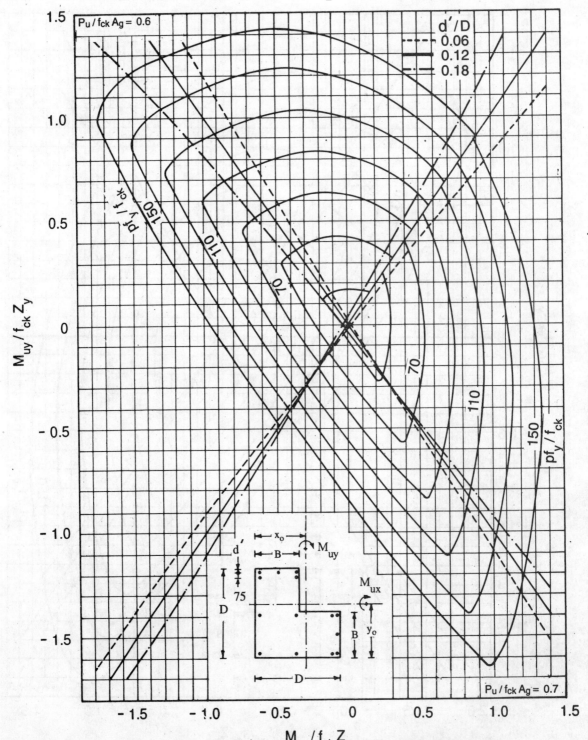

Chart L1.3 Design curves for L1 section of column for $P_u / f_{ck} A_g$ = 0.6 and 0.7 (1.75 ≤ D / B ≤ 2.25)

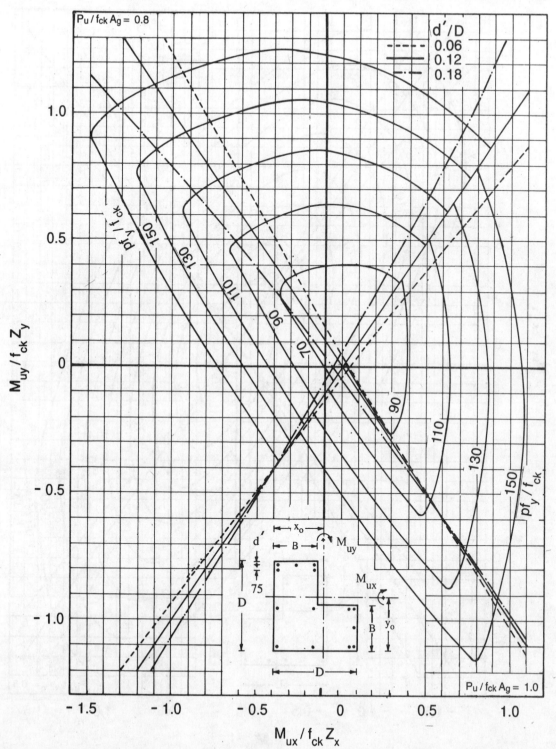

Chart L1.4 Design curves for L1 section of column for $P_u / f_{ck} A_g$ = 0.8 and 1.0 (1.75 ≤ D / B ≤ 2.25)

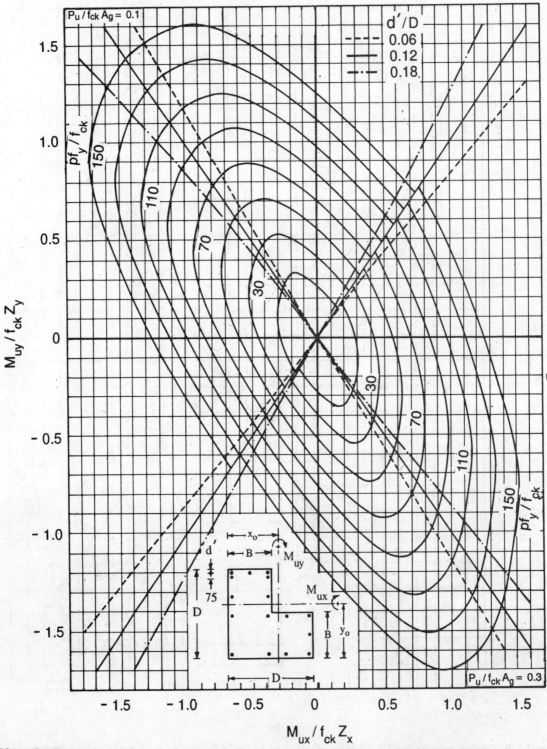

Chart L2.1 Design curves for L2 section of column for $P_u / f_{ck} A_g = 0.1$ and 0.3 $(1.75 \leq D/B \leq 2.25)$

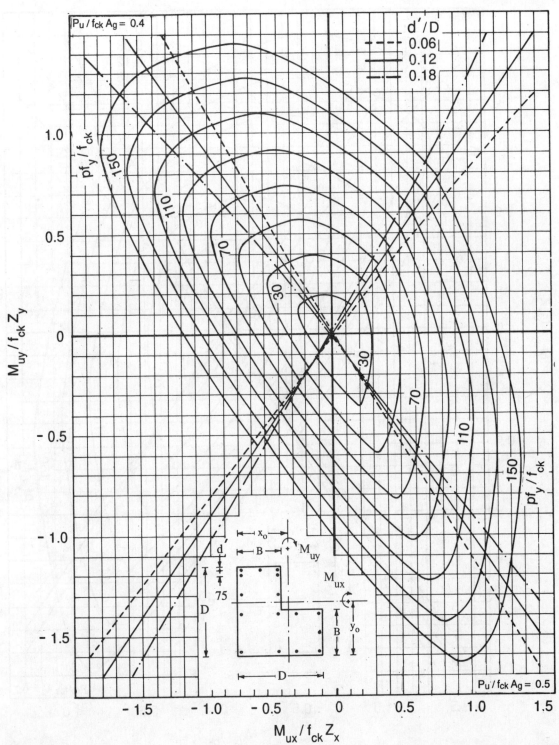

Chart L2.2 Design curves for L2 section of column for $P_u / f_{ck} A_g = 0.4$ and 0.5 $(1.75 \leq D/B \leq 2.25)$

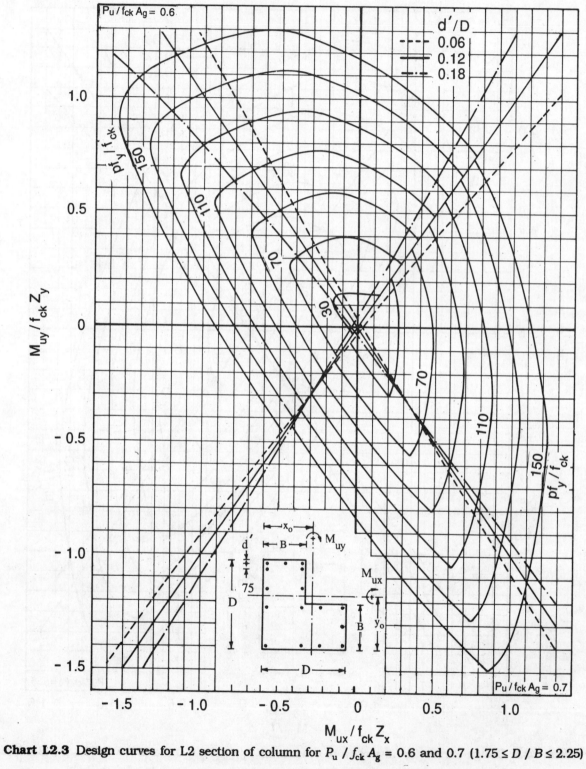

Chart L2.3 Design curves for L2 section of column for $P_u / f_{ck} A_g$ = 0.6 and 0.7 (1.75 ≤ D / B ≤ 2.25)

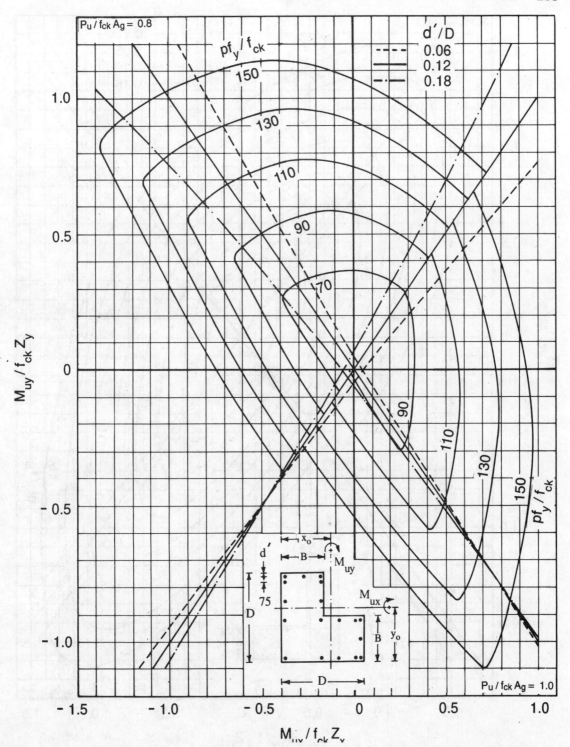

Chart L2.4 Design curves for L2 section of column for $P_u / f_{ck} A_g = 0.8$ and 1.0 $(1.75 \leq D/B \leq 2.25)$

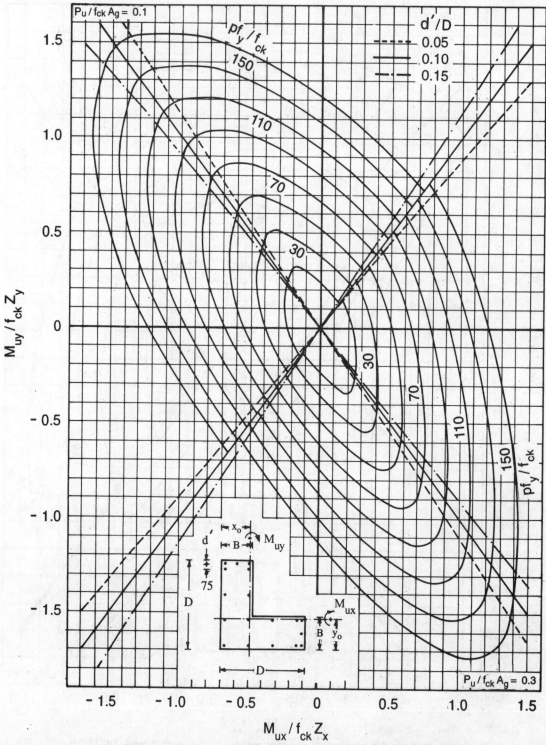

Chart L3.1 Design curves for L3 section of column for $P_u/f_{ck}A_g = 0.1$ and 0.3 $(2.5 \leq D/B \leq 3.0)$

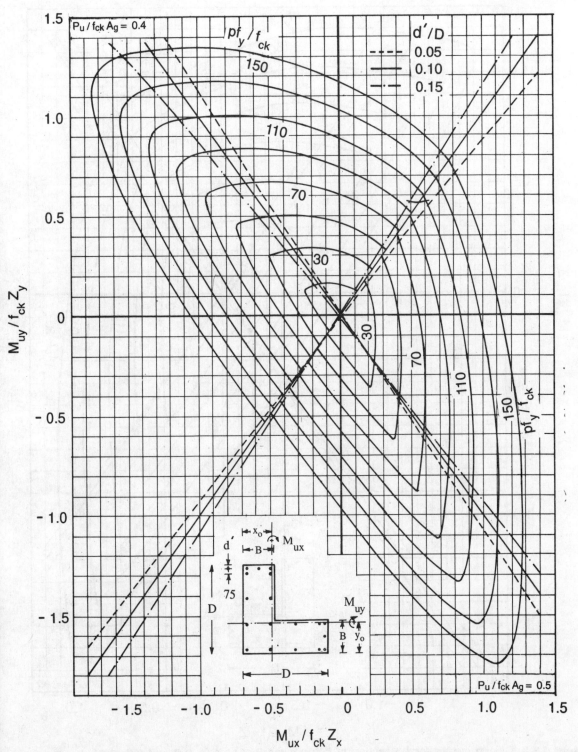

Chart L3.2 Design curves for L3 section of column for $P_u / f_{ck} A_g = 0.4$ and 0.5 ($2.5 \leq D/B \leq 3.0$)

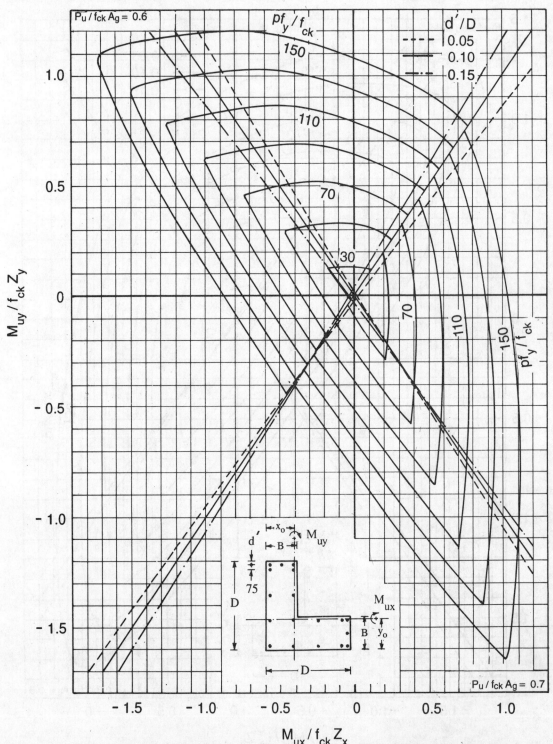

Chart L3.3 Design curves for L3 section of column for $P_u / f_{ck} A_g = 0.6$ and 0.7 ($2.5 \leq D/B \leq 3.0$)

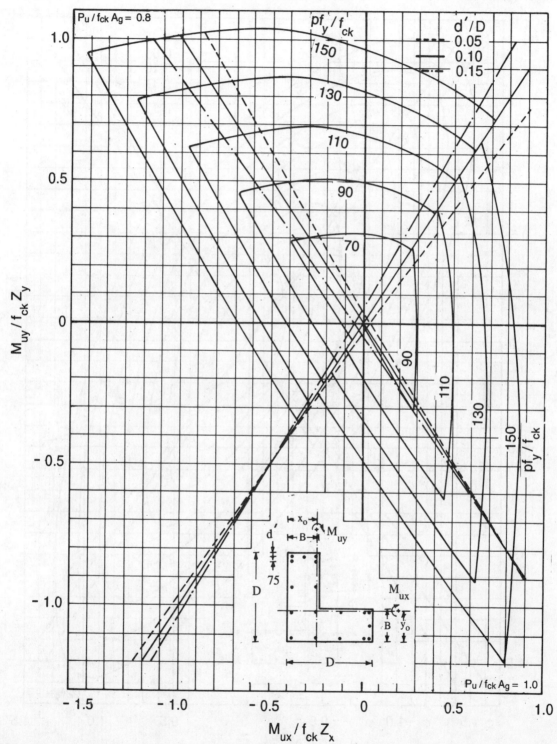

Chart L3.4 Design curves for L3 section of column for $P_u / f_{ck} A_g$ = 0.8 and 1.0 (2.5 ≤ D/B ≤ 3.0)

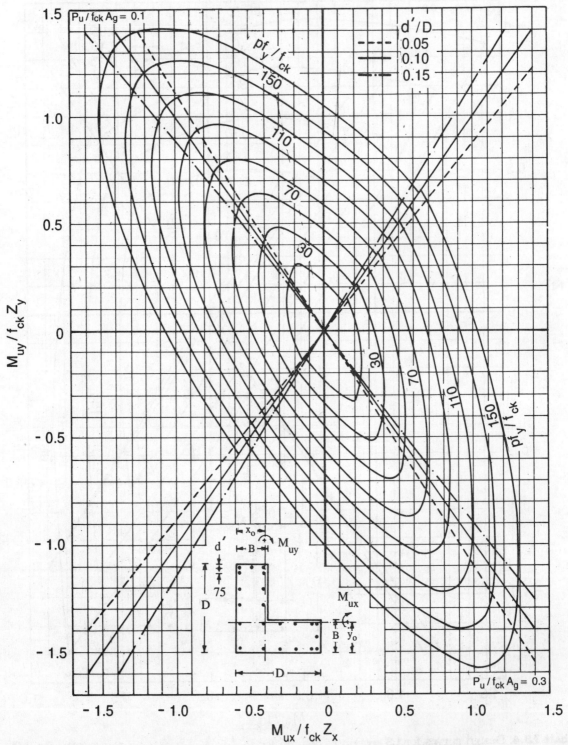

Chart L4.1 Design curves for L4 section of column for $P_u / f_{ck} A_g = 0.1$ and 0.3 $(2.5 \le D/B \le 3.0)$

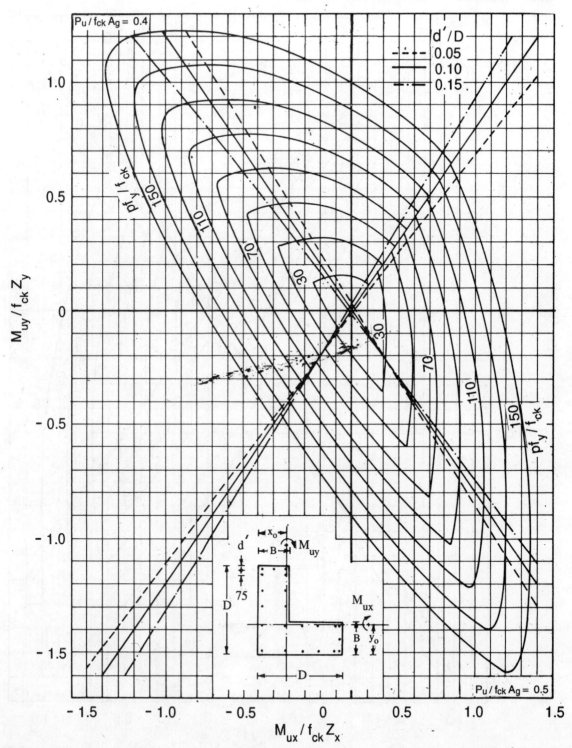

Chart L4.2 Design curves for L4 section of column for $P_u / f_{ck} A_g = 0.4$ and 0.5 ($2.5 \leq D / B \leq 3.0$)

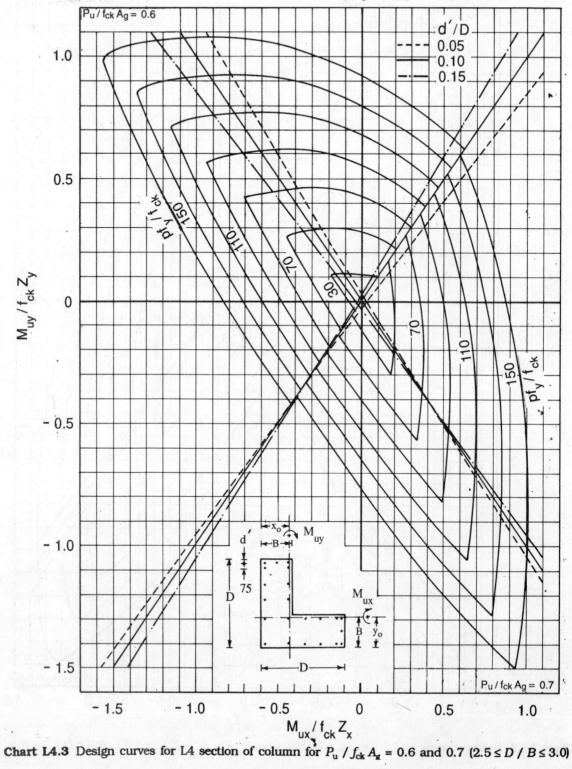

Chart L4.3 Design curves for L4 section of column for $P_u / f_{ck} A_g = 0.6$ and 0.7 ($2.5 \leq D/B \leq 3.0$)

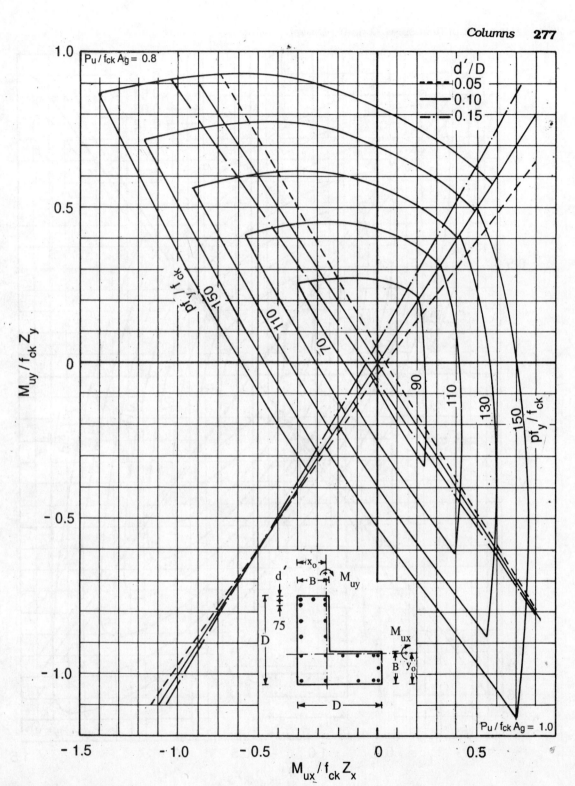

Chart L4.4 Design curves for L4 section of column for $P_u / f_{ck} A_g = 0.8$ and 1.0 $(2.5 \leq D/B \leq 3.0)$

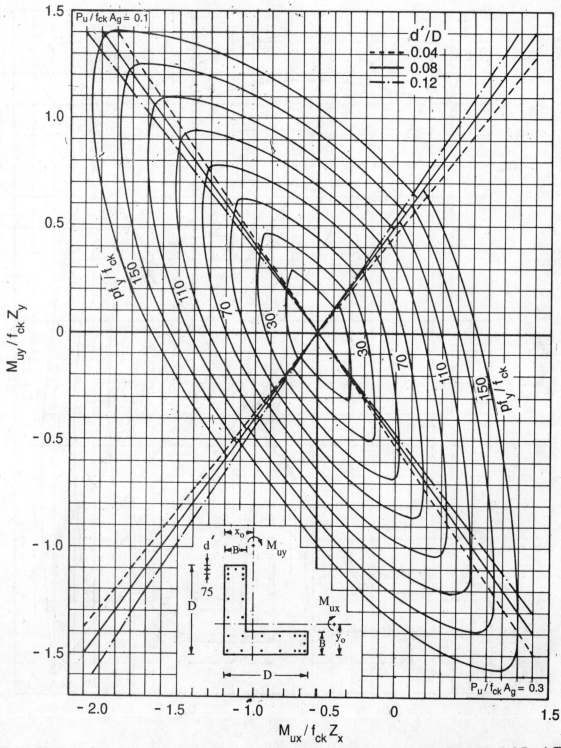

Chart L5.1 Design curves for L5 section of column for $P_u/f_{ck}A_g = 0.1$ and 0.3 $(3.5 \leq D/B \leq 4.5)$

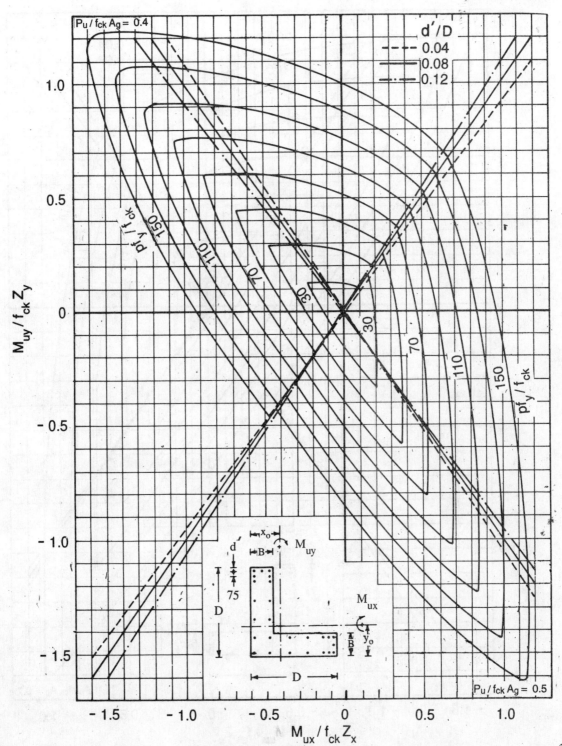

Chart L5.2 Design curves for L5 section of column for $P_u/f_{ck}A_g = 0.4$ and 0.5 ($3.5 \leq D/B \leq 4.5$)

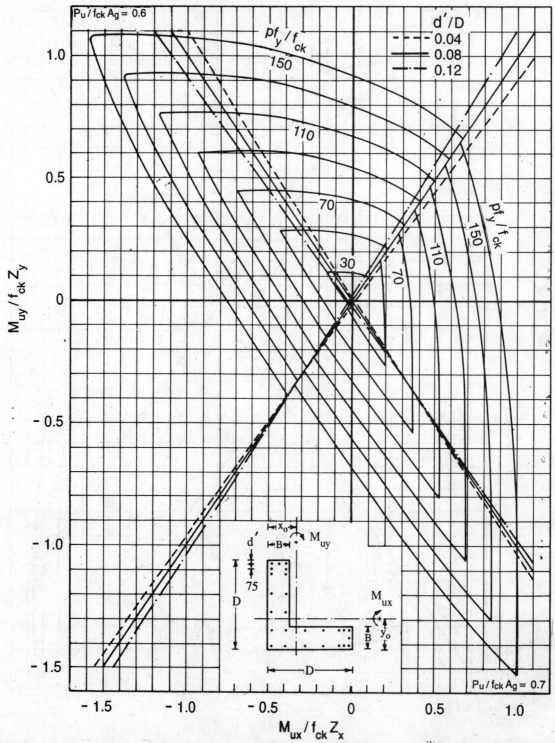

Chart L5.3 Design curves for L5 section of column for $P_u / f_{ck} A_g = 0.6$ and 0.7 ($3.5 \leq D / B \leq 4.5$)

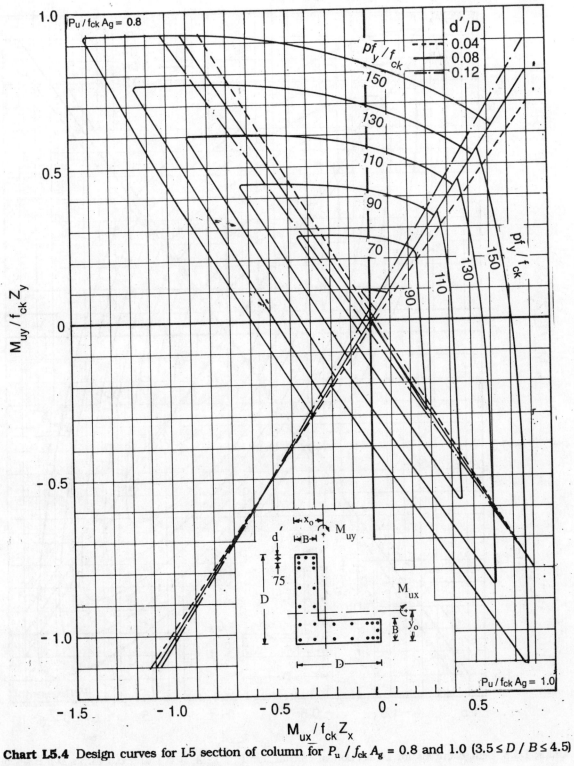

Chart L5.4 Design curves for L5 section of column for $P_u / f_{ck} A_g = 0.8$ and 1.0 $(3.5 \leq D/B \leq 4.5)$

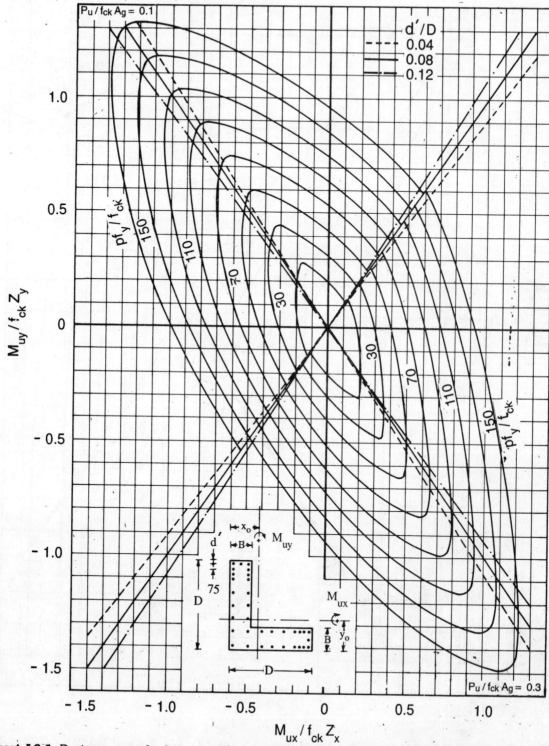

Chart L6.1 Design curves for L6 section of column for $P_u / f_{ck} A_g = 0.1$ and 0.3 ($3.5 \leq D/B \leq 4.5$)

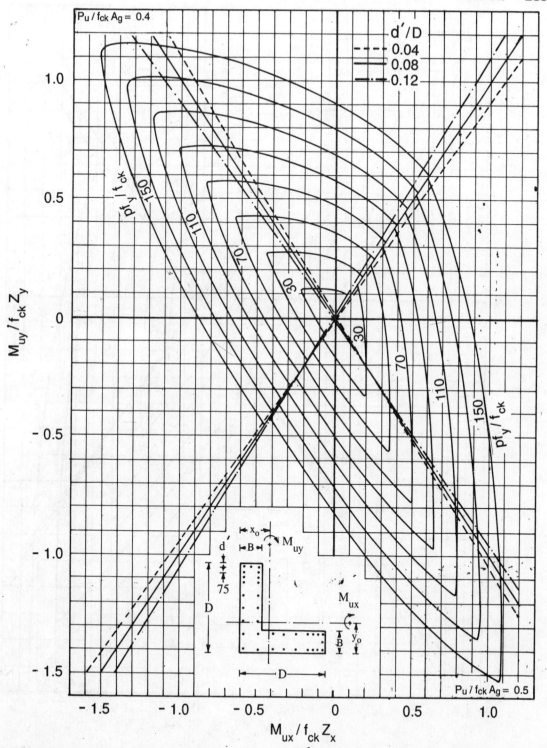

Chart L6.2 Design curves for L6 section of column for $P_u / f_{ck} A_g$ = 0.4 and 0.5 (3.5 ≤ D / B ≤ 4.5)

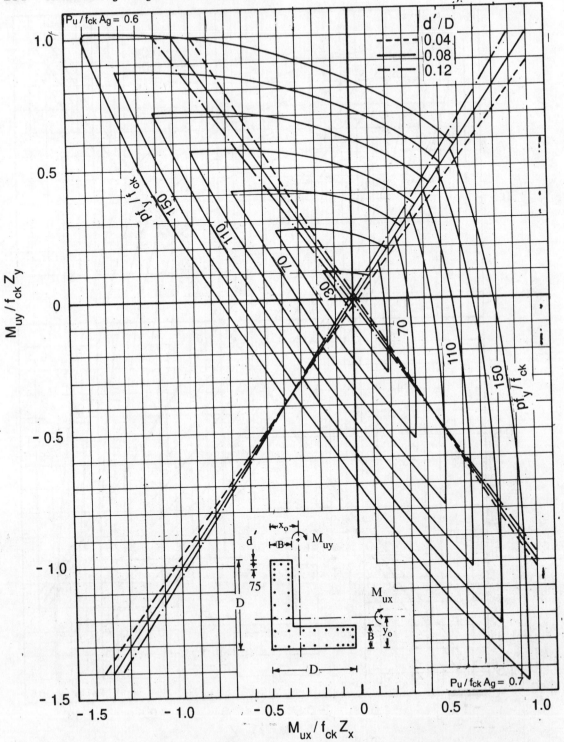

Chart L6.3 Design curves for L6 section of column for $P_u/f_{ck}A_g$ = 0.6 and 0.7 (3.5 ≤ D/B ≤ 4.5)

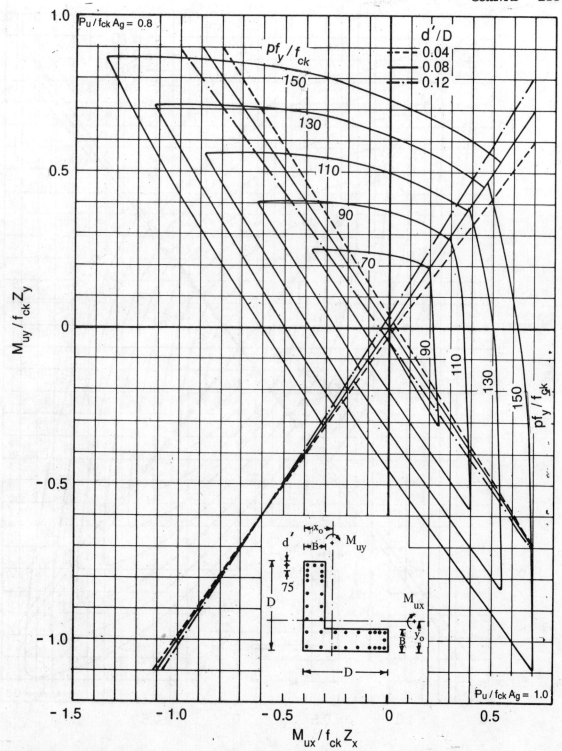

Chart L6.4 Design curves for L6 section of column for $P_u/f_{ck}A_g = 0.8$ and 1.0 ($3.5 \leq D/B \leq 4.5$)

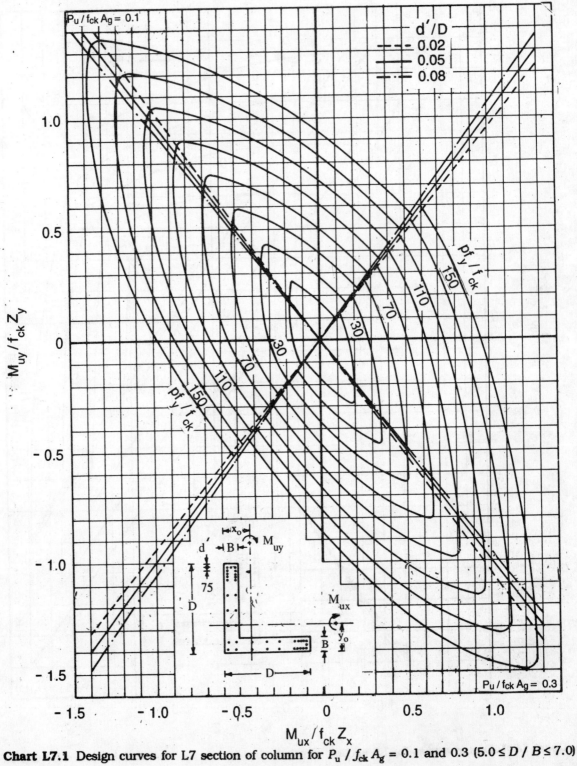

Chart L7.1 Design curves for L7 section of column for $P_u/f_{ck}A_g = 0.1$ and 0.3 ($5.0 \leq D/B \leq 7.0$)

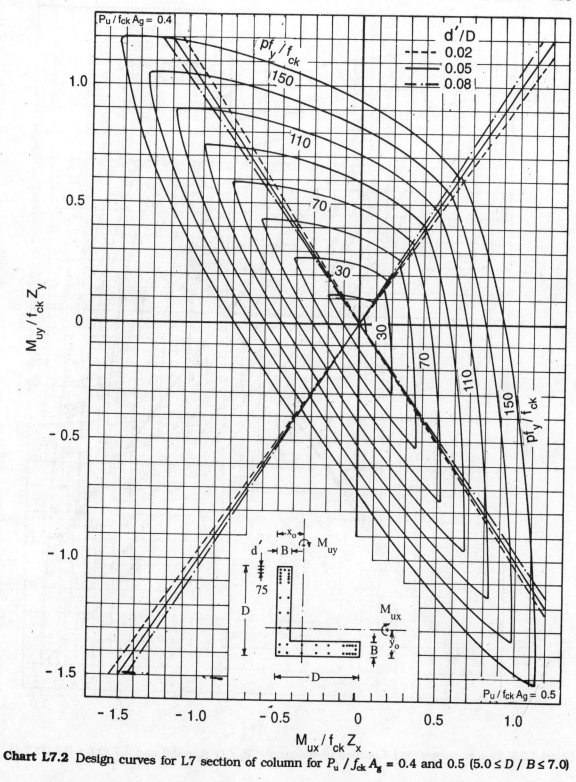

Chart L7.2 Design curves for L7 section of column for $P_u/f_{ck}A_g = 0.4$ and 0.5 ($5.0 \leq D/B \leq 7.0$)

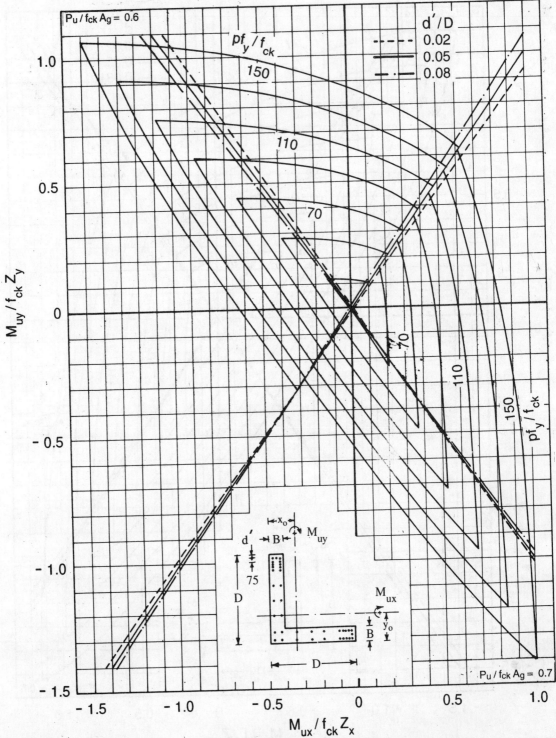

Chart L7.3 Design curves for L7 section of column for $P_u / f_{ck} A_g = 0.6$ and 0.7 ($5.0 \leq D/B \leq 7.0$)

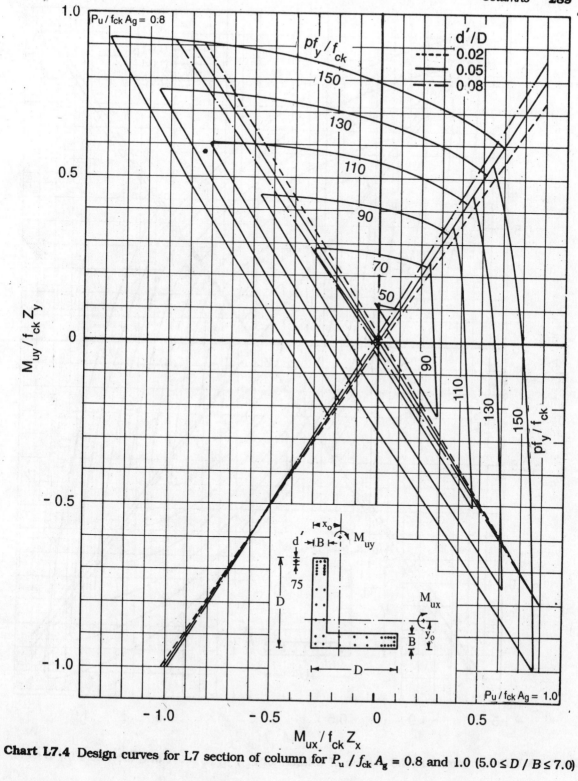

Chart L7.4 Design curves for L7 section of column for $P_u/f_{ck}A_g = 0.8$ and 1.0 ($5.0 \le D/B \le 7.0$)

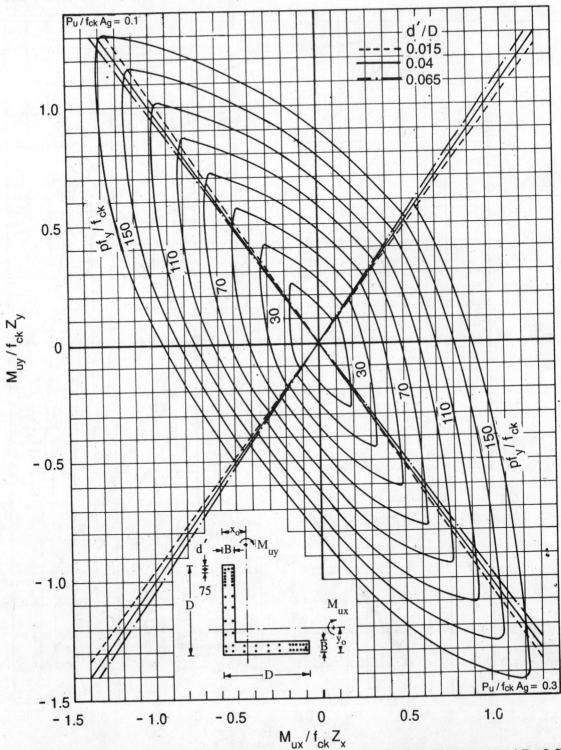

Chart L8.1 Design curves for L8 section of column for $P_u/f_{ck}A_g = 0.1$ and 0.3 ($5.0 \leq D/B \leq 9.0$)

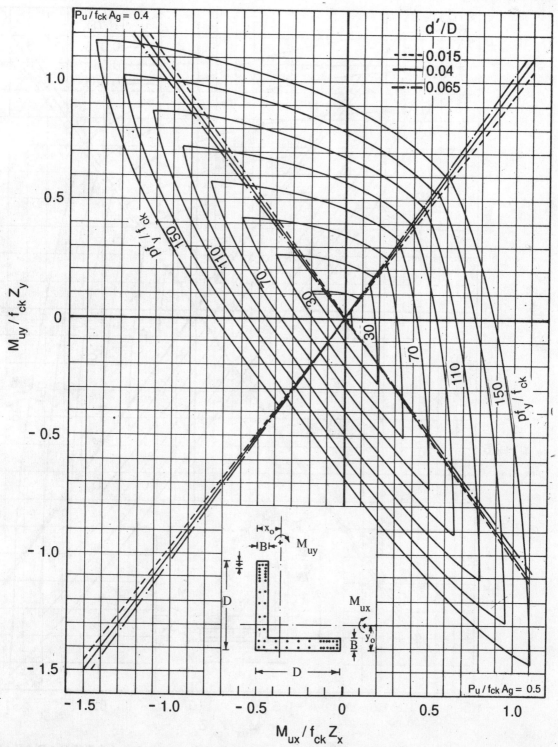

Chart L8.2 Design curves for L8 section of column for $P_u / f_{ck} A_g = 0.4$ and 0.5 $(5.0 \leq D/B \leq 9.0)$

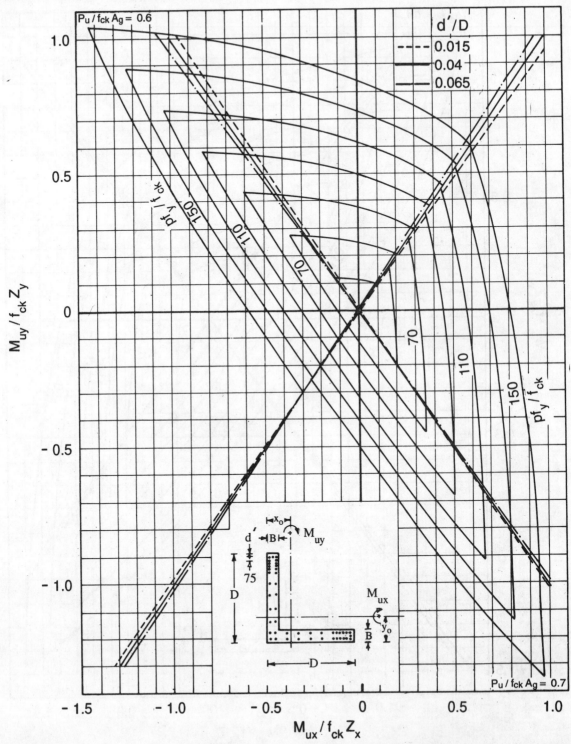

Chart I.8.3 Design curves for L8 section of column for $P_u / f_{ck} A_g$ = 0.6 and 0.7 ($5.0 \leq D/B \leq 9.0$)

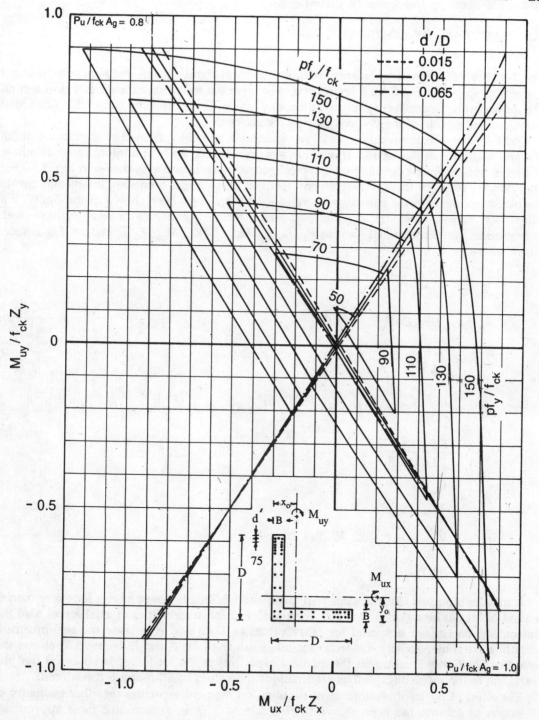

Chart L8.4 Design curves for L8 section of column for $P_u / f_{ck} A_g = 0.8$ and 1.0 $(5.0 \leq D/B \leq 9.0)$

5.6 T-SECTION OF COLUMNS

The T-section of column is geometrically unsymmetrical about x-x axis and symmetrical about y-y axis as shown in Fig. 5.33. Interaction curves for bixial moments for such a typical section are shown in the figure. It is symmetrical about y-y axis. Therefore, only half of the interaction curves on either side of the y-y axis may be considered.

The T-section of column may have large variation in geometry defined by the ratio of depth (D) and width (B) of the section. It may have many possible ways of reinforcement detailing. Different reinforcement detailing for a typical geometry of T-section is shown in Fig. 5.34. For studying the effect of these reinforcement detailing on moment capacities, interaction curves for biaxial moments for a particular value of axial force have been shown in the figure. For interaction curves, moments M_{ux} and M_{uy}, axial force P_u and area of steel A_s have been expressed in non-dimensional form as $P_u/f_{ck}A_g$, $M_{ux}/f_{ck}Z_x$, $M_{uy}/f_{ck}Z_y$ and pf_y/f_{ck} respectively where,

$$p = \frac{100 A_s}{A_g}$$

$$A_g = (2D - B)B$$

$$Z_x = \frac{I_{xx}}{y_o}$$

$$I_{xx} = \frac{BD^3}{12} + DB(y_o - 0.5B)^2 + \frac{B(D-B)^3}{12} + (D-B)B\left(\frac{D+B}{2} - y_o\right)^2$$

$$y_o = \frac{DB + D^2 - B^2}{2(2D - B)}$$

$$Z_y = \frac{I_{yy}}{x_o}$$

$$I_{yy} = \frac{BD^3}{12} + (D-B)\frac{B^3}{12}$$

$$x_o = 0.5 D$$

It may be observed that the same reinforcement detailing is not optimum for every combination of biaxial moments. It has also been observed for other values of axial force. Also the interaction curves are not valid for other values of D/B and B because the reinforcement detailing are dimensionally dissimilar for other values of D/B and B. Figure 5.35 shows the effect of D/B and B values on the moment capacities of the section. It is observed that the variation in the moment capacities are small for the values of D/B and B considered.

The effect of different reinforcement detailing on moment capacities for other geometry of T-section of column has been shown in Figs 5.36 to 5.39. In general, the most appropriate reinforcement detailing may be evolved based on maximum area under the interaction curve. However, it may not be always possible to adopt a particular type of reinforcement detailing

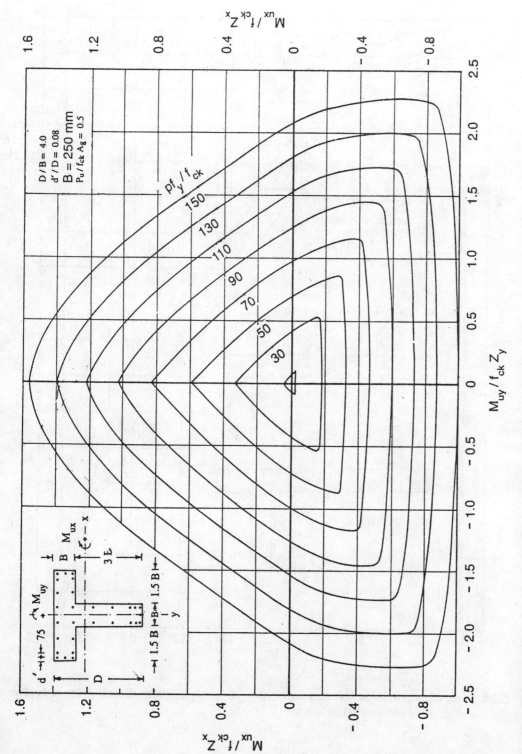

Fig. 5.33 Interaction curves for biaxial moments for a typical T section of column

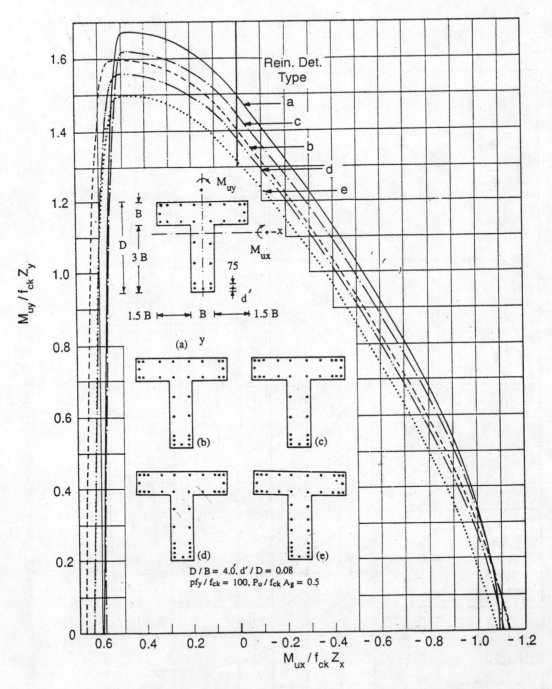

Fig. 5.34 Interaction curves for different types of reinforcement detailing for a typical T section of column ($D/B = 4.0$)

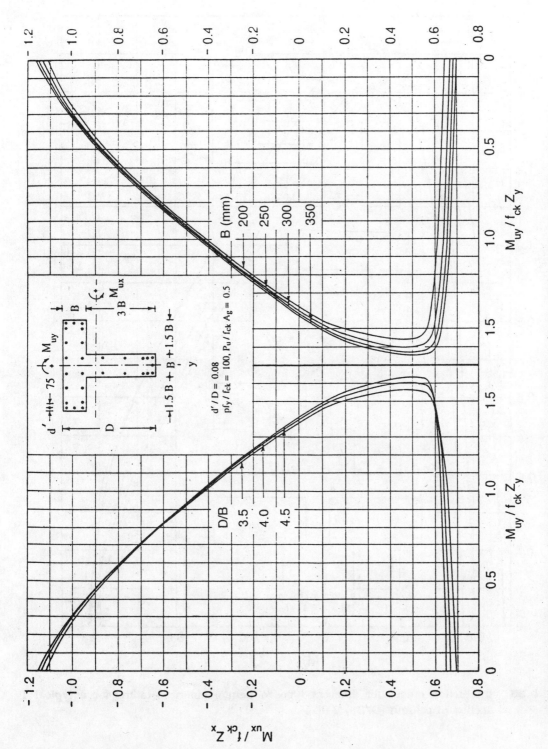

Fig. 5.35 Interaction curves for different values of D/B and B for a typical T section of column

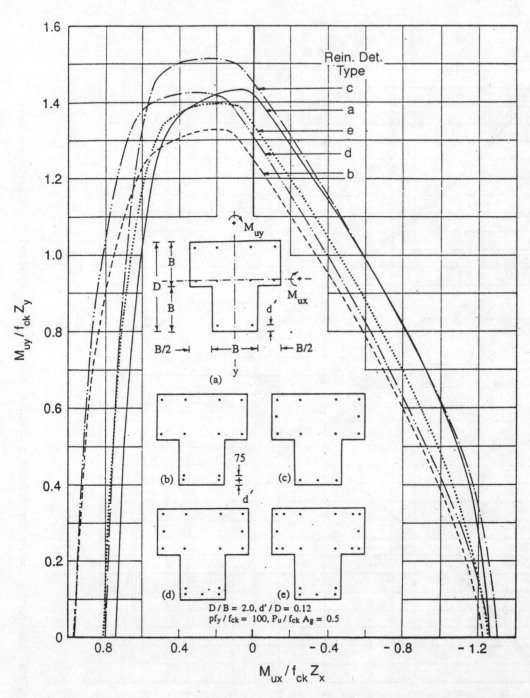

Fig. 5.36 Interaction curves for different types of reinforcement detailing for a typical T section of column ($D/B = 2.0$)

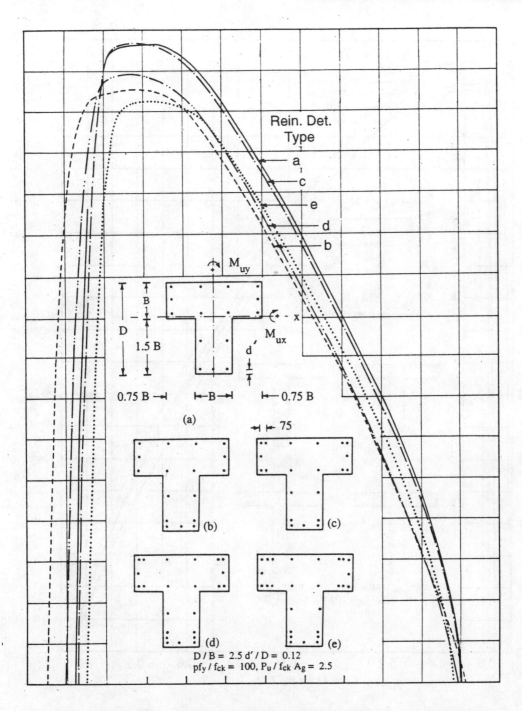

Fig. 5.37 Interaction curves for different types of reinforcement detailing for a typical T section of column ($D/B = 2.5$)

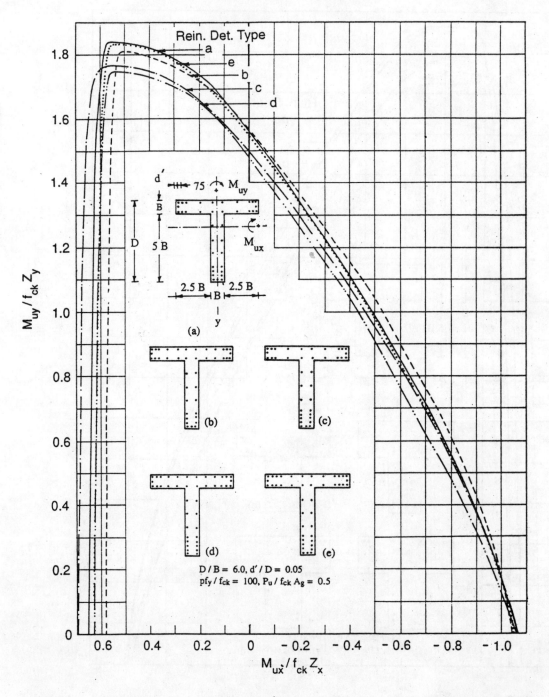

Fig. 5.38 Interaction curves for different types of reinforcement detailing for a typical T section of column ($D/B = 6.0$)

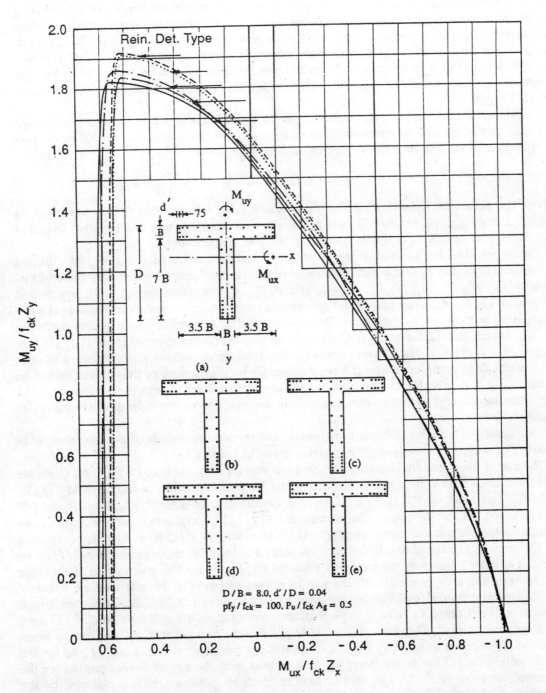

Fig. 5.39 Interaction curves for different types of reinforcement detailing for a typical T section of column ($D/B = 8$)

because of cross-sectional dimensions, area of steel, requirements for reinforcement detailing etc. Therefore, design charts have been prepared for several possible reinforcement detailing as shown in Fig. 5.40. It also shows some alternate reinforcement detailing which has approximately same moment capacity as that of the proposed reinforcement detailing. It also shows a range of values of D/B and B which can be considered for the use of design charts because of reasonably close values of moment capacities for D/B and B values within their recommended range.

The values of axial load, P_{ub} corresponding to the condition of maximum compression strain of 0.0035 in the extreme compression fibre of concrete and tension strain of 0.002 in the outermost layer of tension steel can be determined by Eq. 5.18 as,

$$P_{ub} = q_c f_{ck} A_g + q_s p A_g$$

where the values of q_c and q_s for P_{ub} about x-axis (P_{ubx}) and y-axis (P_{uby}) are given in Table 5.8. For P_{ubx}, the values of q_c and q_s are given for both positive and negative values of M_{ux} causing compression in the flange and in the web respectively.

The design chart has been obtained by plotting biaxial moments M_{ux} and M_{uy} for different values of area of steel A_s and a particular value of axial force P_u expressed in non-dimensional form as $M_{ux}/f_{ck} Z_x$, $M_{uy}/f_{ck} Z_y$, $p f_y/f_{ck}$ and $P_u/f_{ck} A_g$ respectively (Fig. 5.41), where p is defined as $100 A_s/A_g$. Only half of the symmetrical interaction curves about the axis of symmetry have been shown in the figure. The values of $P_u/f_{ck} A_g$ considered for design charts are 0.1, 0.3, 0.4, 0.5, 0.6, 0.7, 0.8 and 1.0. Design curves for two consecutive values of $P_u/f_{ck} A_g$ are shown on one chart. It also shows curves for three values of ratio of effective cover to reinforcement (d') to depth of section (D) appropriate for each geometry of the column section as given in Fig. 5.40. The interaction curves have been plotted for intermediate value of d'/D. For other values of d'/D, necessary corrections are made with the help of d'/D curves as explained below.

The design charts for different geometry and reinforcement detailing are shown in Charts T1.1 to T8.4. The index of the charts is given in Table 5.9.

The use of the charts for design is simple as explained with the help of Fig. 5.41. Consider that the area of steel is required to be determined for positive values of moments $M_{ux}/f_{ck} Z_x$ and $M_{uy}/f_{ck} Z_y$ and for $P_u/f_{ck} A_g = 0.1$ and 0.3. On design charts for $P_u/f_{ck} A_g = 0.1$ and 0.3, point a is plotted for the given value of moment $M_{ux}/f_{ck} Z_x$. On the vertical line drawn from point a, point b is plotted corresponding to the given value of d'/D from the interpolation of d'/D curves. A horizontal line is drawn from point b to intersect the intermediate d'/D curve at point c. Point c indicates the modified value of $M_{ux}/f_{ck} Z_x$ for the given value of d'/D for determining the area of steel with the use of curves plotted for the intermediate value of d'/D. Similarly, point a' is plotted for the given value of moment $M_{uy}/f_{ck} Z_y$. On the horizontal line drawn from point a', point b' is plotted corresponding to the given value of d'/D from the interpolation of d'/D curves. A vertical line is drawn from point b' to intersect the intermediate d'/D curve at point c'. Point c' indicates the modified value of $M_{uy}/f_{ck} Z_y$ for the given value of d'/D for determining the area of steel with the use of curves plotted for the intermediate value of d'/D. The area of steel is given by point d which is obtained by the intersection of vertical line drawn from point c and horizontal line drawn from point c'.

The use of the charts is explained with the help of following examples.

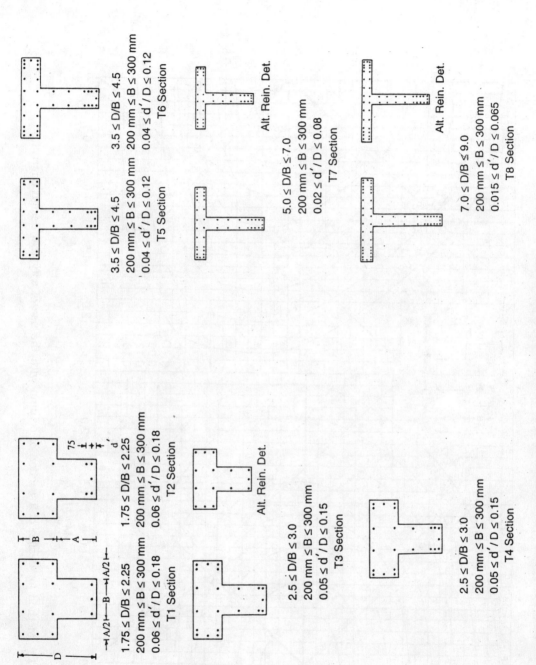

Fig. 5.40 Reinforcement detailing for different geometry of T section of column

304 *Handbook of Reinforced Concrete Design*

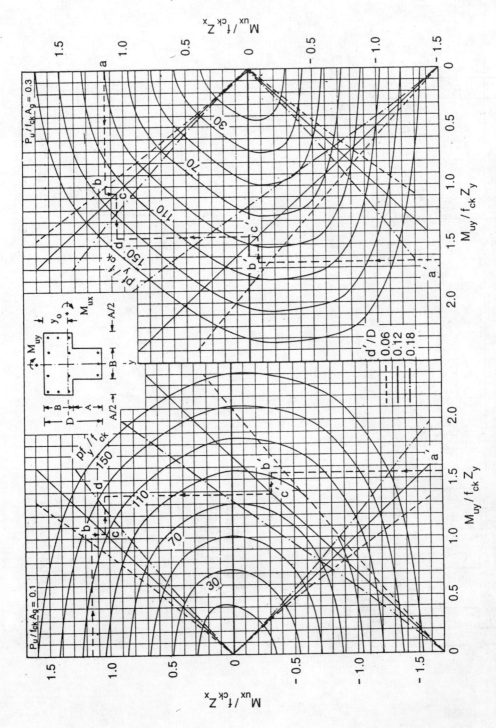

Fig. 5.41 Method for design of T section of column with the use of a typical design chart T1-1

Table 5.8 Values of q_c and q_s for T-shape of column

Cross-sections	Values of $\frac{d'}{D}$	Values of q_c	Values of q_c and q_s for P_{ubx} for positive moment M_{ux}									
			Values of q_s for concrete and steel of grades										
			M20			M25			M30		
			Fe 415	Fe 500	Fe 550	Fe 415	Fe 500	Fe 550	Fe 415	Fe 500	Fe 550
T1 Section	0.0600	0.1512	−0.3509	−0.3313	−0.3074	−0.3575	−0.3379	−0.3140	−0.3641	−0.3445	−0.3206
	0.0900	0.1448	−0.5716	−0.5605	−0.5394	−0.5768	−0.5657	−0.5446	−0.5819	−0.5708	−0.5497
	0.1200	0.1375	−0.8139	−0.8136	−0.7990	−0.8191	−0.8188	−0.8041	−0.8242	−0.8239	−0.8093
	0.1500	0.1319	−1.0815	−1.0898	−1.0769	−1.0867	−1.0950	−1.0821	−1.0918	−1.1001	1.0872
	0.1800	0.1260	−1.3581	1.3862	−1.3856	−1.3633	−1.3913	−1.3908	−1.3684	−1.3965	−1.3959
T2 Section	0.0600	0.1512	0.1342	0.2234	0.2778	0.1255	0.2147	0.2691	0.1169	0.2060	0.2604
	0.0900	0.1448	−0.0679	0.0044	0.0510	−0.0753	−0.0031	0.0435	−0.0828	−0.0105	0.0361
	0.1200	0.1375	−0.2931	−0.2432	−0.2209	−0.3005	−0.2506	−0.2283	−0.3079	−0.2580	−0.2357
	0.1500	0.1319	−0.5487	−0.5401	−0.5290	−0.5560	−0.5474	−0.5363	−0.5633	−0.5547	−0.5435
	0.1800	0.1260	−0.8443	−0.8687	−0.8682	−0.8514	−0.8757	−0.8752	−0.8584	−0.8828	−0.8823
T3 Section	0.0500	0.1334	−0.4312	−0.4071	−0.3787	−0.4381	−0.4140	−0.3856	−0.4451	−0.4210	−0.3926
	0.0750	0.1299	−0.6481	−0.6290	−0.6056	−0.6548	−0.6358	−0.6123	−0.6615	−0.6425	−0.6190
	0.1000	0.1264	−0.8733	−0.8669	−0.8501	−0.8797	−0.8733	−0.8565	−0.8862	−0.8798	−0.8629
	0.1250	0.1229	−1.1033	−1.1231	−1.1132	−1.1093	−1.1291	−1.1192	−1.1153	−1.1352	−1.1252
	0.1500	0.1194	−1.3053	−1.3950	−1.4027	−1.3108	−1.4005	−1.4082	−1.3163	−1.4061	−1.4138

(Contd.)

Table 5.8 (Contd.)

Values of q_c and q_s for P_{ubx} for positive moment M_{ux}

Values of q_s for concrete and steel of grades

Cross-sections	Values of $\frac{d'}{D}$	Values of q_c	M20 Fe 415	M20 Fe 500	M20 Fe 550	M25 Fe 415	M25 Fe 500	M25 Fe 550	M30 Fe 415	M30 Fe 500	M30 Fe 550
T4 Section	0.0500	0.1334	−0.1717	−0.1119	−0.0708	−0.1796	−0.1198	−0.0788	−0.1876	−0.1278	−0.0867
	0.0750	0.1299	−0.3805	−0.3328	−0.3098	−0.3882	−0.3404	−0.3174	−0.3958	−0.3480	−0.3250
	0.1000	0.1264	−0.6002	−0.5844	−0.5690	−0.6074	−0.5915	−0.5762	−0.6145	−0.5987	−0.5833
	0.1250	0.1229	−0.8380	−0.8561	−0.8470	−0.8446	−0.8627	−0.8536	−0.8512	−0.8693	−0.8603
	0.1500	0.1194	−1.0641	−1.1461	−1.1531	−1.0705	−1.1525	−1.1595	−1.0769	−1.1589	−1.1659
T5 Section	0.0400	0.1260	−0.6624	−0.6397	−0.6093	−0.6693	−0.6466	−0.6162	−0.6762	−0.6535	−0.6231
	0.0600	0.1234	−0.8408	−0.8284	−0.8049	−0.8474	−0.8351	−0.8115	−0.8540	−0.8417	−0.8181
	0.0800	0.1208	−1.0218	−1.0313	−1.0098	−1.0283	−1.0378	−1.0163	−1.0347	−1.0443	−1.0227
	0.1000	0.1182	−1.1419	−1.2367	−1.2286	−1.1482	−1.2430	−1.2349	−1.1545	−1.2493	−1.2412
	0.1200	0.1155	−1.2526	−1.4008	−1.4584	−1.2587	−1.4070	−1.4645	−1.2648	−1.4131	−1.4706
T6 Section	0.0400	0.1260	−0.7195	−0.7010	−0.6762	−0.7262	−0.7077	−0.6829	−0.7329	−0.7144	−0.6895
	0.0600	0.1234	−0.8975	−0.8915	−0.8778	−0.9041	−0.8980	−0.8843	−0.9106	−0.9046	−0.8909
	0.0800	0.1208	−1.0784	−1.0985	−1.0916	−1.0848	−1.1048	−1.0980	−1.0912	−1.1112	−1.1044
	0.1000	0.1182	−1.1981	−1.3146	−1.3189	−1.2043	−1.3208	−1.3251	−1.2104	−1.3269	−1.3312
	0.1200	0.1155	−1.3144	−1.4876	−1.5562	−1.3203	−1.4935	−1.5621	−1.3262	−1.4994	−1.5680

(Contd.)

Table 5.8 (*Contd.*)

Cross-sections	Values of $\dfrac{d'}{D}$	Values of q_c	Values of q_c and q_s for P_{ubx} for positive moment M_{ux}								
			Values of q_s for concrete and steel of grades								
			M20			M25			M30		
			Fe 415	Fe 500	Fe 550	Fe 415	Fe 500	Fe 550	Fe 415	Fe 500	Fe 550
T7 Section	0.0200	0.1228	−0.7731	−0.7485	−0.7162	−0.7802	−0.7556	−0.7233	−0.7873	−0.7627	−0.7303
	0.0350	0.1209	−0.9048	−0.8904	−0.8597	−0.9117	−0.8973	−0.8666	−0.9187	−0.9043	−0.8735
	0.0500	0.1191	−0.9938	−1.0340	−1.0113	−1.0005	−1.0408	−1.0181	−1.0073	−1.0475	−1.0249
	0.0650	0.1172	−1.0718	−1.1696	−1.1686	−1.0783	−1.1761	−1.1752	−1.0849	−1.1827	−1.1817
	0.0800	0.1153	−1.1488	−1.2783	−1.3216	−1.1553	−1.2847	−1.3280	−1.1617	−1.2911	−1.3344
T8 Section	0.0150	0.1207	−0.9855	−0.9805	−0.9547	−0.9922	−0.9872	−0.9613	−0.9988	−0.9938	−0.9680
	0.0275	0.1192	−1.0553	−1.0983	−1.0769	−1.0618	−1.1048	−1.0834	−1.0683	−1.1113	−1.0899
	0.0400	0.1176	−1.1181	−1.2191	−1.2037	−1.1246	−1.2256	−1.2102	−1.1310	−1.2320	−1.2166
	0.0525	0.1161	−1.1826	−1.3051	−1.3340	−1.1889	−1.3115	−1.3403	−1.1953	−1.3178	−1.3467
	0.0650	0.1146	−1.2406	−1.3853	−1.4373	−1.2468	−1.3916	−1.4435	−1.2531	−1.3978	−1.4498

(*Contd.*)

Table 5.8 (Contd.)

Values of q_c and q_s for P_{ubx} for negative moment M_{ux}

Values of q_s for concrete and steel of grades

Cross-sections	Values of $\dfrac{d'}{D}$	Values of q_c	M20			M25			M30		
			Fe 415	Fe 500	Fe 550	Fe 415	Fe 500	Fe 550	Fe 415	Fe 500	Fe 550
T1 Section	0.0600	0.2808	1.3518	1.5311	1.6295	1.3366	1.5159	1.6143	1.3214	1.5008	1.5991
	0.0900	0.2734	1.4019	1.5673	1.6591	1.3864	1.5518	1.6436	1.3708	1.5362	1.6280
	0.1200	0.2669	1.4550	1.6027	1.6834	1.4392	1.5869	1.6676	1.4233	1.5711	1.6517
	0.1500	0.2587	1.4996	1.6317	1.7049	1.4835	1.6156	1.6887	1.4673	1.5994	1.6726
	0.1800	0.2508	1.5430	1.6630	1.7110	1.5266	1.6466	1.6946	1.5101	1.6302	1.6782
T2 Section	0.0600	0.2808	0.8723	1.0277	1.1129	0.8591	1.0145	1.0998	0.8460	1.0014	1.0866
	0.0900	0.2734	0.9234	1.0667	1.1463	0.9099	1.0533	1.1328	0.8965	1.0398	1.1194
	0.1200	0.2669	0.9777	1.1057	1.1756	0.9639	1.0920	1.1619	0.9502	1.0782	1.1482
	0.1500	0.2587	1.0251	1.1396	1.2030	1.0112	1.1257	1.1891	0.9972	1.1117	1.1751
	0.1800	0.2508	1.0722	1.1762	1.2178	1.0580	1.1620	1.2036	1.0437	1.1478	1.1894
T3 Section	0.0500	0.3049	1.4709	1.7328	1.8418	1.4558	1.7178	1.8268	1.4408	1.7027	1.8117
	0.0750	0.3020	1.4976	1.7810	1.8953	1.4823	1.7657	1.8801	1.4671	1.7505	1.8649
	0.1000	0.2986	1.5257	1.8107	1.9491	1.5103	1.7953	1.9337	1.4949	1.7799	1.9183
	0.1250	0.2948	1.5474	1.8357	1.9811	1.5319	1.8201	1.9656	1.5164	1.8046	1.9500
	0.1500	0.2906	1.5615	1.8519	2.0074	1.5459	1.8362	1.9918	1.5302	1.8205	1.9761

(Contd.)

Table 5.8 (Contd.)

Cross-sections	Values of $\dfrac{d'}{D}$	Values of q_c	Values of q_c and q_s for P_{ubx} for negative moment M_{ux}									
			Values of q_s for concrete and steel of grades										
			M20			M25			M30		
			Fe 415	Fe 500	Fe 550	Fe 415	Fe 500	Fe 550	Fe 415	Fe 500	Fe 550
T4 Section	0.0500	0.3049	1.2677	1.5068	1.6063	1.2534	1.4926	1.5921	1.2392	1.4784	1.5779
	0.0750	0.3020	1.2995	1.5582	1.6627	1.2851	1.5438	1.6483	1.2707	1.5294	1.6339
	0.1000	0.2986	1.3331	1.5933	1.7196	1.3185	1.5787	1.7051	1.3039	1.5641	1.6905
	0.1250	0.2948	1.3612	1.6244	1.7572	1.3465	1.6097	1.7425	1.3319	1.5950	1.7278
	0.1500	0.2906	1.3829	1.6480	1.7900	1.3681	1.6332	1.7752	1.3533	1.6184	1.7605
T5 Section	0.0400	0.3129	1.5548	1.8818	2.0633	1.5397	1.8666	2.0482	1.5246	1.8515	2.0331
	0.0600	0.3155	1.5789	1.9071	2.0889	1.5637	1.8919	2.0737	1.5484	1.8767	2.0585
	0.0800	0.3088	1.6017	1.9279	2.1130	1.5864	1.9126	2.0977	1.5710	1.8973	2.0824
	0.1000	0.3108	1.6247	1.9479	2.1303	1.6093	1.9325	2.1148	1.5939	1.9170	2.0994
	0.1200	0.3039	1.6484	1.9682	2.1465	1.6328	1.9527	2.1309	1.6173	1.9371	2.1154
T6 Section	0.0400	0.3129	1.6414	1.9828	2.1703	1.6259	1.9673	2.1549	1.6105	1.9519	2.1394
	0.0600	0.3155	1.6669	2.0096	2.1974	1.6513	1.9940	2.1818	1.6357	1.9784	2.1662
	0.0800	0.3088	1.6910	2.0316	2.2228	1.6754	2.0159	2.2072	1.6597	2.0003	2.1915
	0.1000	0.3108	1.7115	2.0528	2.2412	1.6957	2.0370	2.2255	1.6800	2.0213	2.2098
	0.1200	0.3039	1.7318	2.0743	2.2585	1.7160	2.0585	2.2427	1.7002	2.0427	2.2269

(Contd.)

Table 5.8 (Contd.)

Values of q_c and q_s for P_{ubx} for negative moment M_{ux}

Cross-sections	Values of $\dfrac{d'}{D}$	Values of q_c	Values of q_s for concrete and steel of grades									
			M20			M25			M30		
			Fe 415	Fe 500	Fe 550	Fe 415	Fe 500	Fe 550	Fe 415	Fe 500	Fe 550
T7 Section	0.0200	0.3201	1.6115	1.9725	2.1788	1.5963	1.9573	2.1636	1.5810	1.9420	2.1483
	0.0350	0.3301	1.6269	1.9923	2.1998	1.6116	1.9770	2.1845	1.5963	1.9616	2.1692
	0.0500	0.3250	1.6397	2.0109	2.2179	1.6244	1.9955	2.2025	1.6090	1.9802	2.1871
	0.0650	0.3198	1.6519	2.0290	2.2342	1.6365	2.0136	2.2188	1.6210	1.9982	2.2034
	0.0800	0.3147	1.6643	2.0417	2.2503	1.6488	2.0262	2.2348	1.6333	2.0107	2.2193
T8 Section	0.0150	0.3294	1.7234	2.1107	2.3357	1.7077	2.0949	2.3200	1.6920	2.0792	2.3043
	0.0275	0.3253	1.7336	2.1255	2.3534	1.7179	2.1097	2.3376	1.7021	2.0940	2.3219
	0.0400	0.3211	1.7441	2.1407	2.3659	1.7283	2.1249	2.3501	1.7125	2.1091	2.3344
	0.0525	0.3169	1.7534	2.1561	2.3771	1.7376	2.1403	2.3613	1.7218	2.1245	2.3455
	0.0650	0.3325	1.7627	2.1698	2.3885	1.7468	2.1539	2.3727	1.7310	2.1381	2.3568

(Contd.)

Table 5.8 (Contd.)

Cross-sections	Values of $\frac{d'}{D}$	Values of q_c	Values of q_s for concrete and steel of grades								
			M20			M25			M30		
			Fe 415	Fe 500	Fe 550	Fe 415	Fe 500	Fe 550	Fe 415	Fe 500	Fe 550
T1 Section	0.0600	0.2133	0.5756	0.6925	0.7313	0.5646	0.6815	0.7203	0.5536	0.6705	0.7093
	0.0900	0.2030	0.5322	0.5778	0.6137	0.5216	0.5672	0.6031	0.5110	0.5565	0.5925
	0.1200	0.1963	0.4187	0.4539	0.4833	0.4086	0.4438	0.4733	0.3986	0.4338	0.4633
	0.1500	0.1864	0.2901	0.3168	0.3445	0.2809	0.3076	0.3353	0.2718	0.2984	0.3261
	0.1800	0.1768	0.1501	0.1705	0.1860	0.1421	0.1625	0.1779	0.1340	0.1544	0.1698
T2 Section	0.0600	0.2133	0.6282	0.7476	0.7812	0.6173	0.7366	0.7702	0.6063	0.7256	0.7592
	0.0900	0.2030	0.5833	0.6227	0.6539	0.5728	0.6121	0.6433	0.5622	0.6016	0.6327
	0.1200	0.1963	0.4579	0.4884	0.5139	0.4480	0.4784	0.5040	0.4380	0.4685	0.4940
	0.1500	0.1864	0.3179	0.3409	0.3649	0.3088	0.3319	0.3559	0.2998	0.3229	0.3469
	0.1800	0.1768	0.1660	0.1837	0.1971	0.1583	0.1760	0.1893	0.1505	0.1682	0.1815
T3 Section	0.0500	0.2171	0.7015	0.7904	0.8464	0.6909	0.7798	0.8357	0.6802	0.7691	0.8251
	0.0750	0.2060	0.6184	0.7022	0.7532	0.6082	0.6921	0.7431	0.5980	0.6819	0.7329
	0.1000	0.1952	0.5293	0.6058	0.6502	0.5197	0.5963	0.6407	0.5102	0.5867	0.6311
	0.1250	0.1848	0.4346	0.4996	0.5370	0.4258	0.4908	0.5283	0.4170	0.4820	0.5195
	0.1500	0.1795	0.3280	0.3804	0.4054	0.3200	0.3723	0.3973	0.3119	0.3643	0.3893

(Contd.)

Table 5.8 (Contd.)

Cross-sections	Values of $\frac{d'}{D}$	Values of q_c	Values of q_c and q_s for P_{uby}									
			Values of q_s for concrete and steel of grades										
			M20			M25			M30		
			Fe 415	Fe 500	Fe 550	Fe 415	Fe 500	Fe 550	Fe 415	Fe 500	Fe 550
T4 Section	0.0500	0.2171	0.7423	0.8234	0.8745	0.7317	0.8129	0.8640	0.7211	0.8023	0.8534
	0.0750	0.2060	0.6531	0.7296	0.7762	0.6430	0.7195	0.7661	0.6329	0.7094	0.7560
	0.1000	0.1952	0.5578	0.6277	0.6682	0.5483	0.6182	0.6588	0.5389	0.6088	0.6493
	0.1250	0.1848	0.4566	0.5160	0.5502	0.4480	0.5074	0.5416	0.4395	0.4988	0.5330
	0.1500	0.1795	0.3437	0.3915	0.4144	0.3359	0.3837	0.4065	0.3281	0.3759	0.3987
T5 Section	0.0400	0.2179	0.8062	0.8960	0.9519	0.7953	0.8851	0.9410	0.7844	0.8742	0.9301
	0.0600	0.2134	0.7407	0.8245	0.8701	0.7301	0.8138	0.8594	0.7194	0.8031	0.8488
	0.0800	0.2010	0.6707	0.7427	0.7808	0.6604	0.7324	0.7705	0.6501	0.7221	0.7603
	0.1000	0.1892	0.5954	0.6506	0.6866	0.5857	0.6408	0.6768	0.5759	0.6310	0.6670
	0.1200	0.1850	0.5076	0.5545	0.5854	0.4984	0.5454	0.5762	0.4893	0.5362	0.5671
T6 Section	0.0400	0.2179	0.8165	0.9169	0.9713	0.8056	0.9060	0.9604	0.7947	0.8951	0.9495
	0.0600	0.2134	0.7538	0.8423	0.8894	0.7431	0.8317	0.8787	0.7325	0.8210	0.8681
	0.0800	0.2010	0.6852	0.7588	0.8042	0.6750	0.7486	0.7939	0.6647	0.7383	0.7837
	0.1000	0.1892	0.6044	0.6714	0.7117	0.5946	0.6616	0.7019	0.5848	0.6518	0.6921
	0.1200	0.1850	0.5174	0.5776	0.6073	0.5083	0.5685	0.5981	0.4991	0.5593	0.5890

(Contd.)

Table 5.8 (*Contd.*)

Cross-sections	Values of $\frac{d'}{D}$	Values of q_c	Values of q_c and q_s for P_{uby}							
			Values of q_s for concrete and steel of grades							
			M20			M25			M30		
			Fe 415	Fe 500	Fe 550	Fe 415	Fe 500	Fe 550	Fe 415	Fe 500	Fe 550
T7 Section	0.0200	0.2413	0.8645	0.9711	1.0348	0.8529	0.9595	1.0232	0.8413	0.9479	1.0116
	0.0350	0.2253	0.8212	0.9208	0.9805	0.8097	0.9094	0.9691	0.7983	0.8979	0.9577
	0.0500	0.2243	0.7749	0.8660	0.9222	0.7637	0.8548	0.9110	0.7526	0.8437	0.8999
	0.0650	0.2107	0.7263	0.8093	0.8619	0.7155	0.7984	0.8511	0.7046	0.7876	0.8402
	0.0800	0.1971	0.6702	0.7483	0.7995	0.6598	0.7378	0.7891	0.6494	0.7274	0.7786
T8 Section	0.0150	0.2391	0.8907	0.9974	1.0642	0.8789	0.9856	1.0524	0.8671	0.9738	1.0407
	0.0275	0.2361	0.8554	0.9571	1.0216	0.8439	0.9455	1.0100	0.8323	0.9340	0.9984
	0.0400	0.2236	0.8163	0.9144	0.9749	0.8050	0.9030	0.9636	0.7936	0.8916	0.9522
	0.0525	0.2253	0.7757	0.8704	0.9253	0.7646	0.8593	0.9142	0.7535	0.8482	0.9031
	0.0650	0.2080	0.7335	0.8225	0.8741	0.7227	0.8118	0.8633	0.7120	0.8010	0.8525

Table 5.9 Index for Charts T1.1 to T8.4

Reinforcement Detailing	Chart No.	$P_u/f_{ck} A_g$	Page No.
T1 Section	T1.1	0.1 and 0.3	322
	T1.2	0.4 and 0.5	323
	T1.3	0.6 and 0.7	324
	T1.4	0.8 and 1.0	325
T2 Section	T2.1	0.1 and 0.3	326
	T2.2	0.4 and 0.5	327
	T2.3	0.6 and 0.7	328
	T2.4	0.8 and 1.0	329
T3 Section	T3.1	0.1 and 0.3	330
	T3.2	0.4 and 0.5	331
	T3.3	0.6 and 0.7	332
	T3.4	0.8 and 1.0	333
T4 Section	T4.1	0.1 and 0.3	334
	T4.2	0.4 and 0.5	335
	T4.3	0.6 and 0.7	336
	T4.4	0.8 and 1.0	337
T5 Section	T5.1	0.1 and 0.3	338
	T5.2	0.4 and 0.5	339
	T5.3	0.6 and 0.7	340
	T5.4	0.8 and 1.0	341
T6 Section	T6.1	0.1 and 0.3	342
	T6.2	0.4 and 0.5	343
	T6.3	0.6 and 0.7	344
	T6.4	0.8 and 1.0	345
T7 Section	T7.1	0.1 and 0.3	346
	T7.2	0.4 and 0.5	347
	T7.3	0.6 and 0.7	348
	T7.4	0.8 and 1.0	349
T8 Section	T8.1	0.1 and 0.3	350
	T8.2	0.4 and 0.5	351
	T8.3	0.6 and 0.7	352
	T8.4	0.8 and 1.0	353

Example 5.9: Design a biaxially eccentrically loaded T-shaped column section for the following data:

Column section: $\dfrac{D}{B}$ = 4

B = 250 mm

D = 1000 mm

Ultimate axial load, P_u = 4000 kN

Ultimate moments, M_{ux} = 750 kN.m

M_{uy} = 750 kN.m

Grade of concrete : M20

Grade of steel : Fe 415

Solution

Consider 28 mm φ bars with clear cover of 40 mm.

∴ Effectives cover, d' = 40 + 14 = 54 mm

and d'/D = 54/1000 = 0.054

Area of the section,

$$A_g = (2D - B)B = (2 \times 1000 - 250) \times 250 = 437500 \text{ mm}^2$$

Centre of gravity of the section,

$$x_o = 0.5\,D = 0.5 \times 1000 = 500 \text{ mm}$$

$$y_o = \frac{DB + D^2 - B^2}{2(2D - B)} = \frac{1000 \times 250 + 1000^2 - 250^2}{2 \times (2 \times 1000 - 250)} = 339.286 \text{ mm}$$

Moment of intertia about x and y axes of the section,

$$I_{xx} = \frac{DB^3}{12} + DB(y_o - 0.5B)^2 + \frac{B(D-B)^3}{12} + (D-B)B\left(\frac{D+B}{2} - y_o\right)^2$$

$$= \frac{1000 \times 250^3}{12} + 1000 \times 250 \times (339.286 - 0.5 \times 250)^2 + \frac{250 \times (1000 - 250)^3}{12}$$

$$+ (1000 - 250) \times 250 \times \left(\frac{1000 + 250}{2} - 339.286\right)^2$$

$$= 36876.74 \times 10^6 \text{ mm}^4$$

$$I_{yy} = \frac{BD^3}{12} + \frac{(D-B)B^3}{12}$$

$$= \frac{250 \times 1000^3}{12} + \frac{(1000-250) \times 250^3}{12}$$

$$= 21809.90 \times 10^6 \text{ mm}^4$$

Section modulus about x and y axes,

$$Z_x = \frac{I_{xx}}{y_o} = \frac{36876.74 \times 10^6}{339.2857} = 10868.925 \times 10^4 \text{ mm}^3$$

$$Z_y = \frac{I_{yy}}{x_o} = \frac{21809.90 \times 10^6}{500.0} = 4361.979 \times 10^4 \text{ mm}^3$$

Compute

$$\frac{P_u}{f_{ck} A_g} = \frac{4000 \times 10^3}{20 \times 437500} = 0.4571$$

$$\frac{M_{ux}}{f_{ck} Z_x} = \frac{750 \times 10^6}{20 \times 10868.925 \times 10^4} = 0.345$$

$$\frac{M_{uy}}{f_{ck} Z_y} = \frac{750 \times 10^6}{20 \times 4361.979 \times 10^4} = 0.86$$

Area of steel is determined from the linear interpolation of area of steel determined for $P_u/f_{ck} A_g = 0.4$ and 0.5 from Chart T5.2. The use of chart for determining the area of steel has been explained with the help of Fig. 5.41. Area of steel for $f_{ck} A_g = 0.4$ and 0.5 are,

For $\dfrac{P_u}{f_{ck} A_g} = 0.4$, $\dfrac{p f_y}{f_{ck}} = 73.5$

For $\dfrac{P_u}{f_{ck} A_g} = 0.5$, $\dfrac{p f_y}{f_{ck}} = 84.5$

∴ For $\dfrac{P_u}{f_{ck} A_g} = 0.4571$, $\dfrac{p f_y}{f_{ck}} = 73.5 + \dfrac{84.5 - 73.5}{0.5 - 0.4} \times (0.4571 - 0.4) = 79.781$

∴ $$p = \frac{79.781 \times 20}{415} = 3.8449$$

∴ $$A_s = \frac{3.8449 \times 437500}{100} = 16821.44 \text{ mm}^2$$

The diameter of reinforcement bars is chosen to provide 29 bars in accordance with recommended reinforcement detailing. Accordingly 29×28 mm ϕ bars ($A_s = 17856$ mm^2) is provided as shown in Fig. Ex. 5.9.

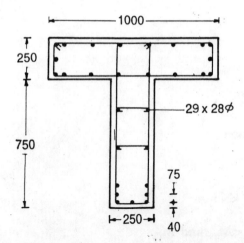

Fig. Ex. 5.9 Reinforcement detailing

Example 5.10: Design a biaxial eccentrically loaded braced T-shaped column deforming in single curvature for the following data.

Cross-section: $\frac{D}{B}$	=	2
B	=	250 mm
D	=	500 mm
Ultimate axial load, P_u	=	2000 kN
Ultimate moment about x-axis,		
at bottom, M_{ux1}	=	275.0 kN.m
at top, M_{ux2}	=	125.0 kN.m
Ultimate moment about y-axis		
at bottom, M_{uy1}	=	150.0 kN
at top, M_{uy2}	=	100.0 kN.m
Unsupported length of column, l	=	9m
Effective length about x-axis, l_{ex}	=	6.9 m
Effective length about y-axis, l_{ey}	=	7.95 m
Grade of concrete	:	M 25
Grade of steel	:	Fe 415

Solution

Moment due to minumum of eccentricities:

$$e_{min} = e_{x,min} = e_{y,min}$$

$$= \frac{l}{500} + \frac{D}{30} \text{ or 20 mm whichever is more}$$

$$= \frac{9000}{500} + \frac{500}{30} \text{ or 20 mm whichever is more}$$

$$= 34.67 \text{ mm}$$

∴ $$M_{uxe} = M_{uye} = P_u e_{min} = 2000 \times 34.67 \text{ kN.mm}$$

$$= 69.34 \text{ kN.m} < M_{ux} \text{ and } M_{uy}$$

Check for short/long column:

$$l_{ex}/D = 6900/500 = 13.8 > 12$$

$$l_{ey}/D = 7950/500 = 15.91 > 12$$

Hence it is a slender column about both axes of bending.
Additional moments due to slenderness effects,

$$M_{ux} = P_u e_y = P_u \frac{D}{2000}\left(\frac{l_{ey}}{D}\right)^2$$

$$= 2000 \times \frac{500}{2000} \times 15.91^2 \text{ kN.mm} = 126.564 \text{ kN.m}$$

and $$M_{uy} = P_u e_x = P_u \frac{D}{2000}\left(\frac{l_{ex}}{D}\right)^2$$

$$= 2000 \times \frac{500}{2000} \times 13.8^2 \text{ kN.mm} = 95.22 \text{ kN.m}$$

Additional moments M_{ux} and M_{uy} are reduced by multiplying with a factor K given by,

$$K = \frac{P_{u2} - P_u}{P_{u2} - P_{ub}}$$

where P_{uz} and P_{ub} are determined for the assumed area of longitudinal reinforcement and its distribution. Consider percentage of reinforcement, $p = 5.0$ provided as reinforcement detailing type 1 (Fig. 5.41) with clear cover of 40 mm. Consider 28 mm ϕ bars.

∴ Effectives cover, $d' = 40 + 28/2 = 54$ mm

and $$d'/D = 54/500 = 0.108$$

Area of the section,

$$A_g = (2D - B)B = (2 \times 500 - 250) \times 250 = 187500 \text{ mm}^2$$

Centre of gravity of the section,

$$x_o = 0.5 D = 0.5 \times 500 = 250 \text{ mm}$$

$$y_o = \frac{DB + D^2 - D^2}{2(2D - B)} = \frac{500 \times 250 + 500^2 - 250^2}{2 \times (2 \times 500 - 250)} = 208.333 \text{ mm}$$

Moment of intertia about x and y-axes of the section,

$$I_{xx} = \frac{DB^3}{12} + DB(y_o - 0.5B)^2 + \frac{B(D-B)^3}{12} + (D-B)B\left(\frac{D+B}{2} - y_o\right)^2$$

$$= \frac{500 \times 250^3}{12} + 500 \times 250 \times (208.333 - 0.5 \times 250)^2 + \frac{250 \times (500-250)^3}{12}$$

$$+ (500 - 250) \times \left(\frac{500 + 250}{2} - 208.333\right)^2$$

$$= 3580.719 \times 10^6 \text{ mm}^4$$

$$I_{yy} = \frac{BD^3}{12} + \frac{(D-B)B^3}{12}$$

$$= \frac{250 \times 500^3}{12} + \frac{(500 - 250) \times 250^3}{12}$$

$$= 2929.68 \times 10^6 \text{ mm}^4$$

Section modulus about x and y axes of the section,

$$Z_x = \frac{3580.719 \times 10^6}{208.33} = 1718.75 \text{ mm}^3$$

and

$$Z_y = \frac{2929.68 \times 10^6}{250} = 1171.87 \times 10^4 \text{ mm}^2$$

Compute $P_{uz} = 0.466 f_{ck} A_g + (0.75 f_y - 0.446 f_{ck}) p A_g / 100$

$$= 0.446 \times 25 \times 187500 + (0.75 \times 415 - 0.446 \times 25) \times 0.05 \times 187500 \text{ N} = 4904.063 \text{ kN}$$

and $P_{ub} = q_c f_{ck} A_g + q_s p A_g$

where q_c and q_s are determined from Table 5.6. as,

$$q_{cx} = 0.1448 + \frac{0.1375 - 0.1448}{0.03} \times 0.018 = 0.1404$$

$$q_{sx} = -0.0753 + \frac{-0.3005 - 0.0753}{0.03} \times 0.018 = -0.2104$$

$$q_{cy} = 0.2030 + \frac{0.1963 - 0.2030}{0.03} \times 0.018 = 0.199$$

$$q_{sy} = 0.5728 + \frac{0.4480 - 0.5728}{0.03} \times 0.018 = 0.498$$

$\therefore \quad P_{ubx} = 0.1404 \times 25 \times 187500 - 0.2104 \times 5 \times 187500 \text{ N} = 460.875 \text{ kN}$

$P_{uby} = 0.199 \times 25 \times 187500 + 0.498 \times 5 \times 187500 \text{ N} = 1399.69 \text{ kN}$

and
$$K_x = \frac{4904.063 - 2000}{4904.063 - 460.875} = 0.6536$$

$$K_y = \frac{4904.063 - 2000}{4904.063 - 1399.69} = 0.8287$$

Reduced additional moments are,

$$M'_{ax} = K_x M_{ax} = 0.6536 \times 126.563 = 82.72 \text{ kN.m}$$

and
$$M'_{ay} = K_y M_{ay} = 0.8287 \times 95.063 = 78.78 \text{ kN.m}$$

The ultimate design moments are,

$$M_{ux} = M'_{ux} + M'_{ax}$$

and
$$M_{uy} = M'_{uy} + M'_{ay}$$

where $M'_{ux} = 0.6 M_{ux, max} + 0.4 M_{uy, min}$

$= 0.6 \times 275 + 0.4 \times 125$

$= 215 \text{ kN.m} > 0.4 M_{ux, max}$

and $M'_{uy} = 0.6 M_{uy, max} + 0.4 M_{uy, min}$

$= 0.6 \times 150 + 0.4 \times 100$

$= 130 \text{ kN.m} > 0.4 M_{uy, max}$

$\therefore \quad M_{ux} = 215 + 82.72 = 297.72 \text{ kN.m}$

and $\quad M_{uy} = 130 + 78.78 = 208.78 \text{ kN.m}$

Compute $\quad \dfrac{P_u}{f_{ck} A_g} = \dfrac{2000 \times 10^3}{25 \times 187500} = 0.4267$

$$\frac{M_{ux}}{f_{ck} Z_x} = \frac{297.72 \times 10^6}{25 \times 1718.77 \times 10^4} = 0.693$$

$$\frac{M_{uy}}{f_{ck} Z_y} = \frac{208.78 \times 10^6}{25 \times 1171.87 \times 10^4} = 0.713$$

Area of steel is determined from the linear interpolation of area of steel determined for $P_u/f_{ck}A_g = 0.4$ and $P_u/f_{ck}A_g = 0.5$ from Chart T 2.2. The use of chart for determining the area of steel has been explained with the help of Fig. 5.41. Area of steel for $P_u/f_{ck}A_g = 0.4$ and 0.5 are,

For $\dfrac{P_u}{f_{ck}A_g} = 0.4$, $\dfrac{pf_y}{f_{ck}} = 88.0$

For $\dfrac{P_u}{f_{ck}A_g} = 0.5$, $\dfrac{pf_y}{f_{ck}} = 98.0$

$\therefore$ For $\dfrac{P_u}{f_{ck}A_g} = 0.4267$, $\dfrac{pf_y}{f_{ck}} = 88.0 + \dfrac{98.0 - 88.0}{0.5 - 0.4} \times (0.4267 - 0.4) = 90.67$

$\therefore \qquad p = 90.67 \times 25 / 415 = 5.462$

and $\qquad A_s = 5.462 \times 187500 / 100 = 10241.25 \text{ mm}^2$

The diameter of reinforcement bars is chosen to provide 15 bars in accordance with recommended reinforcement detailing. Accordingly 9×30 mm $\phi + 6 \times 28$ mm ϕ bars ($A_s = 10056$ mm^2) are provided as shown in Fig. Ex. 5.10. Reinforcement bars of 30 mm are ϕ placed at the edges which shall result in conservative design.

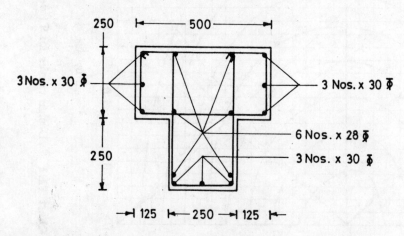

Fig. Ex. 5.10 Reinforcement detailing

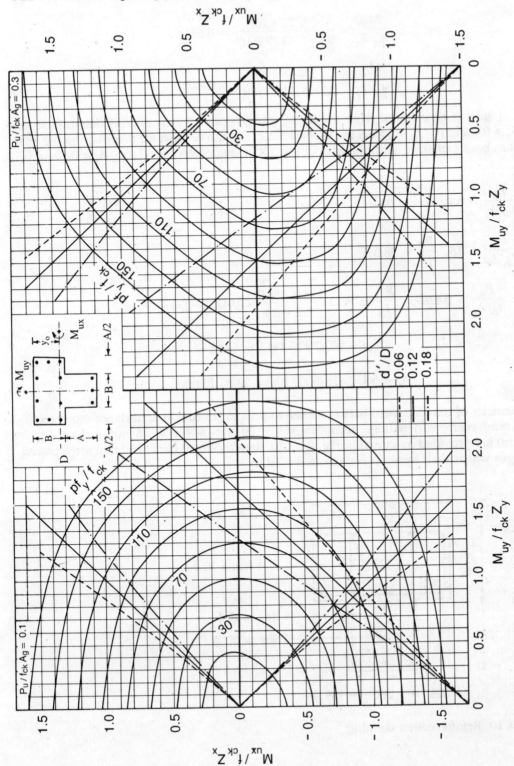

Chart T1.1 Design curves for T1 section of column for $P_u/f_{ck}A_g = 0.1$ and 0.3 $(1.75 \leq D/B \leq 2.25)$

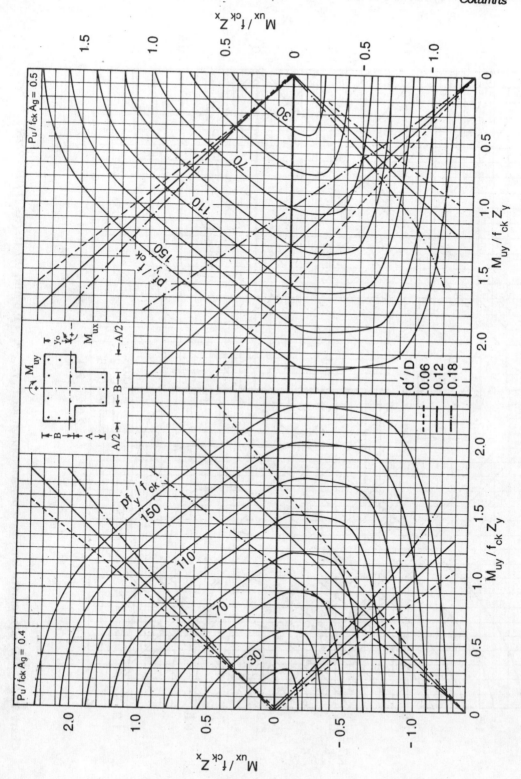

Chart T1.2 Design curves for T1 section of column for $P_u/f_{ck}A_g = 0.4$ and 0.5 ($1.75 \leq D/B \leq 2.25$)

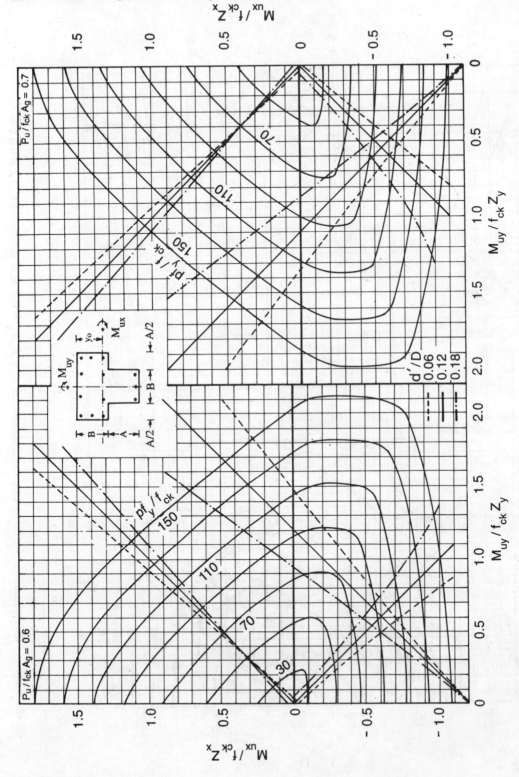

Chart T1.3   Design curves for T1 section of column for $P_u / f_{ck} A_g = 0.6$ and 0.7 ($1.75 \leq D / B \leq 2.25$)

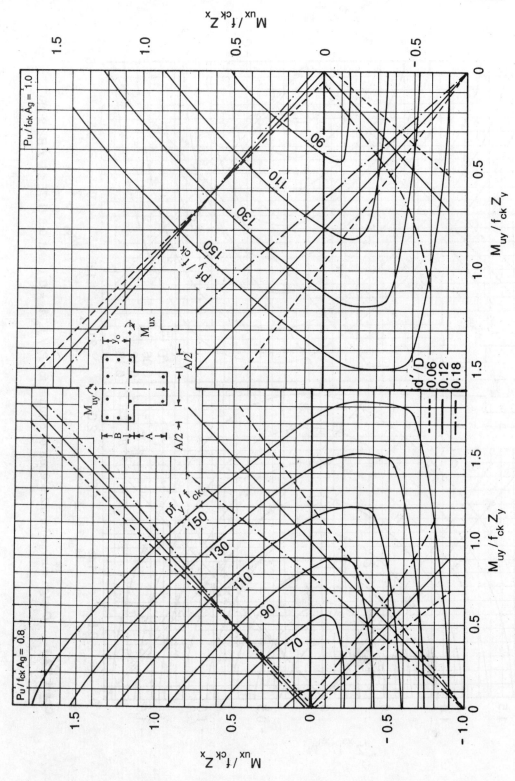

Chart T1.4 Design curves for T1 section of column for $P_u/f_{ck}A_g = 0.8$ and 1.0 ($1.75 \leq D/B \leq 2.25$)

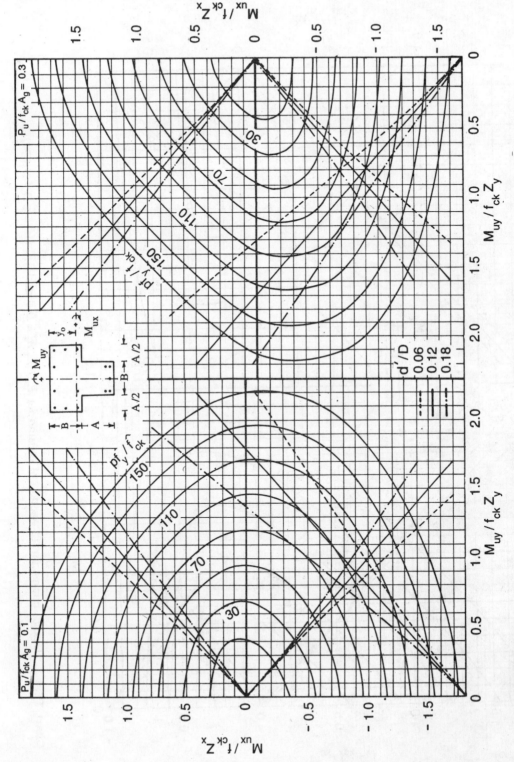

Chart T2.1 Design curves for T2 section of column for $P_u/f_{ck}A_g = 0.1$ and 0.3 $(1.75 \leq D/B \leq 2.25)$

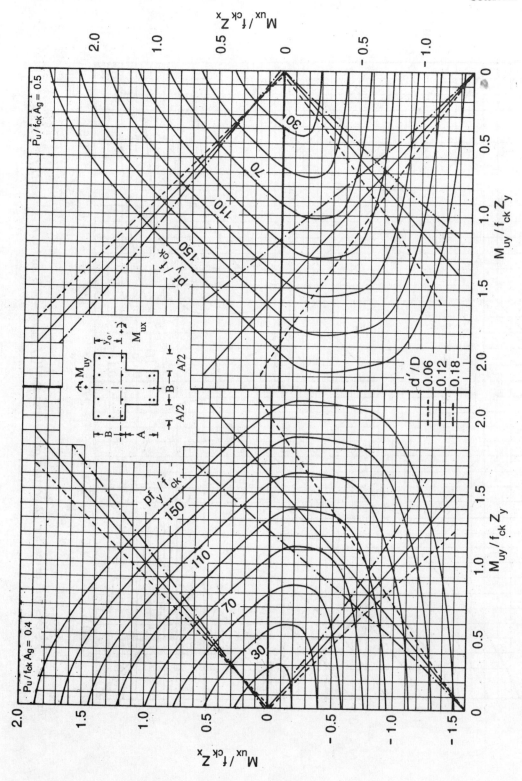

Chart T2.2 Design curves for T2 section of column for $P_u / f_{ck} A_g = 0.4$ and 0.5 ($1.75 \leq D/B \leq 2.25$)

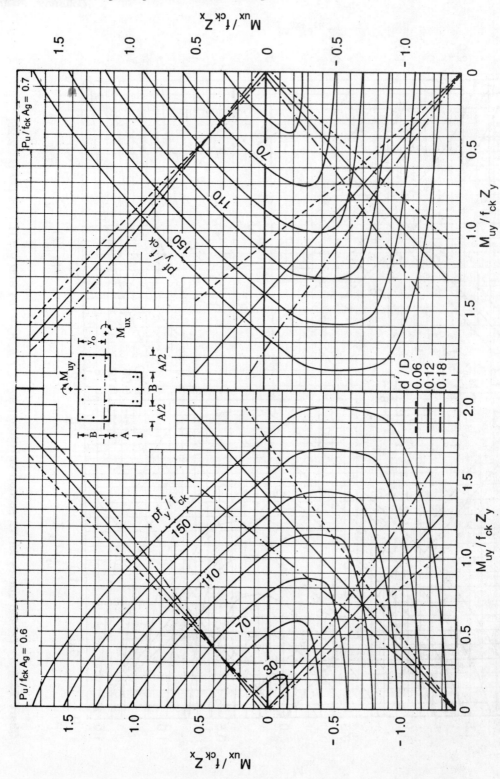

Chart T2.3 Design curves for T2 section of column for $P_u/f_{ck}A_g = 0.6$ and 0.7 $(1.75 \le D/B \le 2.25)$

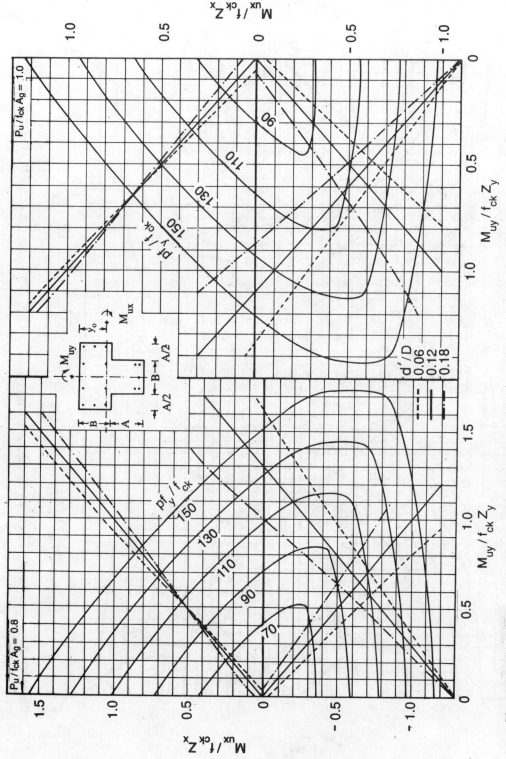

Chart T2.4 Design curves for T2 section of column for $P_u/f_{ck}A_g = 0.8$ and 1.0 $(1.75 \leq D/B \leq 2.25)$

330 Handbook of Reinforced Concrete Design

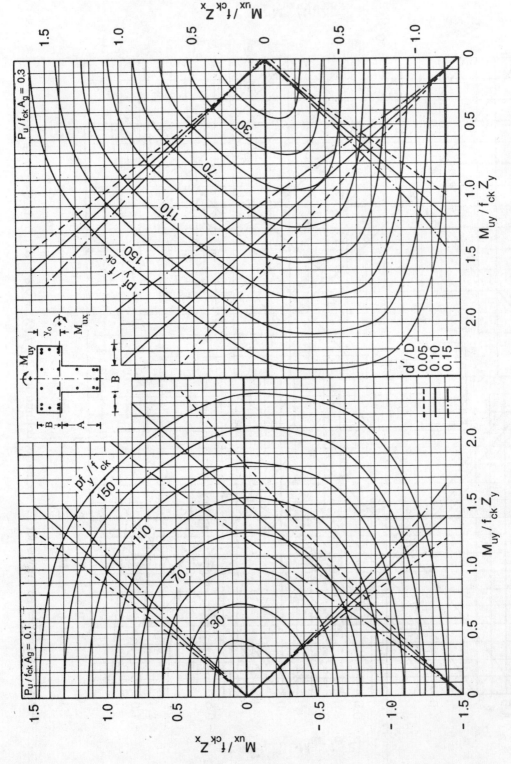

Chart T3.1 Design curves for T3 section of column for $P_u / f_{ck} A_g = 0.1$ and 0.3 $(2.5 \leq D/B \leq 3.0)$

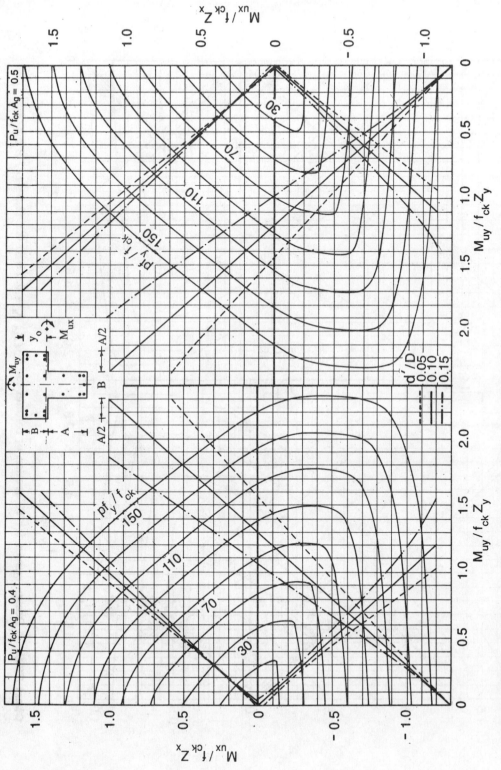

Chart T3.2 Design curves for T3 section of column for $P_u/f_{ck}A_g = 0.4$ and 0.5 ($2.5 \leq D/B \leq 3.0$)

332 Handbook of Reinforced Concrete Design

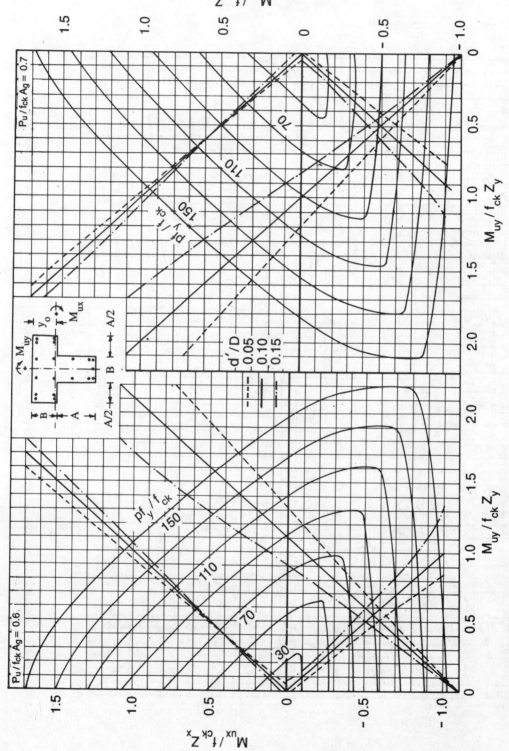

Chart T3.3   Design curves for T3 section of column for $P_u/f_{ck}A_g = 0.6$ and 0.7 $(2.5 \leq D/B \leq 3.0)$

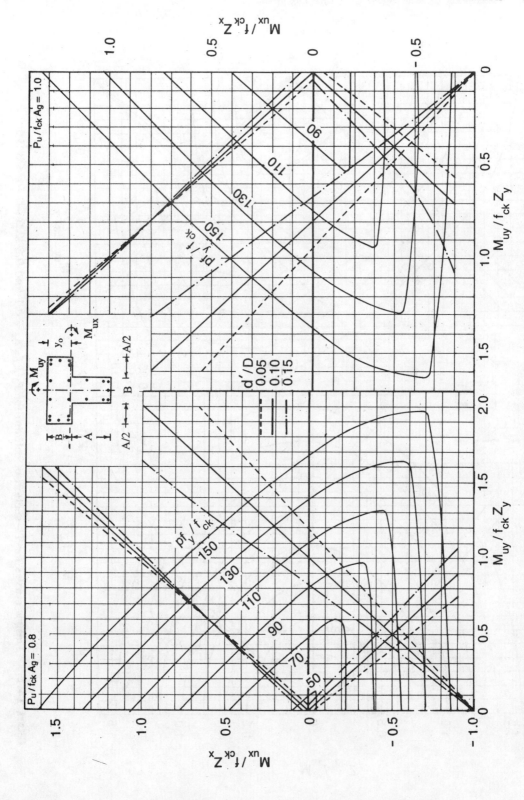

Chart T3.4 Design curves for T3 section of column for $P_u / f_{ck} A_g = 0.8$ and 1.0 $(2.5 \leq D/B \leq 3.0)$

334 Handbook of Reinforced Concrete Design

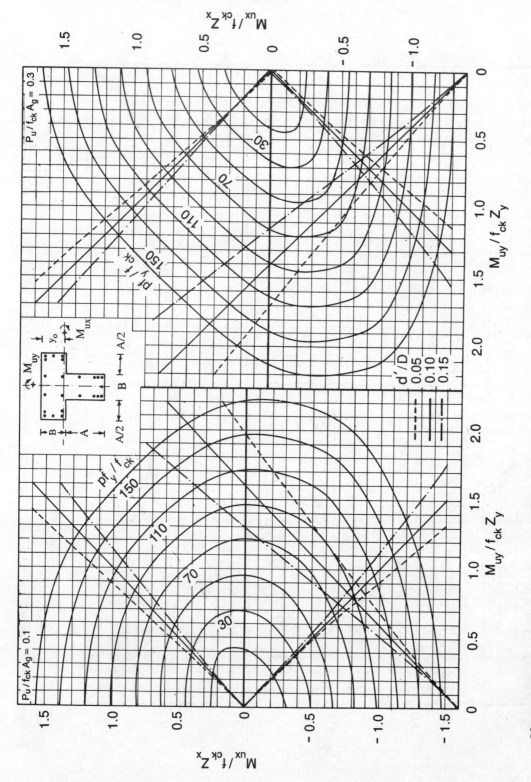

Chart T4.1 Design curves for T4 section of column for $P_u / f_{ck} A_g = 0.1$ and 0.3 $(2.5 \leq D / B \leq 3.0)$

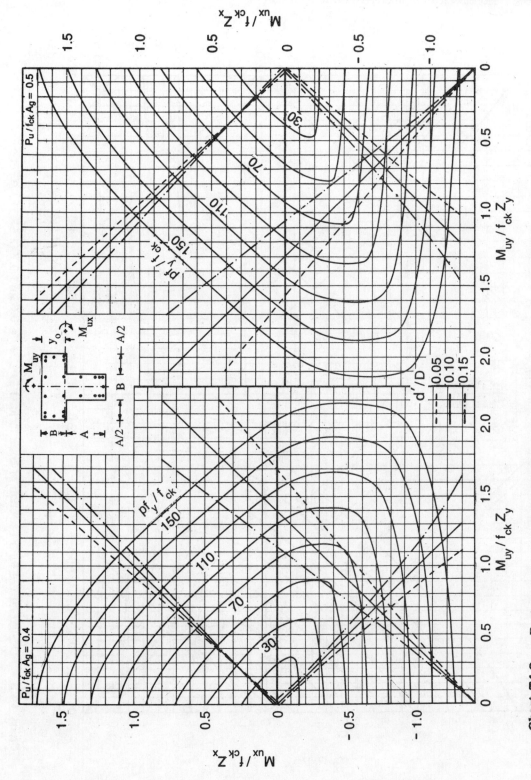

Chart T4.2 Design curves for T4 section of column for $P_u/f_{ck}A_g = 0.4$ and 0.5 $(2.5 \leq D/B \leq 3.0)$

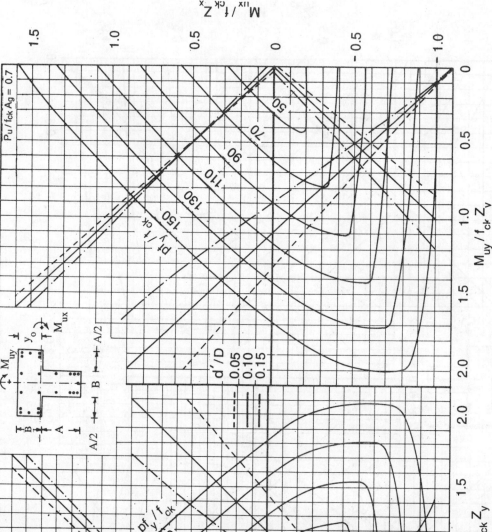

Chart T4.3 Design curves for T4 section of column for $P_u / f_{ck} A_g = 0.6$ and 0.7 $(2.5 \leq D/B \leq 3.0)$

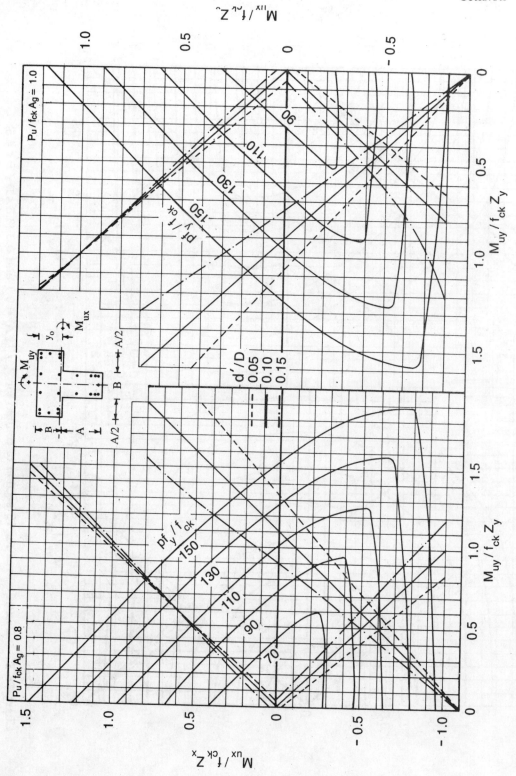

Chart T4.4 Design curves for T4 section of column for $P_u/f_{ck}A_g = 0.8$ and 1.0 ($2.5 \leq D/B \leq 3.0$)

338 Handbook of Reinforced Concrete Design

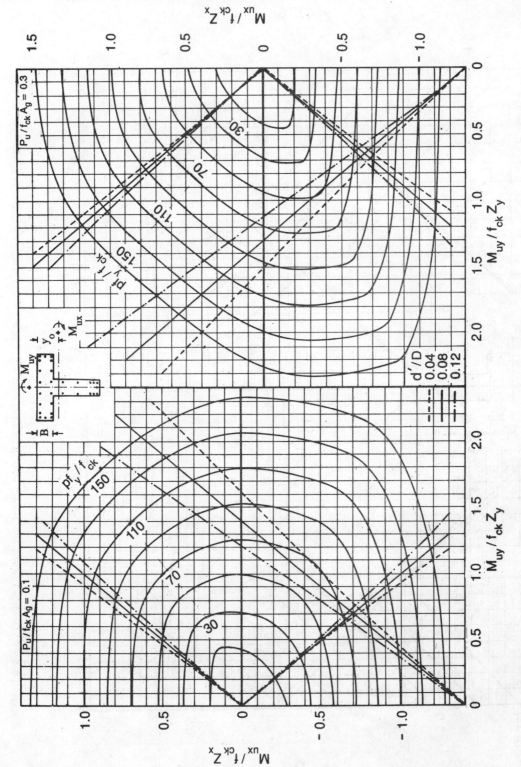

Chart T5.1 Design curves for T5 section of column for $P_u/f_{ck}A_g = 0.1$ and 0.3 ($3.5 \leq D/B \leq 4.5$)

Columns **339**

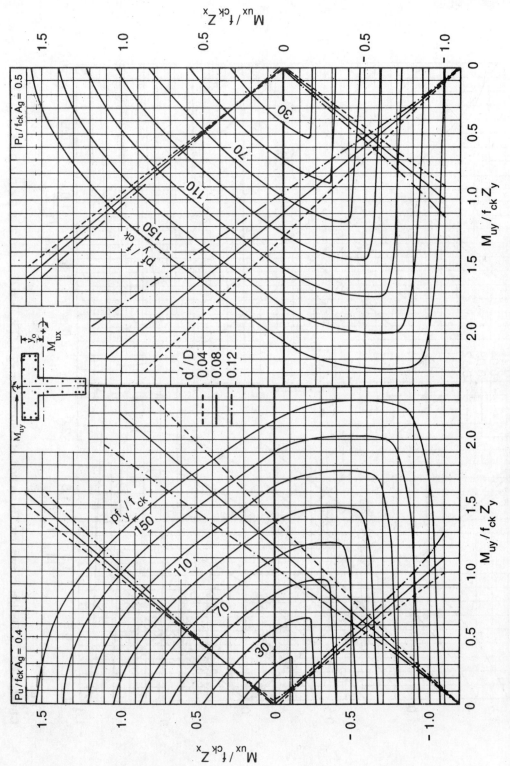

Chart T5.2 Design curves for T5 section of column for $P_u/f_{ck}A_g = 0.4$ and 0.5 $(3.5 \leq D/B \leq 4.5)$

340 Handbook of Reinforced Concrete Design

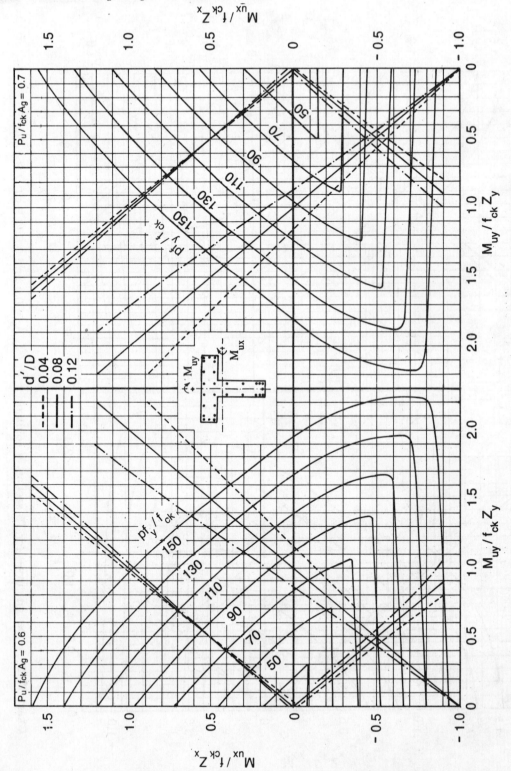

Chart T5.3   Design curves for T5 section of column for $P_u/f_{ck}A_g$ = 0.6 and 0.7 ($3.5 \leq D/B \leq 4.5$)

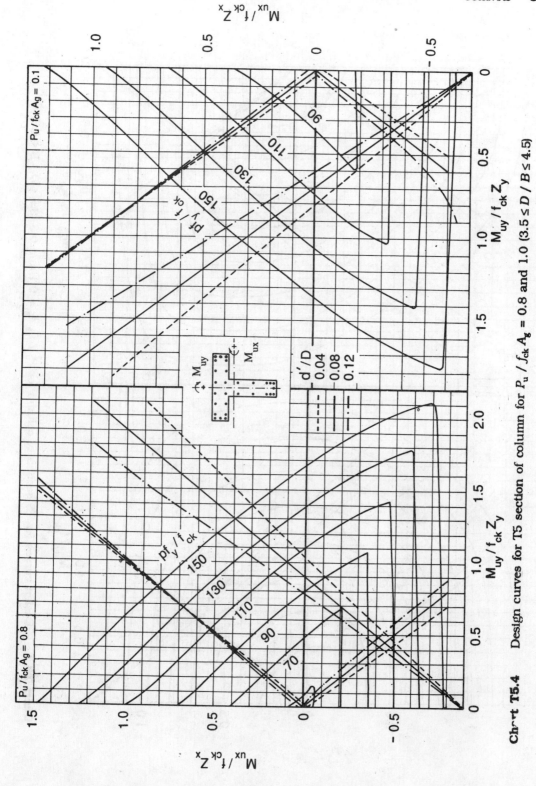

Chart T5.4 Design curves for T5 section of column for $P_u/f_{ck}A_g = 0.8$ and 1.0 $(3.5 \le D/B \le 4.5)$

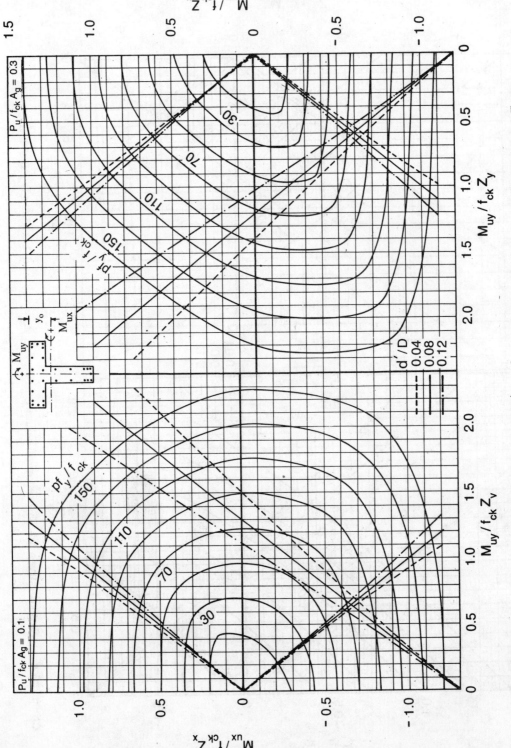

Chart T6.1 Design curves for T6 section of column for $P_u/f_{ck}A_g = 0.1$ and 0.3 $(3.5 \leq D/B \leq 4.5)$

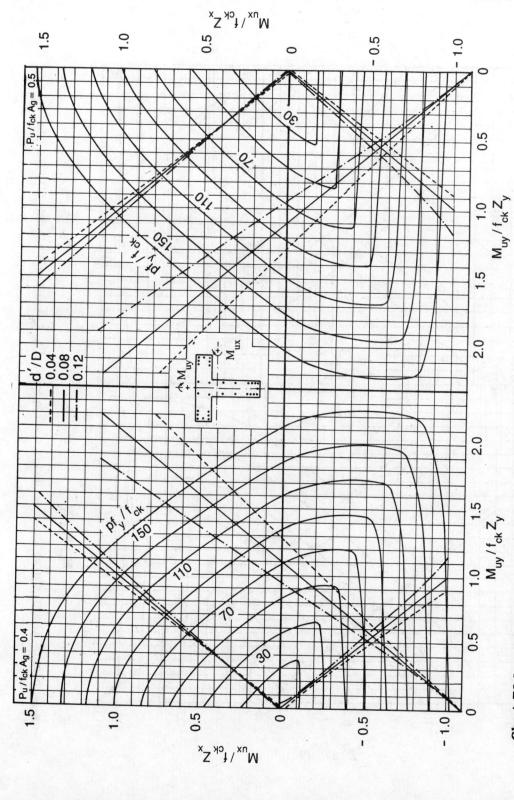

Chart T6.2 Design curves for T6 section of column for $P_u / f_{ck} A_g = 0.4$ and 0.5 ($3.5 \leq D/B \leq 4.5$)

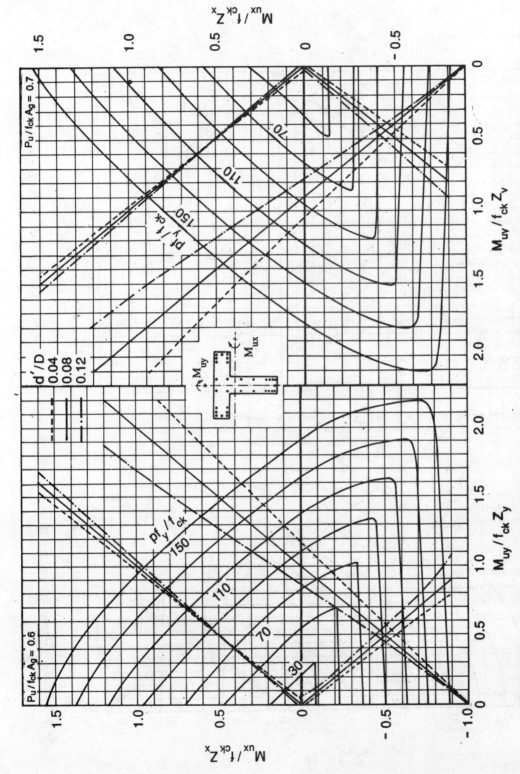

Chart T6.3 Design curves for T6 section of column for $P_u/f_{ck}A_g = 0.6$ and 0.7 ($3.5 \leq D/B \leq 4.5$)

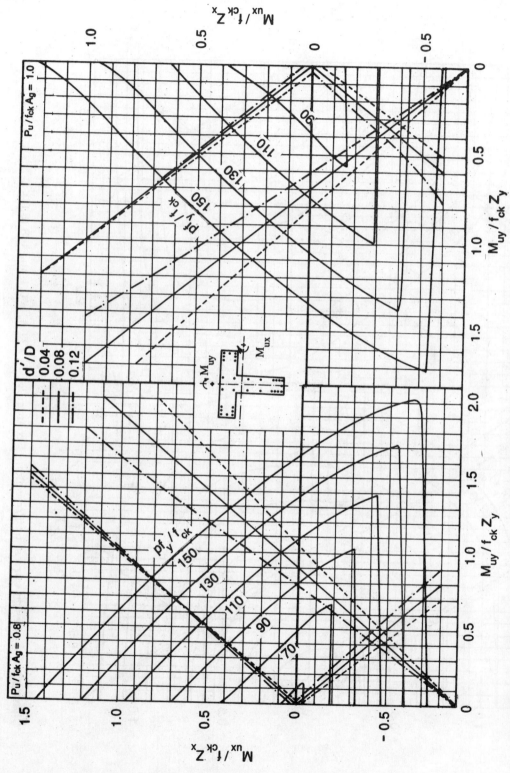

Chart T6.4 Design curves for T6 section of column for $P_u/f_{ck}A_g = 0.8$ and 1.0 ($3.5 \leq D/B \leq 4.5$)

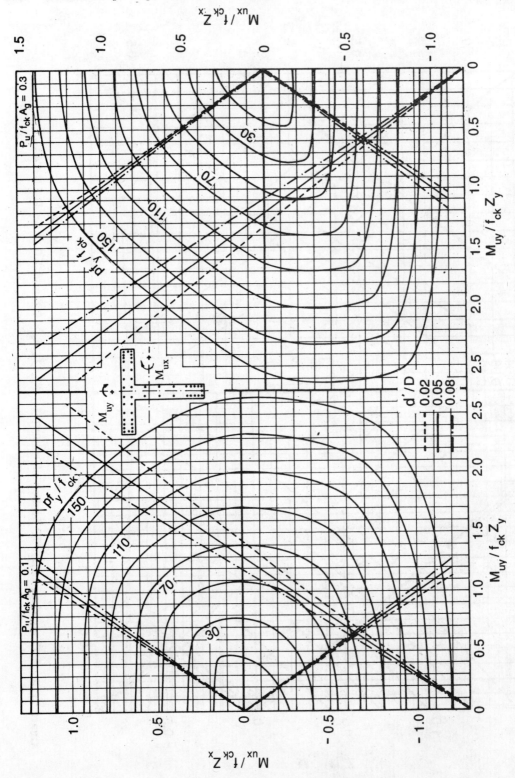

Chart T7.1 Design curves for T7 section of column for $P_u/f_{ck} A_g = 0.1$ and 0.3 $(5.0 \leq D/B \leq 7.0)$

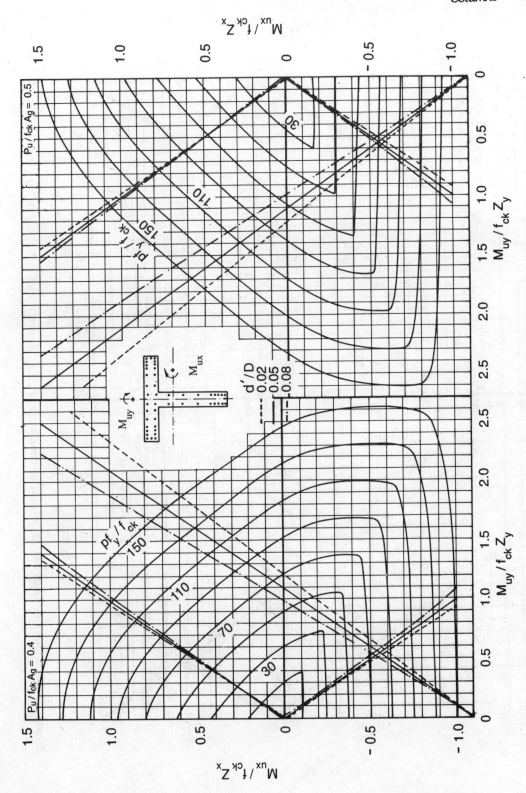

Chart T7.2 Design curves for T7 section of column for $P_u/f_{ck}A_g = 0.4$ and 0.5 $(5.0 \leq D/B \leq 7.0)$

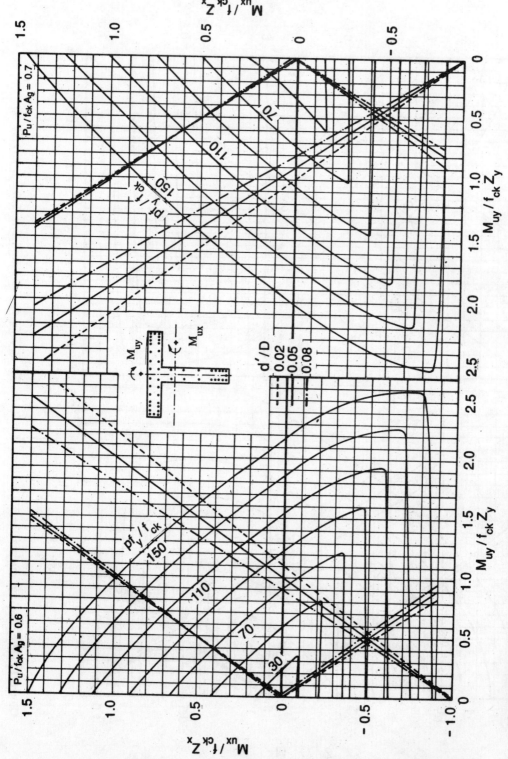

Chart T7.3 Design curves for T7 section of column for $P_u/f_{ck}A_g = 0.6$ and 0.7 ($5.0 \leq D/B \leq 7.0$)

Columns 349

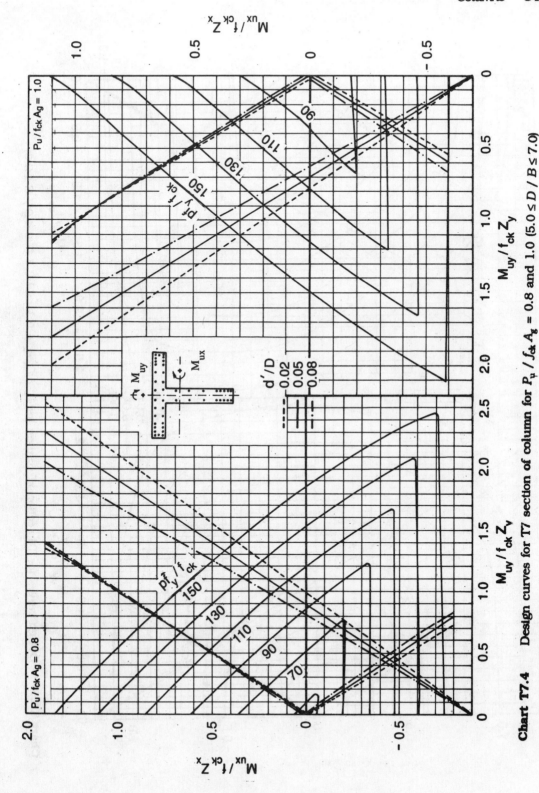

Chart T7.4 Design curves for T7 section of column for $P_u/f_{ck} A_g = 0.8$ and 1.0 $(5.0 \le D/B \le 7.0)$

350 *Handbook of Reinforced Concrete Design*

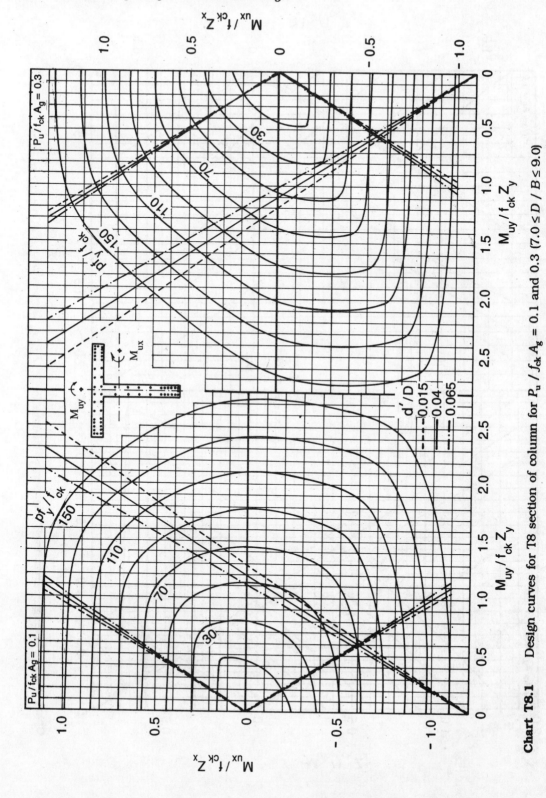

Chart T8.1 Design curves for T8 section of column for $P_u / f_{ck} A_g = 0.1$ and 0.3 $(7.0 \leq D / B \leq 9.0)$

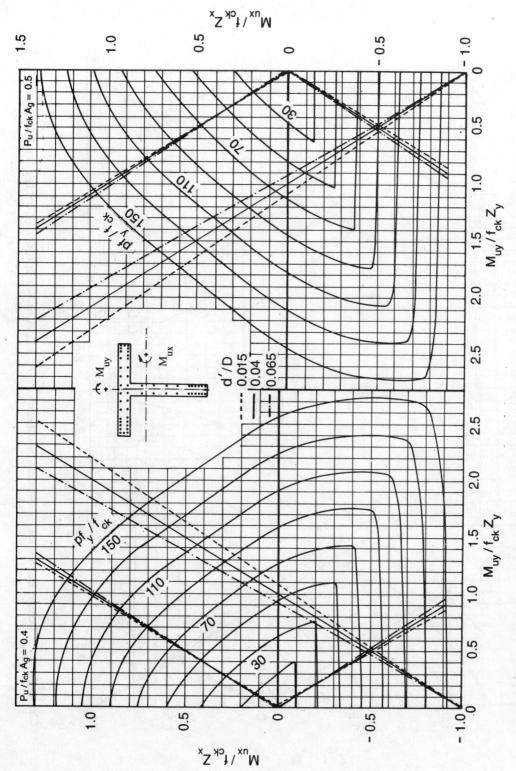

Chart T8.2 Design curves for T8 section of column for $P_u / f_{ck} A_g = 0.4$ and 0.5 $(7.0 \leq D/B \leq 9.0)$

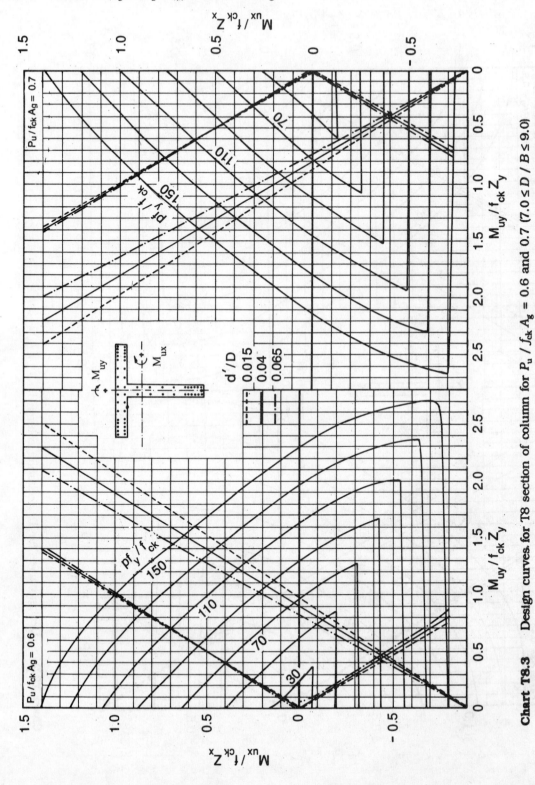

Chart T8.3 Design curves for T8 section of column for $P_u/f_{ck}A_g = 0.6$ and 0.7 $(7.0 \le D/B \le 9.0)$

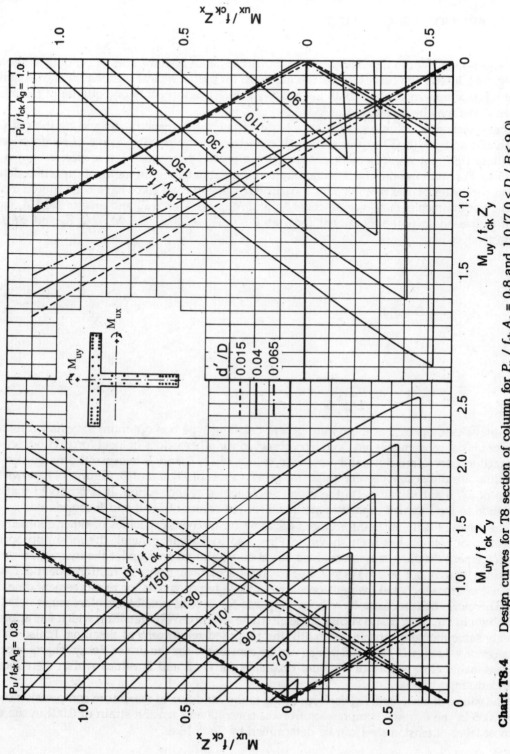

Chart T8.4 Design curves for T8 section of column for $P_u/f_{ck}A_g = 0.8$ and 1.0 $(7.0 \leq D/B \leq 9.0)$

5.7 + SECTION OF COLUMNS

The + section of column is geometrically symmetrical about both the axes of bending as shown in Fig. 5.42. Interaction curves for biaxial moments for such a typical sections are shown in the figure. It is symmetrical about both x-x and y-y axes. Therefore, only one quarter of the interaction curve may be considered.

The + section of column may have a large variation in geometry defined by the ratio of depth (D) and width (B) of the section. It may have many possible ways of reinforcement detailing. Different reinforcement detailing for a typical geometry of + section is shown in Fig. 5.43. For studying the effect of these reinforcement detailing on moment capacities, interaction curves for biaxial moments for a particular value of axial force have been shown in the figure. For interaction curves, moments M_{ux} and M_{uy}, axial force P_u and area of steel A_s have been expressed in non-dimensional form as $P_u/f_{ck}A_g$, $M_{ux}/f_{ck}Z_x$, $M_{uy}/f_{ck}Z_y$ and pf_y/f_{ck} respectively where,

$$p = \frac{100 A_s}{A_g}$$

$$A_g = (2D - B) B$$

$$Z_x, Z_y = \frac{I_{xx}}{y_o} = \frac{I_{yy}}{x_o}$$

$$= B\left(B^2 + D^2 - \frac{B^3}{D}\right) \times \frac{1}{6}$$

It may be observed that the reinforcement detailing type b is optimum for every combination of biaxial moments. It has been also observed for other values of axial force. However the interaction curves are not valid for other values of D/B and B because the reinforcement detailing are dimensionally dissimilar for other values of D/B and B. Figure 5.44 shows the effect of D/B and B values on the moment capacities of the section. It is observed that the variation in the moment capacities are small for the values of D/B and B considered.

The effect of different reinforcement detailing on moment capacities of other geometry of + section of column has been shown in Figs. 5.45 to 5.48. In general, the most appropriate reinforcement detailing may be evolved based on maximum area under the interaction curve. However, it may not be always possible to adopt a particular type of reinforcement detailing because of cross-sectional diemensions, area of steel, requirements for reinforcement detailing etc. Therefore, design charts have been prepared for several possible reinforcement detailing as shown in Fig. 5.49. It also shows some alternate reinforcement detailing which has approximately same moment capacity as that of the proposed reinforcement detailing. It also shows a range of values of D/B and B which can be considered for the use of design charts because of reasonably close values of moment capacities for D/B and B values within their recommended range.

The values of axial load, P_{ub} corresponding to the condition of maximum compression strain of 0.0035 in the extreme compression fibre of concrete and tension strain of 0.002 in the outermost layer of tension steel can be determined by Eq. 5.18 as,

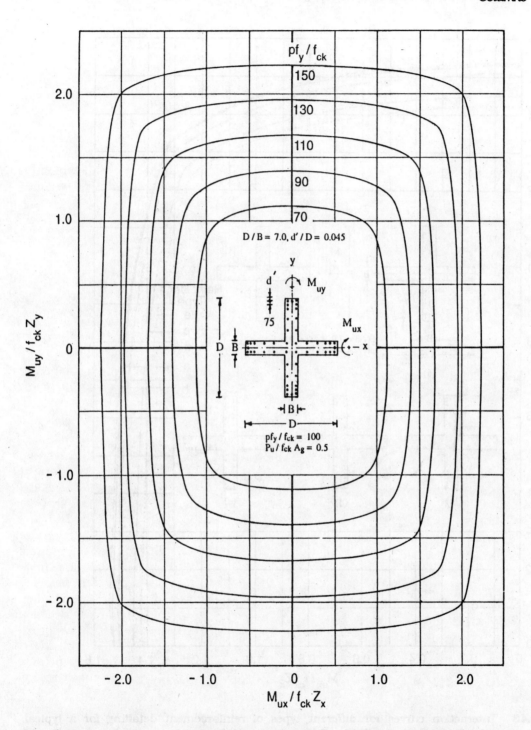

Fig. 5.42 Interaction curves for biaxial moments for a typical + section of column

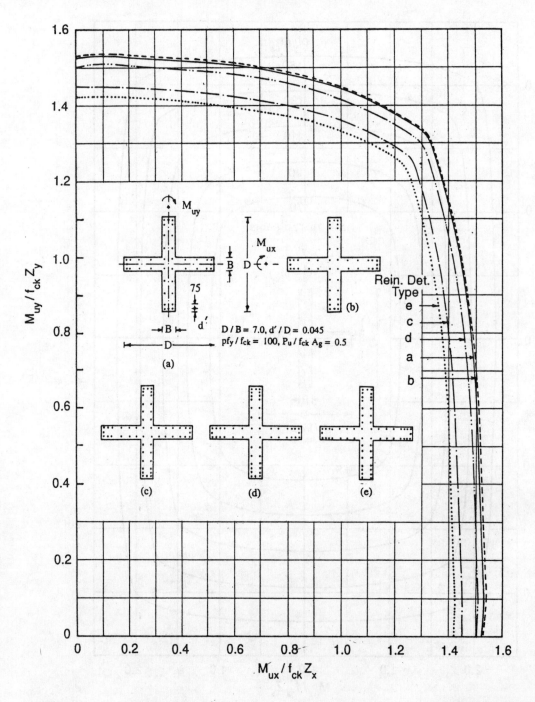

Fig. 5.43 Interaction curves for different types of reinforcement detailing for a typical + section of column ($D/B = 7.0$)

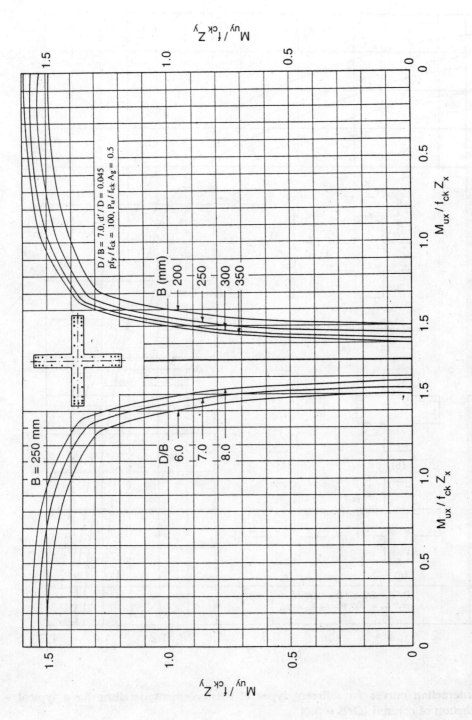

Fig. 5.44 Interaction curves for different values of D/B and B for a typical + section of column

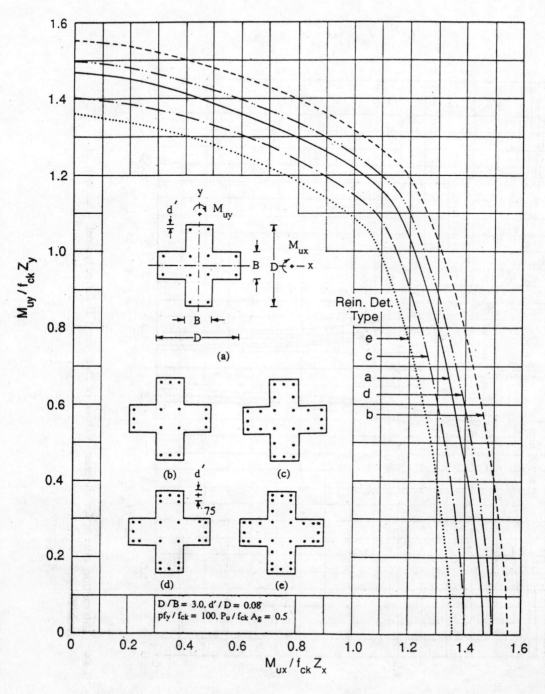

Fig. 5.45 Interaction curves for different types of reinforcement detailing for a typical + section of column ($D/B = 3.0$)

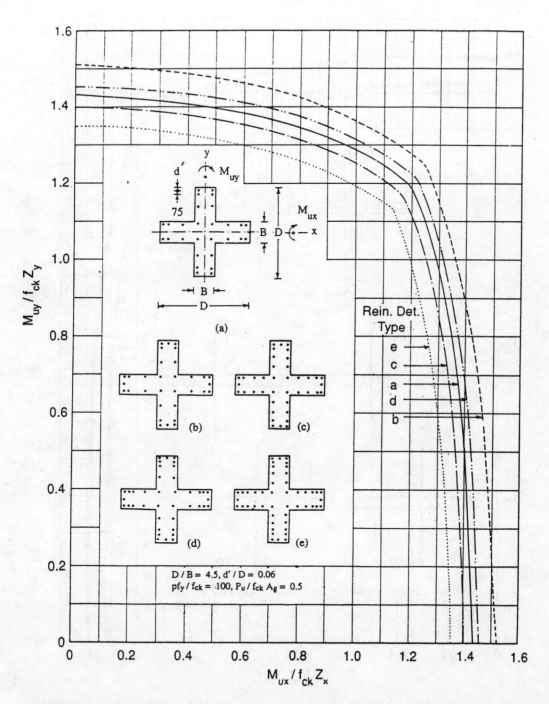

Fig. 5.46 Interaction curves for different types of reinforcement detailing for a typical + section of column ($D/B = 4.5$)

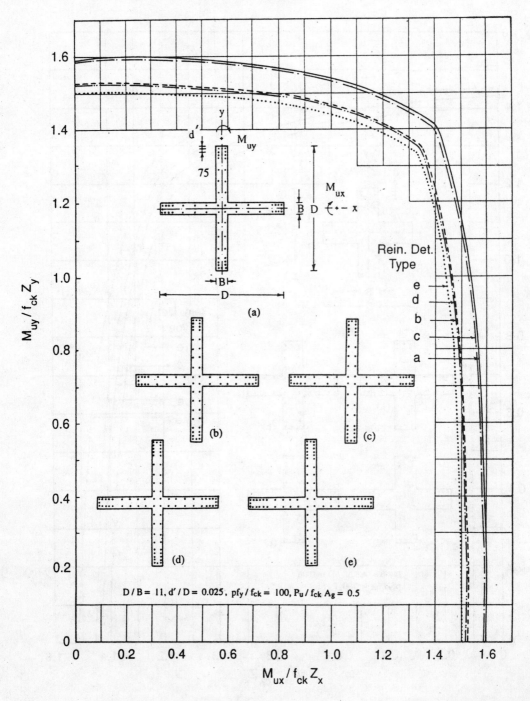

Fig. 5.47 Interaction curves for different types of reinforcement detailing for a typical + section of column ($D/B = 11$)

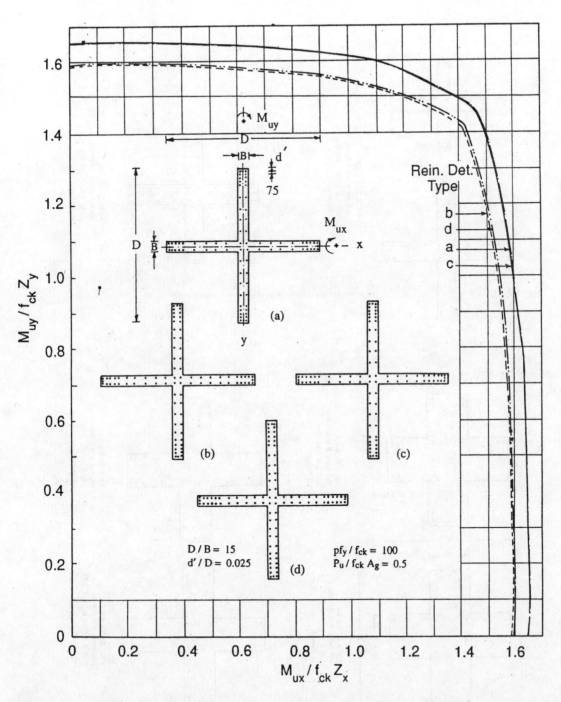

Fig. 5.48 Interaction curves for different types of reinforcement detailing for a typical + section of column ($D/B = 15$)

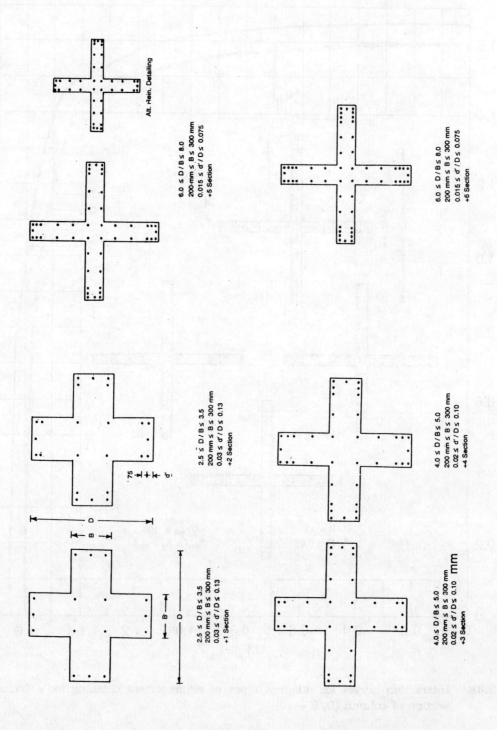

(Contd.)

Columns 363

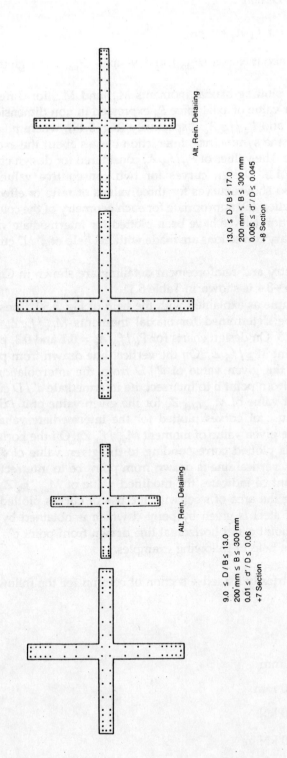

Fig. 5.49 Reinforcement detailing for different geometry of + section of column

$$P_{ub} = q_c f_{ck} A_g + q_s p A_g$$

where the values of q_c and q_s for P_{ub} about x-axis (P_{ubx}) and y-axis (P_{uby}) are given in Table 5.10.

The design chart has been obtained by plotting biaxial moments M_{ux} and M_{uy} for different values of area of steel A_s and a particular value of axial force P_u expressed in non-dimensional form as $M_{ux}/f_{ck} Z_x$, $M_{uy}/f_{ck} Z_y$, pf_y/f_{ck} and $P_u/f_{ck} A_g$ respectively (Fig. 5.49), where p is defined as $100 A_s/A_g$. Only one quarter of the symmetrical interaction curves about the axes of symmetry have been shown in the figure. The values of $P_u/f_{ck} A_g$ considered for design charts are 0.1, 0.3, 0.4, 0.5, 0.6, 0.7, 0.8 and 1.0. Design curves for two consequtive values of $P_u/f_{ck} A_g$ are shown on one chart. It also shows curves for three values of ratio of effective cover to reinforcement (d') to depth of section (D) appropriate for each geometry of the column section as given in Fig. 5.50. The interaction curves have been plotted for intermediate value of d'/D. For other values of d'/D, necessary corrections are made with the help of d'/D curves as explained below.

The design charts for different geometry and reinforcement detailing are shown in Charts +1.1 to +8.4. The index of Charts +1.1 to +8.4 is shown in Table 5.11.

The use of the charts for design is simple as explained with the help of Fig. 5.50. Consider that the area of steel is required to be determined for biaxial moments $M_{ux}/f_{ck} Z_x$ and $M_{uy}/f_{ck} Z_y$ and for $P_u/f_{ck} A_g = 0.1$ and 0.3. On design charts for $P_u/f_{ck} A_g = 0.1$ and 0.3, point a is plotted for the given value of moment $M_{ux}/f_{ck} Z_x$. On the vertical line drawn from point a, point b is plotted corresponding to the given value of d'/D from the interpolation of d'/D curves. A horizontal line is drawn from point b to intersect the intermediate d'/D curve at point c. Point c indicates the modified value of $M_{ux}/f_{ck} Z_x$ for the given value of d'/D for determining the area of steel with the use of curves plotted for the intermediate value of d'/D. Similarly, point aé is plotted for the given value of moment $M_{uy}/f_{ck} Z_y$. On the horizontal line drawn from point aé, point bé is plotted corresponding to the given value of d'/D from the interpolation of d'/D curves. A vertical line is drawn from point bé to intersect the intermediate d'/D curve at point cé. Point cé indicates the modified value of $M_{uy}/f_{ck} Z_y$ for the given value of d'/D for determining the area of steel with the use of curves plotted for intermediate value of d'/D. The area of steel is given by point d which is obtained by the intersection of vertical line drawn from point c and horizontal line drawn from point c'.

The use of charts is explained with the help of following examples.

Example 5.11: Design a biaxial eccentrically loaded + section of column for the following data:

Column section: $\dfrac{D}{B}$ = 4

B = 250 mm

D = 1000 mm

Ultimate axial load, P_u = 4000 kN

Ultimate moments, M_{ux} = 1000 kN.m

Table 5.10 Values of q_c and q_s for +-shape of column

Cross-sections	Values of $\frac{d'}{D}$	Values of q_c	Values of q_c and q_s for P_{ubx} and P_{uby} Values of q_s for concrete and steel of grades								
			M20			M25			M30		
			Fe 415	Fe 500	Fe 550	Fe 415	Fe 500	Fe 550	Fe 415	Fe 500	Fe 550
+1 Section	0.0300	0.2233	0.8422	0.8871	0.9263	0.8313	0.8762	0.9155	0.8205	0.8653	0.9046
	0.0550	0.2114	0.7480	0.7924	0.8252	0.7377	0.7821	0.8149	0.7273	0.7717	0.8045
	0.0800	0.2058	0.6493	0.6894	0.7192	0.6396	0.6797	0.7095	0.6299	0.6700	0.6998
	0.1050	0.1944	0.5445	0.5776	0.6040	0.5354	0.5685	0.5949	0.5263	0.5594	0.5858
	0.1300	0.1836	0.4316	0.4579	0.4814	0.4232	0.4496	0.4730	0.4149	0.4412	0.4647
+2 Section	0.0300	0.2233	0.8232	0.9038	0.9567	0.8124	0.8930	0.9459	0.8016	0.8822	0.9351
	0.0550	0.2114	0.7349	0.8129	0.8610	0.7246	0.8025	0.8507	0.7142	0.7922	0.8403
	0.0800	0.2058	0.6423	0.7149	0.7600	0.6325	0.7052	0.7502	0.6228	0.6955	0.7405
	0.1050	0.1944	0.5424	0.6090	0.6472	0.5332	0.5997	0.6380	0.5240	0.5905	0.6288
	0.1300	0.1836	0.4355	0.4921	0.5253	0.4269	0.4835	0.5167	0.4184	0.4750	0.5082
+3 Section	0.0200	0.2340	0.9099	0.9929	1.0492	0.8986	0.9815	1.0379	0.8872	0.9702	1.0265
	0.0400	0.2202	0.8439	0.9225	0.9732	0.8328	0.9113	0.9621	0.8216	0.9002	0.9510
	0.0600	0.2067	0.7738	0.8465	0.8870	0.7630	0.8357	0.8762	0.7522	0.8250	0.8654
	0.0800	0.2023	0.6989	0.7599	0.7946	0.6885	0.7496	0.7843	0.6782	0.7393	0.7739
	0.1000	0.1896	0.6184	0.6644	0.6976	0.6087	0.6547	0.6878	0.5989	0.6450	0.6781

(Contd.)

Table 5.10 (*Contd.*)

Values of q_c and q_s for P_{ubx} and P_{uby}

Values of q_s for concrete and steel of grades

Cross-sections	Values of $\frac{d'}{D}$	Values of q_c	M20 Fe 415	M20 Fe 500	M20 Fe 550	M25 Fe 415	M25 Fe 500	M25 Fe 550	M30 Fe 415	M30 Fe 500	M30 Fe 550
+4 Section	0.0200	0.2340	0.9222	1.0143	1.0721	0.9108	1.0030	1.0608	0.8995	0.9917	1.0495
	0.0400	0.2202	0.8562	0.9430	0.9920	0.8451	0.9319	0.9809	0.8340	0.9208	0.9698
	0.0600	0.2067	0.7873	0.8633	0.9062	0.7766	0.8526	0.8954	0.7658	0.8418	0.8846
	0.0800	0.2023	0.7118	0.7761	0.8162	0.7015	0.7658	0.8059	0.6912	0.7555	0.7956
	0.1000	0.1896	0.6253	0.6834	0.7209	0.6156	0.6737	0.7112	0.6059	0.6639	0.7015
+5 Section	0.0150	0.2439	0.9547	1.0401	1.0930	0.9429	1.0283	1.0811	0.9310	1.0164	1.0693
	0.0300	0.2261	0.9062	0.9843	1.0354	0.8946	0.9727	1.0238	0.8830	0.9611	1.0122
	0.0450	0.2266	0.8552	0.9259	0.9745	0.8439	0.9146	0.9632	0.8326	0.9033	0.9519
	0.0600	0.2129	0.7976	0.8655	0.9106	0.7867	0.8545	0.8996	0.7757	0.8436	0.8887
	0.0750	0.2112	0.7378	0.8013	0.8427	0.7273	0.7908	0.8322	0.7168	0.7803	0.8217
+6 Section	0.0150	0.2439	0.9590	1.0486	1.1046	0.9471	1.0368	1.0928	0.9353	1.0250	1.0810
	0.0300	0.2261	0.9111	0.9938	1.0474	0.8995	0.9822	1.0358	0.8879	0.9706	1.0243
	0.0450	0.2266	0.8581	0.9366	0.9875	0.8469	0.9253	0.9762	0.8356	0.9140	0.9649
	0.0600	0.2129	0.8014	0.8765	0.9216	0.7905	0.8656	0.9106	0.7796	0.8547	0.8997
	0.0750	0.2112	0.7427	0.8125	0.8529	0.7322	0.8020	0.8424	0.7218	0.7916	0.8319

(Contd.)

Table 5.10 (*Contd.*)

Values of q_c and q_s for P_{ubx} and P_{uby}

Values of q_s for concrete and steel of grades

Cross-sections	Values of $\dfrac{d'}{D}$	Values of q_c	M20			M25			M30		
			Fe 415	Fe 500	Fe 550	Fe 415	Fe 500	Fe 550	Fe 415	Fe 500	Fe 550
+7 Section	0.0100	0.2418	0.9741	1.0630	1.1203	0.9621	1.0509	1.1082	0.9500	1.0389	1.0961
	0.0225	0.2295	0.9327	1.0193	1.0722	0.9208	1.0074	1.0603	0.9090	0.9956	1.0485
	0.0350	0.2265	0.8900	0.9726	1.0227	0.8783	0.9609	1.0111	0.8667	0.9493	0.9995
	0.0475	0.2334	0.8460	0.9229	0.9712	0.8347	0.9116	0.9599	0.8234	0.9003	0.9486
	0.0600	0.2194	0.8000	0.8718	0.9179	0.7891	0.8608	0.9069	0.7781	0.8499	0.8960
+8 Section	0.0050	0.2404	0.9899	1.0802	1.1375	0.9776	1.0680	1.1252	0.9654	1.0557	1.1130
	0.0150	0.2565	0.9581	1.0449	1.1007	0.9461	1.0328	1.0886	0.9340	1.0207	1.0765
	0.0250	0.2265	0.9255	1.0088	1.0625	0.9136	0.9969	1.0506	0.9017	0.9850	1.0387
	0.0350	0.2242	0.8918	0.9713	1.0229	0.8801	0.9597	1.0112	0.8685	0.9480	0.9996
	0.0450	0.2377	0.8559	0.9328	0.9811	0.8445	0.9214	0.9697	0.8331	0.9101	0.9584

(*Contd.*)

368 Handbook of Reinforced Concrete Design

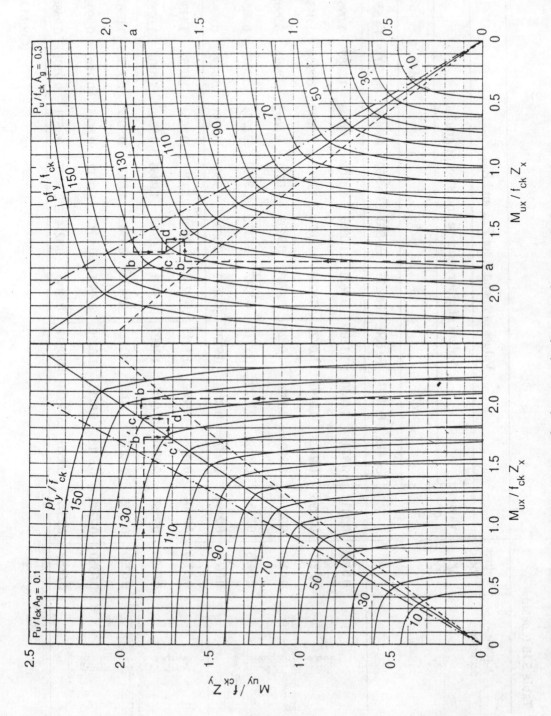

Fig. 5.50 Method for design of + section of column with the use of a typical chart +1-1

Table 5.11 Index for Charts +1.1 to +8.4

Reinforcement Detailing	Chart No.	$P_u / f_{ck} A_g$	Page No.
+1 Section	+1.1	0.1 and 0.3	372
	+1.2	0.4 and 0.5	373
	+1.3	0.6 and 0.7	374
	+1.4	0.8 and 1.0	375
+2 Section	+2.1	0.1 and 0.3	376
	+2.2	0.4 and 0.5	377
	+2.3	0.6 and 0.7	378
	+2.4	0.8 and 1.0	379
+3 Section	+3.1	0.1 and 0.3	380
	+3.2	0.4 and 0.5	381
	+3.3	0.6 and 0.7	382
	+3.4	0.8 and 1.0	383
+4 Section	+4.1	0.1 and 0.3	384
	+4.2	0.4 and 0.5	385
	+4.3	0.6 and 0.7	386
	+4.4	0.8 and 1.0	387
+5 Section	+5.1	0.1 and 0.3	388
	+5.2	0.4 and 0.5	389
	+5.3	0.6 and 0.7	390
	+5.4	0.8 and 1.0	391
+6 Section	+6.1	0.1 and 0.3	392
	+6.2	0.4 and 0.5	393
	+6.3	0.6 and 0.7	394
	+6.4	0.8 and 1.0	395
+7 Section	+7.1	0.1 and 0.3	396
	+7.2	0.4 and 0.5	397
	+7.3	0.6 and 0.7	398
	+7.4	0.8 and 1.0	399
+8 Section	+8.1	0.1 and 0.3	400
	+8.2	0.4 and 0.5	401
	+8.3	0.6 and 0.7	402
	+8.4	0.8 and 1.0	403

370 Handbook of Reinforced Concrete Design

$$M_{uy} = 1200 \text{ kN.m}$$

Grade of concrete : M25

Grade of steel : Fe 415

Solution

Consider 28 mm φ bars with clear cover of 40 mm

∴ Effectives cover, $d' = 40 + \dfrac{28}{2}$

$$= 54 \text{ mm}$$

and

$$\dfrac{d'}{D} = \dfrac{54}{1000}$$

$$= 0.054$$

Area of the section,

$$A_g = B(2D - B)$$

$$= 250 \times (2 \times 1000 - 250)$$

$$= 437500 \text{ mm}^2$$

Section modulus about x and y axes of the section.

$$Z_x = Z_y = B\left(D^2 + B^2 - \dfrac{B^3}{D}\right) \times \dfrac{1}{6}$$

$$= 250\left(1000^2 + 250^2 - \dfrac{250^3}{1000}\right) \times \dfrac{1}{6}$$

$$= 43619792.0 \text{ mm}^3$$

Compute

$$\dfrac{P_u}{f_{ck} A_g} = \dfrac{4000 \times 10^3}{20 \times 437500}$$

$$= 0.4571$$

$$\dfrac{M_{ux}}{f_{ck} Z_x} = \dfrac{1000 \times 10^6}{20 \times 43619792.0}$$

$$= 1.1463$$

$$\dfrac{M_{uy}}{f_{ck} Z_y} = \dfrac{1200 \times 10^6}{20 \times 43619792.0}$$

$$= 1.376$$

Area of steel is determined from the linear interpolation of area of steel determined for $P_u/f_{ck}A_g = 0.4$ and 0.5 from Chart + 3.2. The use of chart for determining the area of steel has been explained with the help of Fig. 5.50. Area of steel for $P_u/f_{ck}A_g = 0.4$ and 0.5 are,

For $\dfrac{P_u}{f_{ck}A_g} = 0.4,\ \dfrac{pf_y}{f_{ck}} = 92.5$

For $\dfrac{P_u}{f_{ck}A_g} = 0.5,\ \dfrac{pf_y}{f_{ck}} = 102.5$

$\therefore$ For $\dfrac{P_u}{f_{ck}A_g} = 0.4571,\ \dfrac{pf_y}{f_{ck}} = 92.5 + \dfrac{102.5 - 92.5}{0.5 - 0.4} \times (0.4571 - 0.4)$

$$= 98.21$$

$\therefore$
$$p = \dfrac{98.21 \times 20}{415}$$

$$= 4.733$$

and
$$A_s = \dfrac{4.733 \times 437500}{100}$$

$$= 20706.875\ \text{mm}^2$$

The diameter of reinforcement bars is chosen to provide 32 bars in accordance with recommended reinforcement detailing. Accordingly 12×30 mm ϕ + 20×28 mm ϕ bars (A_s = 20797 mm²) are provided as shown in Fig. Ex. 5.11. Reinforcement bars of 30 mm ϕ are placed at the edges which shall result in conservative design.

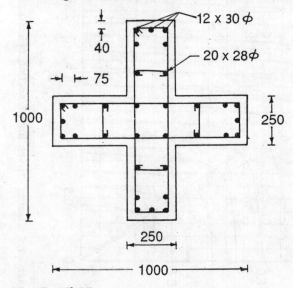

Fig. Ex. 5.11

372 *Handbook of Reinforced Concrete Design*

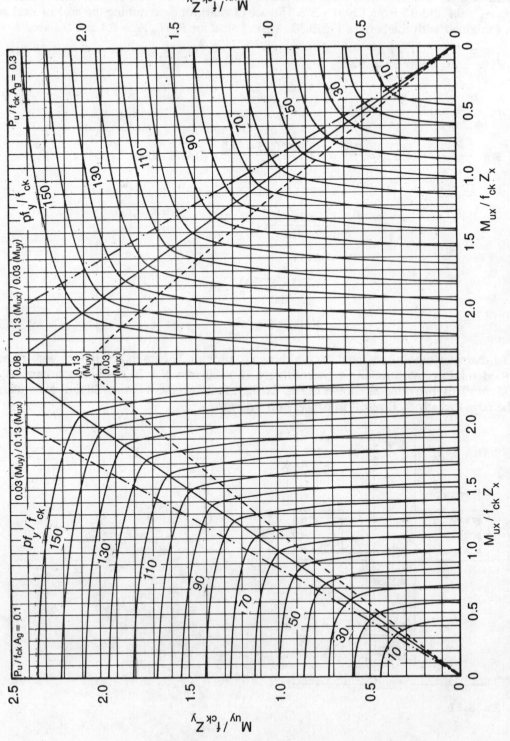

Chart +1.1 Design curves for +1 section of column for $P_u/f_{ck}A_g = 0.1$ and 0.3 $(2.5 \leq D/B \leq 3.5)$

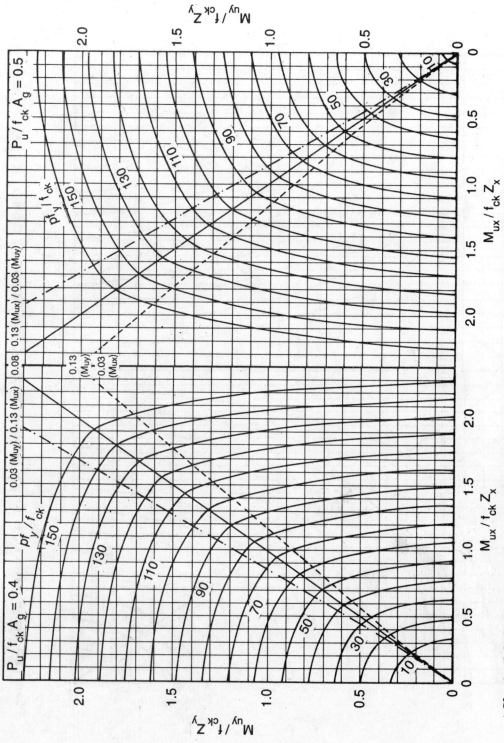

Chart +1.2 Design curves for +1 section of column for $P_u/f_{ck}A_g = 0.4$ and 0.5 $(2.5 \leq D/B \leq 3.5)$

374 *Handbook of Reinforced Concrete Design*

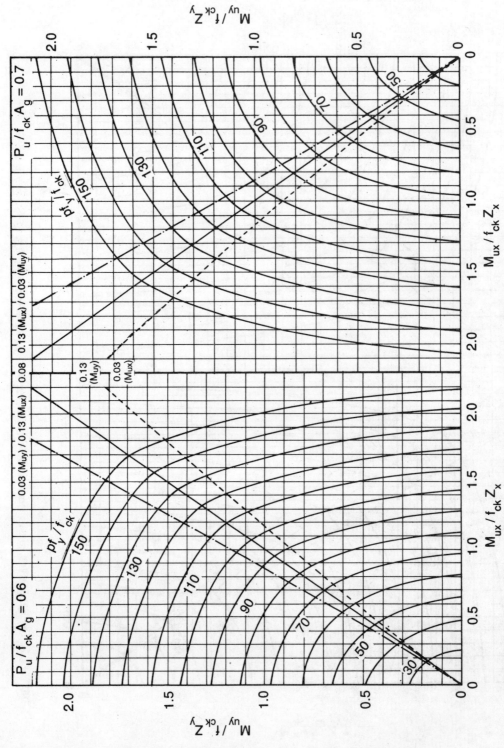

Chart +1.3 Design curves for +1 section of column for $P_u / f_{ck} A_g = 0.6$ and 0.7 $(2.5 \leq D/B \leq 3.5)$

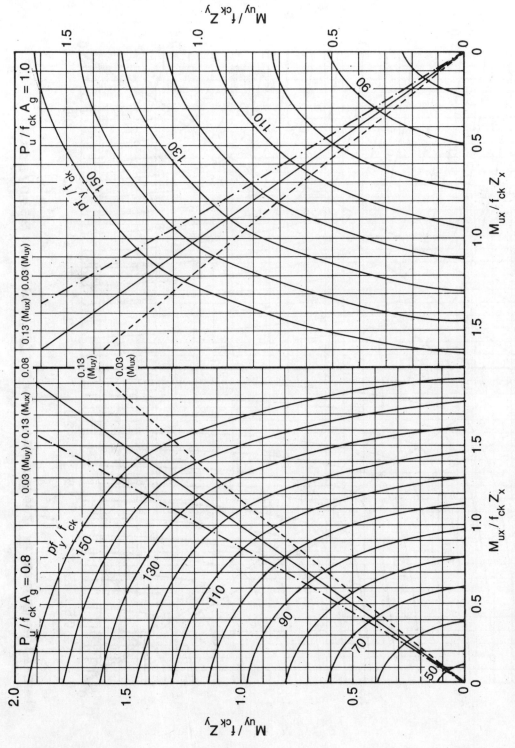

Chart +1.4 Design curves for +1 section of column for $P_u / f_{ck} A_g = 0.8$ and 1.0 $(2.5 \leq D / B \leq 3.5)$

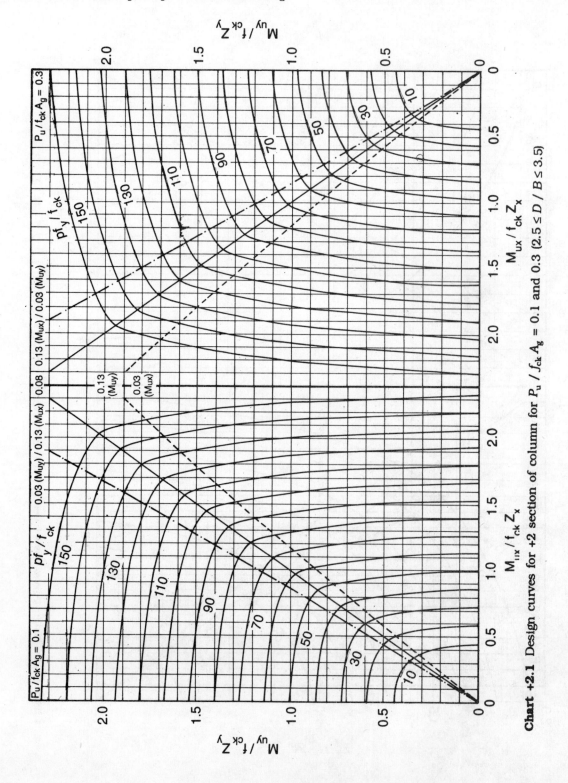

Chart +2.1 Design curves for +2 section of column for $P_u/f_{ck}A_g = 0.1$ and 0.3 $(2.5 \leq D/B \leq 3.5)$

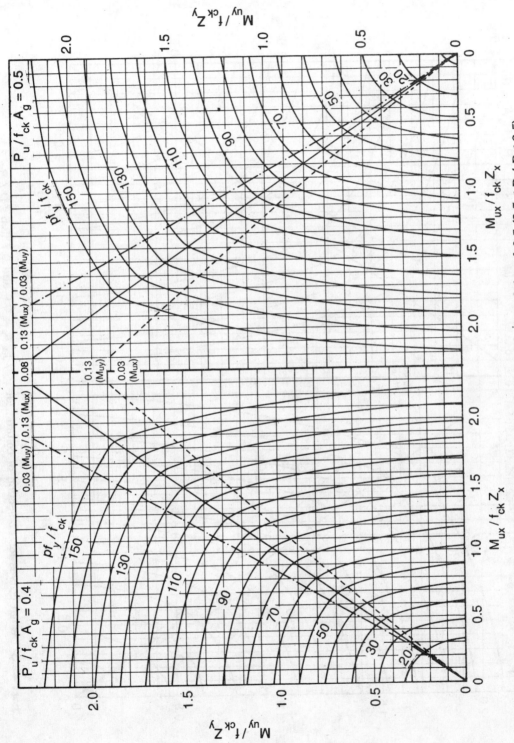

Chart +2.2 Design curves for +2 section of column for $P_u/f_{ck}·A_g = 0.4$ and 0.5 $(2.5 \le D/B \le 3.5)$

378 Handbook of Reinforced Concrete Design

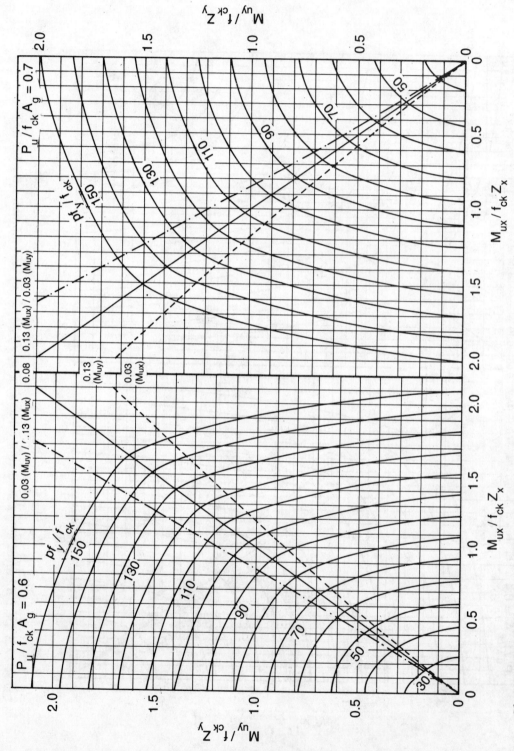

Chart +2.3 Design curves for +2 section of column for $P_u/f_{ck}A_g = 0.6$ and 0.7 $(2.5 \leq D/B \leq 3.5)$

Columns 379

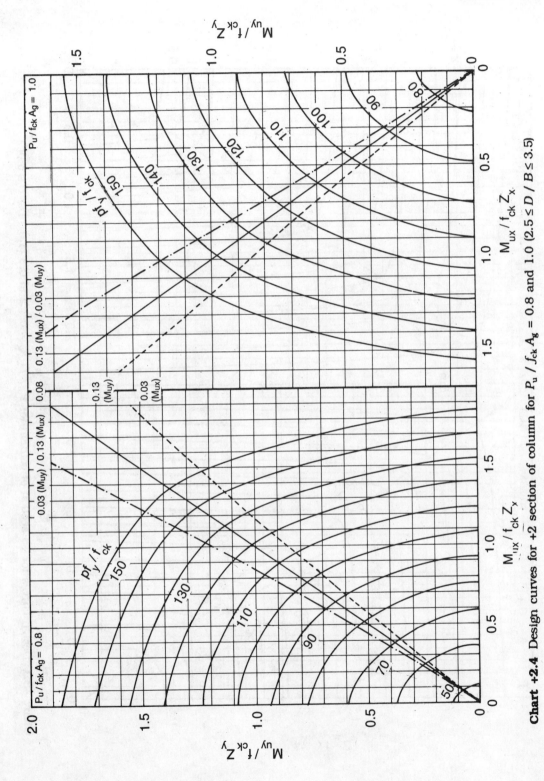

Chart +2.4 Design curves for +2 section of column for $P_u/f_{ck}A_g = 0.8$ and 1.0 $(2.5 \leq D/B \leq 3.5)$

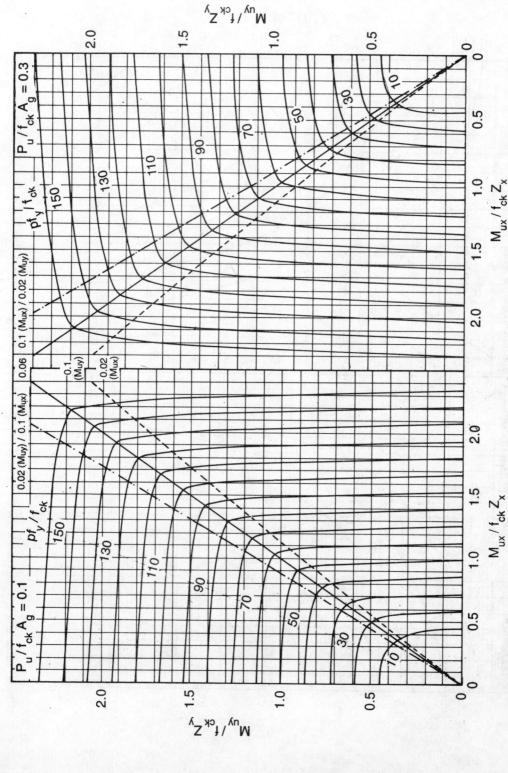

Chart +3.1 Design curves for +3 section of column for $P_u/f_{ck}A_g = 0.1$ and 0.3 ($4.0 \leq D/B \leq 5.0$)

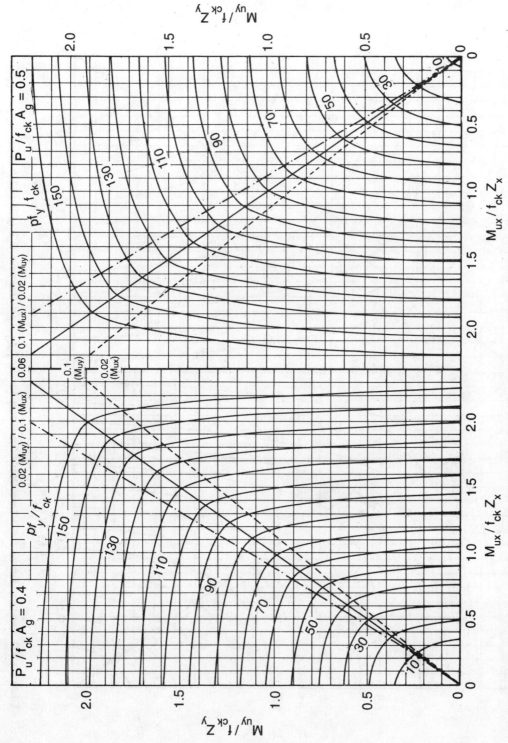

Chart +3.2 Design curves for +3 section of column for $P_u / f_{ck} A_g = 0.4$ and 0.5 $(4.0 \leq D / B \leq 5.0)$

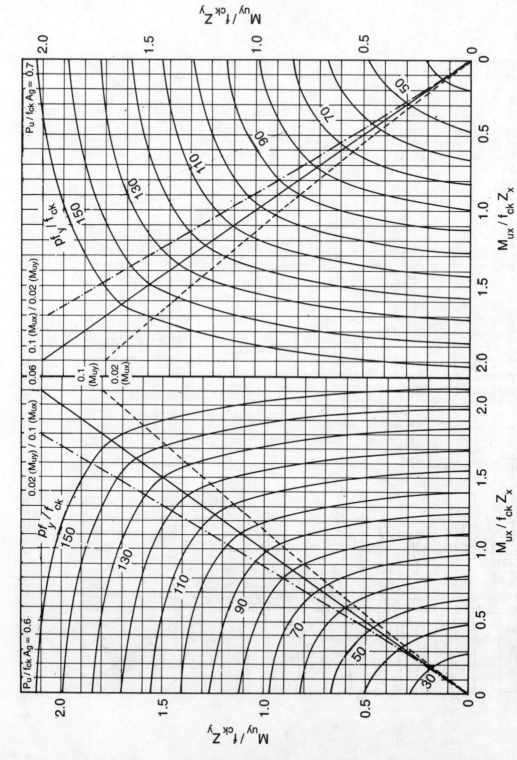

Chart +3.3 Design curves for +3 section of column for $P_u/f_{ck}A_g = 0.6$ and 0.7 ($4.0 \leq D/B \leq 5.0$)

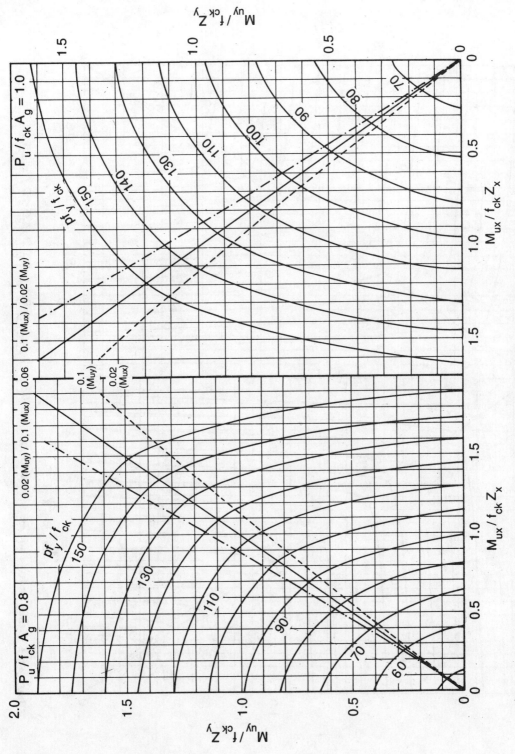

Chart 43.4 Design curves for +3 section of column for $P_u/f_{ck}A_g = 0.8$ and 1.0 ($4.0 \leq D/B \leq 5.0$)

384 Handbook of Reinforced Concrete Design

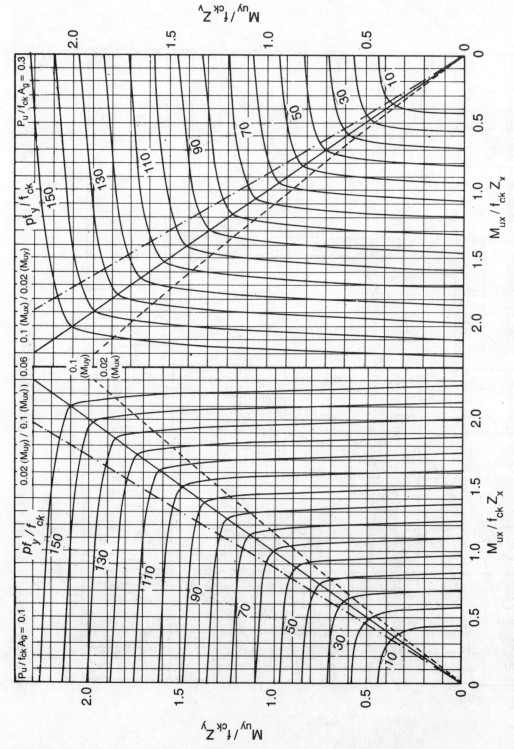

Chart +4.1 Design curves for +4 section of column for $P_u/f_{ck}A_g = 0.1$ and 0.3 ($4.0 \leq D/B \leq 5.0$)

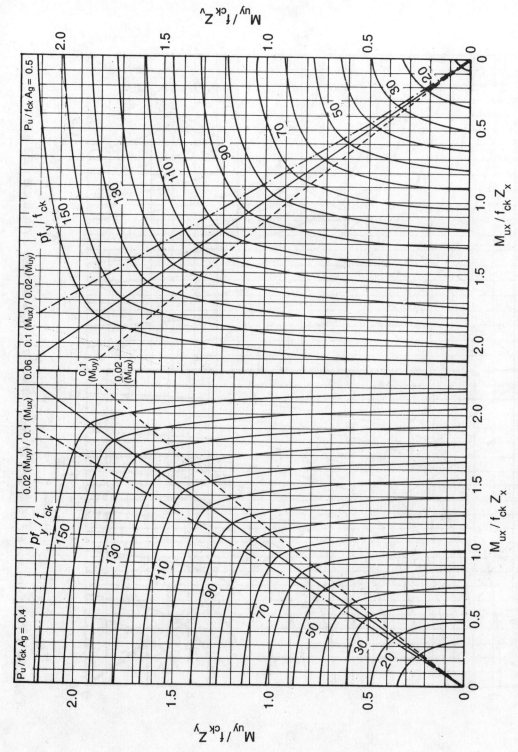

Chart 4.2 Design curves for 4 section of column for $P_u / f_{ck} A_g = 0.4$ and 0.5 $(4.0 \leq D/B \leq 5.0)$

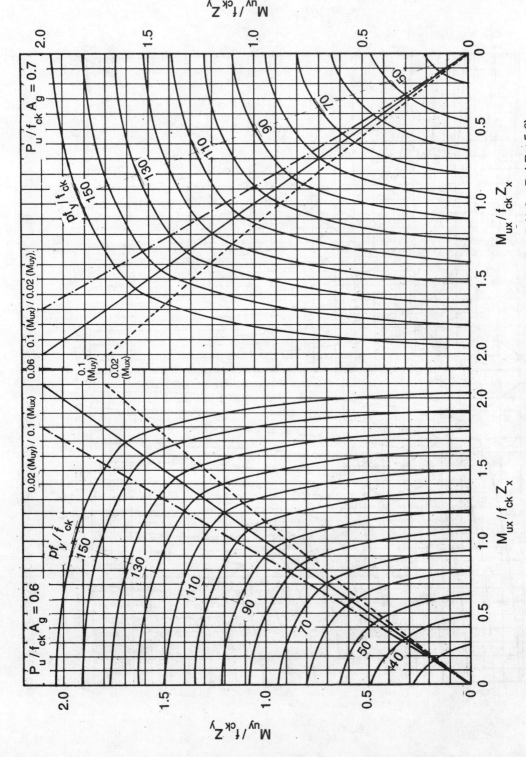

Chart 4.3 Design curves for +4 section of column for $P_u / f_{ck} A_g = 0.6$ and 0.7 $(4.0 \leq D/B \leq 5.0)$

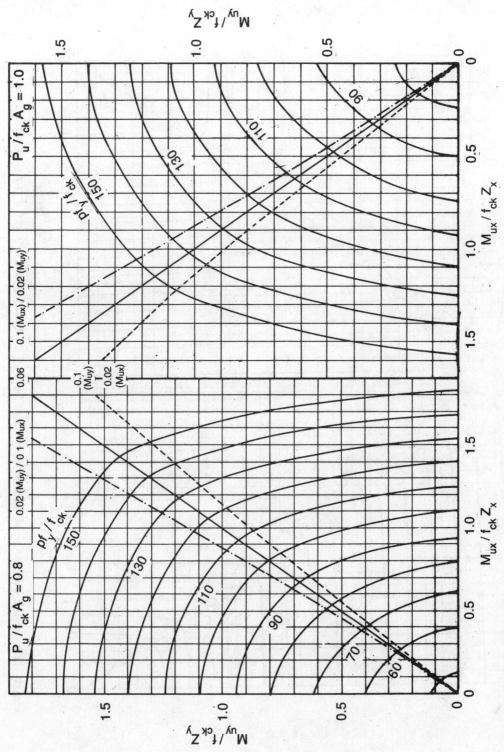

Chart 4.4 Design curves for 4 section of column for $P_u / f_{ck} A_g = 0.8$ and 1.0 $(4.0 \leq D/B \leq 5.0)$

388 Handbook of Reinforced Concrete Design

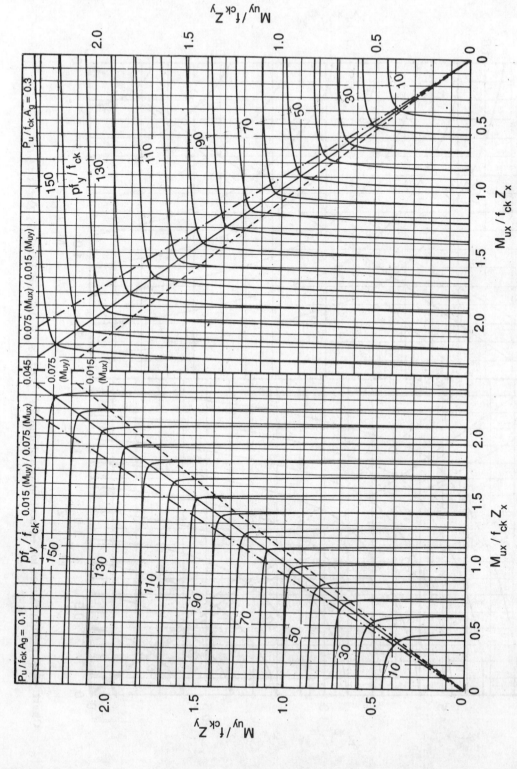

Chart +5.1 Design curves for +5 section of column for $P_u / f_{ck} A_g = 0.1$ and 0.3 $(6.0 \leq D/B \leq 8.0)$

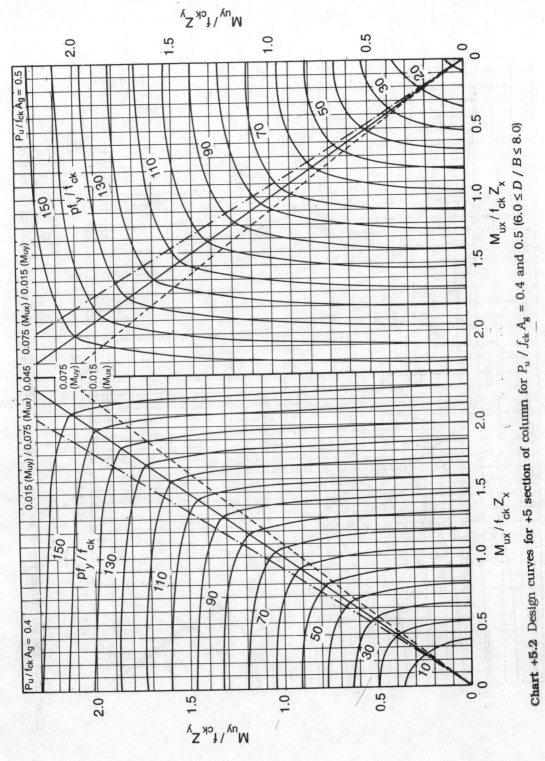

Chart +5.2 Design curves for +5 section of column for $P_u/f_{ck}A_g = 0.4$ and 0.5 $(6.0 \leq D/B \leq 8.0)$

390 Handbook of Reinforced Concrete Design

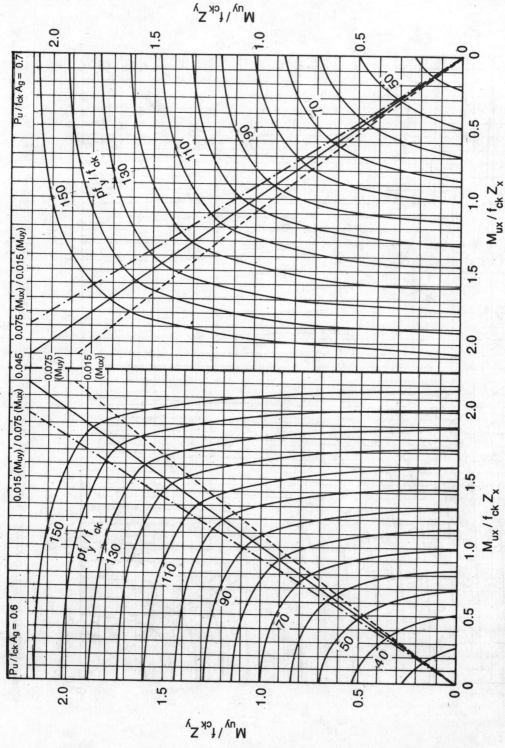

Chart 5.3 Design curves for +b section of column for $P_u/f_{ck}A_g = 0.6$ and 0.7 ($6.0 \leq D/B \leq 8.0$)

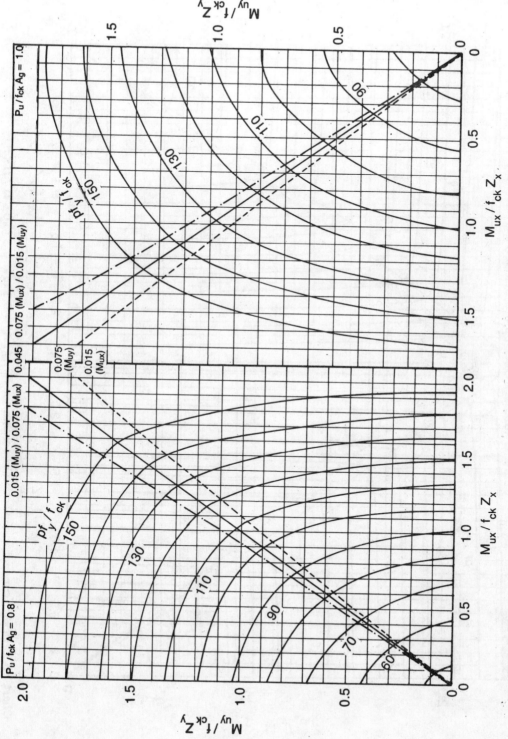

Chart +5.4 Design curves for +5 section of column for $P_u/f_{ck}A_g = 0.8$ and 1.0 ($6.0 \leq D/B \leq 8.0$)

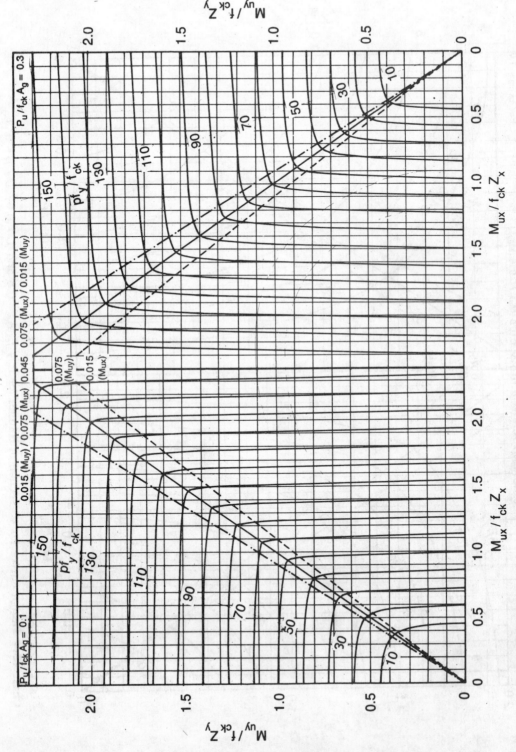

Chart +6.1 Design curves for +6 section of column for $P_u/f_{ck}A_g = 0.1$ and 0.3 ($6.0 \leq D/B \leq 8.0$)

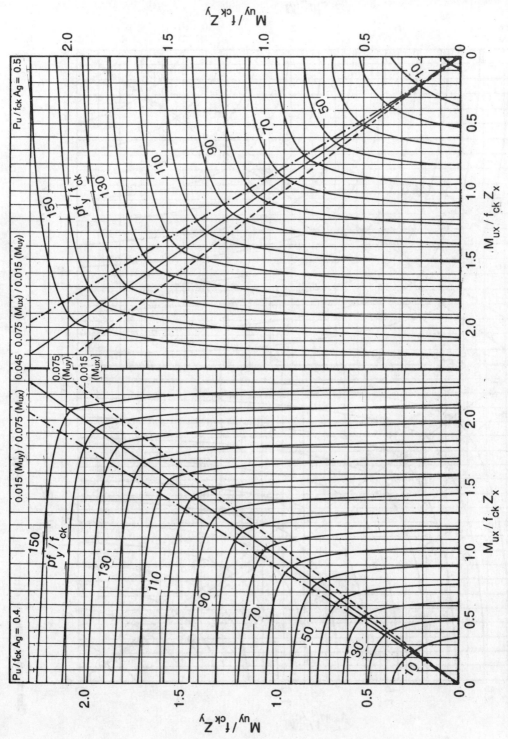

Chart +6.2 Design curves for +6 section of column for $P_u/f_{ck}A_g = 0.4$ and 0.5 $(6.0 \leq D/B \leq 8.0)$

394 Handbook of Reinforced Concrete Design

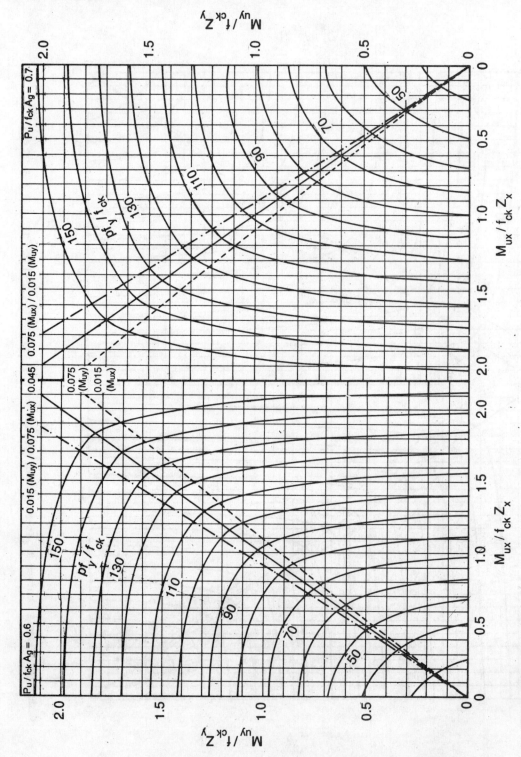

Chart +6.3 Design curves for +6 section of column for $P_u/f_{ck}A_g = 0.6$ and 0.7 $(6.0 \leq D/B \leq 8.0)$

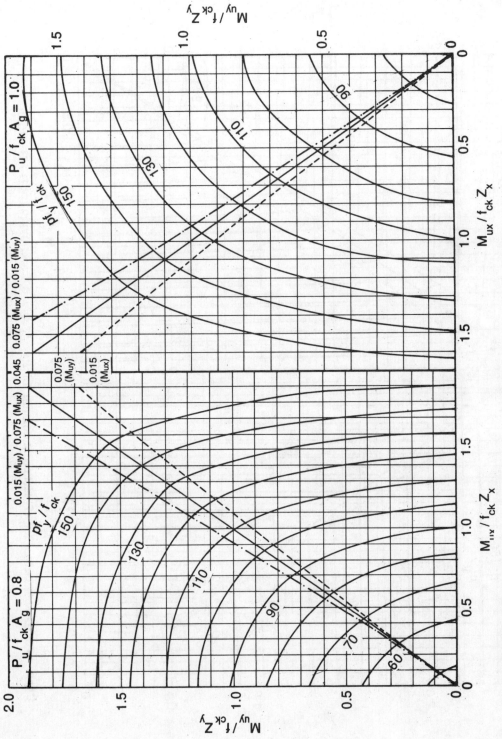

Chart +6.4 Design curves for +6 section of column for $P_u/f_{ck} A_g = 0.8$ and 1.0 $(6.0 \le D/B \le 8.0)$

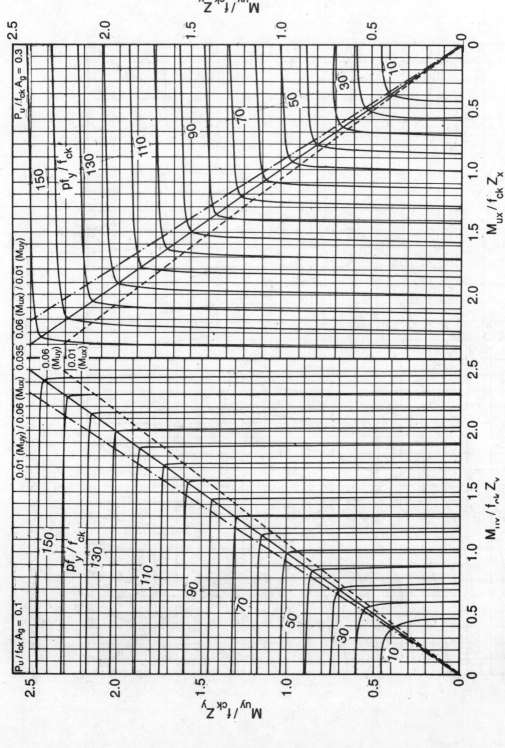

Chart +7.1 Design curves for +7 section of column for $P_u/f_{ck}A_g = 0.1$ and 0.3 ($9.0 \leq D/B \leq 13.0$)

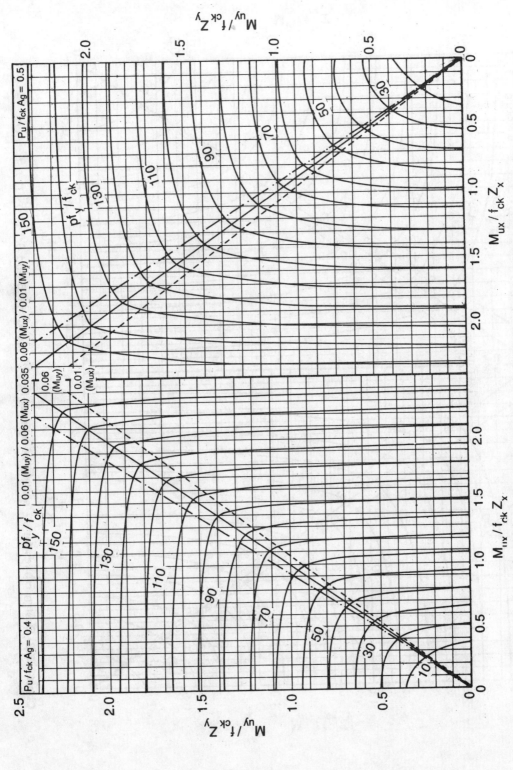

Chart +7.2 Design curves for +7 section of column for $P_u/f_{ck}A_g = 0.4$ and 0.5 $(9.0 \le D/B \le 13.0)$

398 *Handbook of Reinforced Concrete Design*

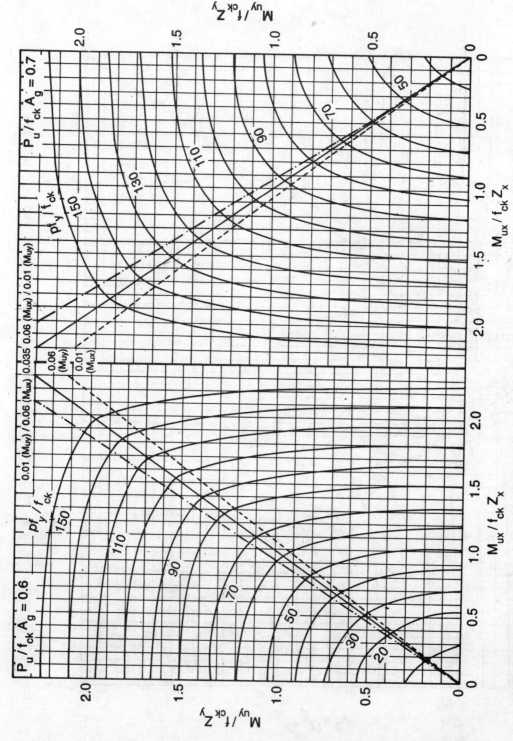

Chart +7.3 Design curves for +7 section of column for $P_u/f_{ck}A_g = 0.6$ and 0.7 $(9.0 \leq D/B \leq 13.0)$

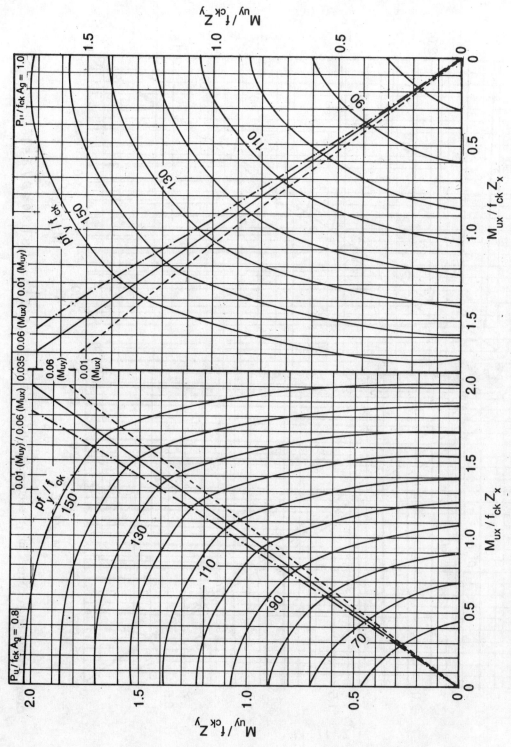

Chart +7.4 Design curves for +7 section of column for $P_u/f_{ck}A_g = 0.8$ and 1.0 $(9.0 \le D/B \le 13.0)$

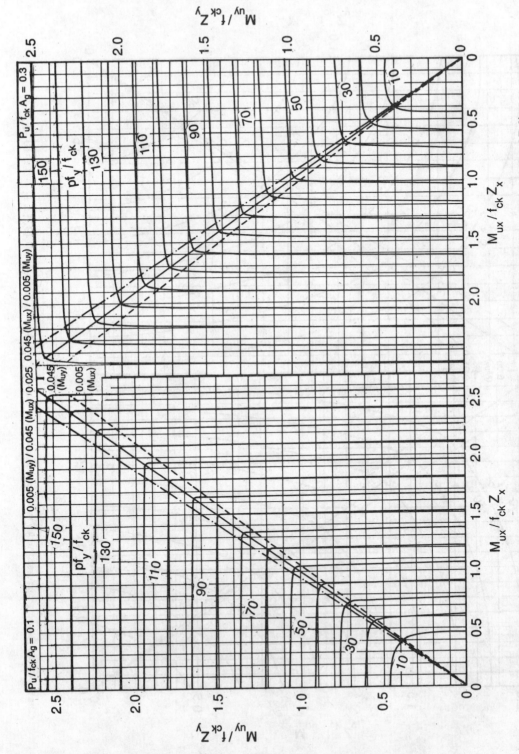

Chart +8.1 Design curves for +8 section of column for $P_u/f_{ck}A_g = 0.1$ and 0.3 $(13.0 \leq D/B \leq 17.0)$

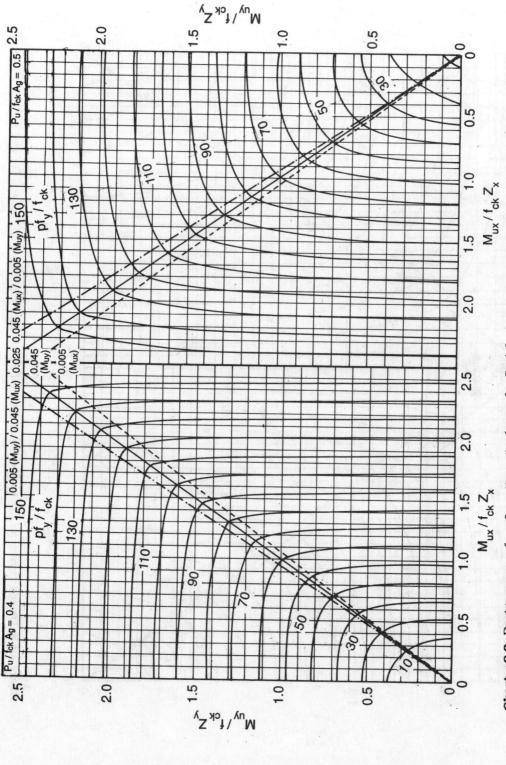

Chart 8.2 Design curves for +8 section of column for $P_u/f_{ck}A_g = 0.4$ and 0.5 $(13.0 \leq D/B \leq 17.0)$

402 Handbook of Reinforced Concrete Design

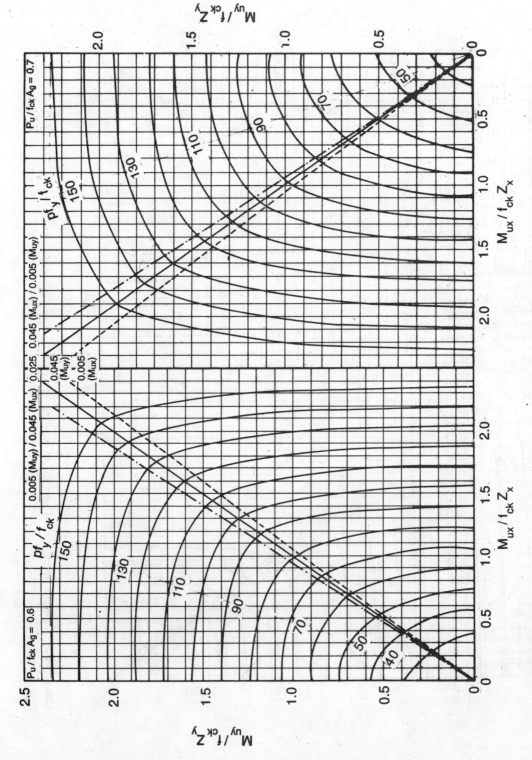

Chart +8.3 Design curves for +8 section of column for $P_u / f_{ck} A_g = 0.6$ and 0.7 $(13.0 \le D/B \le 17.0)$

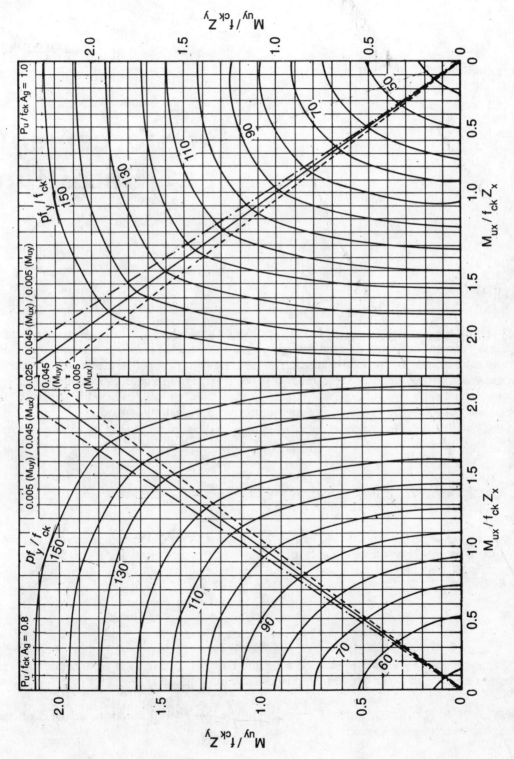

Chart +8.4 Design curves for +8 section of column for $P_u/f_{ck}A_g = 0.8$ and 1.0 $(13.0 \le D/B \le 17.0)$

Chapter 6

Reinforcement Detailing

6.1 INTRODUCTION

Much attention is generally directed towards analysis and design than towards proper detailing. Experience has indicated that the damages occur more due to errors in detailing than due to faulty calculations. In earthquake disasters, a structure with proper reinforcement detailing may show considerable damage but rarely collapse than many structures without proper reinforcement detailing which may totally collapse. Many minor cracks induced in the structure after construction may be avoided if proper reinforcement detailing is made.

In this chapter, the reinforcement detailing of various types of reinforced concrete structural elements is discussed.

6.2 GENERAL DETAILING REQUIREMENTS

The detailing of the reinforcement shall be such that the structural elements satisfy the requirements of its intended purpose satisfactorily. This is achieved by satisfying the detailing requirements as described below.

6.2.1 Cover to Reinforcement

Reinforcement should be provided with adequate cover. The thickness of a nominal cover exclusive of plaster or other decorative finishes shall be as follows.

(i) Ends of reinforcing bars: Cover shall not be less than 25 mm or twice the diameter of such bars whichever is greater.

(ii) Longitudinal bars in beams: Cover shall not be less than 25 mm or the diameter of such bars whichever is greater.

(iii) Tensile, compressive or shear reinforcement in slabs: Cover shall not be less than 15 mm or the diameter of such bars whichever is greater.

(iv) Longitudinal reinforcement in columns: The cover shall not be less than 40 mm or the diameter of such bars whichever is more. In case of columns with a minimum dimension of 200 mm or less, whose reinforcing bars do not exceed a diameter of 12 mm, the cover may be reduced to 25 mm. Increase in the thickness of cover of reinforcement bars shall be provided as described below. However, the additional thickness of cover shall be reduced to half for concrete of grade M25 and above, and the total thickness of cover shall not exceed 75 mm.

(a) *Surface of concrete exposed to the action of harmful chemicals (as in the case of concrete in contact with earth contaminated with such chemicals), acid, vapour, saline atmosphere and sulphurous smoke (as in the case of steam-operated railways)*: The additional cover shall not be less than 15 mm and not greater than 50 mm.

(b) *Reinforced concrete members of marine structures totally immersed in sea water*: Additional cover shall not be less than 40 mm.

(c) *Reinforced concrete members periodically immersed in sea water or subject to sea spray*: The additional cover shall not be less than 50 mm.

6.2.2 Development of Stress in Reinforcement

The stress in the reinforcement at any section shall be developed by an appropriate development length or end anchorage or a combination of both. The development length, L_d is given by:

$$L_d = \frac{\phi \sigma_s}{4\tau_{bd}} \tag{6.1}$$

where ϕ = nominal diameter of the bar

σ_s = stress in the bar at the section under consideration at design load

τ_{bd} = design bond stress

Development length of tension and compression bars based on working and limit state methods of design are given in Tables 6.1 and 6.2 respectively for concrete and steel of different grades.

Table 6.1 Development length for fully stressed tension and compression deformed bars based on working stress method

Permissible stress in steel, σ_{st} (N/mm²)	Development length / ϕ									
	Tension bar				Compression bar					
	Concrete grade				Concrete grade					
	M15	M20	M25	M30	M15	M20	M25	M30	M35	M40
170	51	38	34	31	41	31	27	25	23	21
230	69	52	46	41	55	41	37	33	30	28
275	82	62	55	49	52	39	35	31	29	26
300	90	67	60	54	72	54	48	43	39	36

Table 6.2 Development length for fully stressed tension and compression deformed bars based on limit state method

Characteristic strength of steel, f_y (N/mm²)	Development length / ϕ									
	Tension bar				Compression bar					
	Concrete grade				Concrete grade					
	M15	M20	M25	M30	M15	M20	M25	M30	M35	M40
415	57	47	41	38	47	38	33	31	27	24
480	66	55	47	44	55	44	38	35	31	28
500	68	57	49	46	57	46	39	37	32	29
550	75	63	54	50	60	50	43	40	36	32

The development length of bars under tension also includes the anchorage values of hooks and bends. Standard hook and 90° bend and their anchorage values are given in Fig. 6.1.

In compression, hooks and bends are ineffective and cannot be considered for anchorage. The anchorage length of such bars shall be the length of the straight part plus the projected length of hooks and bends as shown in Fig. 6.2.

The development length of shear reinforcement shall be measured as given below:

(i) Inclined bars in tension zone: Development length shall be considered from the end of the sloping or inclined portion of the bar as shown in Fig. 6.3(a).

(ii) Inclined bars in compression zone: Development length shall be considered from the mid-depth of the beam as shown in Fig. 6.3(b).

(iii) Stirrups: Full development length and anchorage shall be deemed to have been provided when stirrups are provided as shown in Fig. 6.3(c).

Reinforcement Detailing **407**

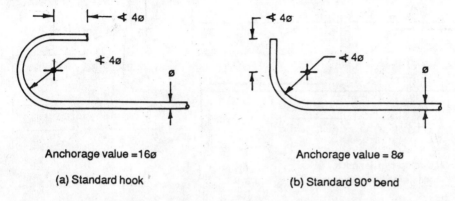

(a) Standard hook — Anchorage value = 16∅

(b) Standard 90° bend — Anchorage value = 8∅

Fig. 6.1 Anchorage value of standard hook and 90° bend

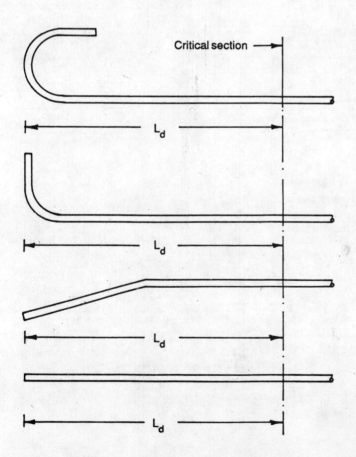

Fig. 6.2 Anchorage length of compression bars

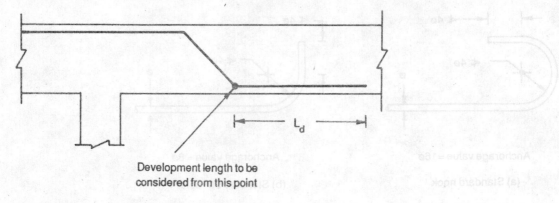

(a) Inclined bar in tension zone

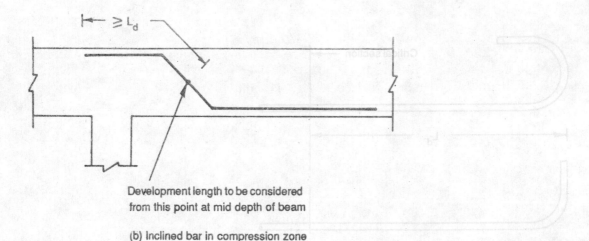

(b) Inclined bar in compression zone

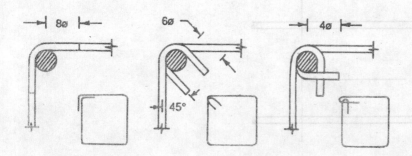

(c) Different ways of anchoring ends of stirrups

Fig. 6.3 Development length of shear reinforcement

6.2.3 Reinforcement Splicing

Splicing is required for transfer of force from one bar to another. Various provisions of reinforcement splicing are discussed below:

(i) Splices of longitudinal bars in flexural members: Splices of reinforcement bars in beams and slabs for their continuity shall be provided at sections away from the section of maximum stress as far as possible and be staggered. The splices are considered to be staggered if the centre to centre distance of the splices is not less than 1.3 times the lap length as shown in Fig. 6.4. It shall not be provided at a section where bending moment is more than 50 per cent of the moment of resistance of the section and not more than half the bars shall be spliced at a section.

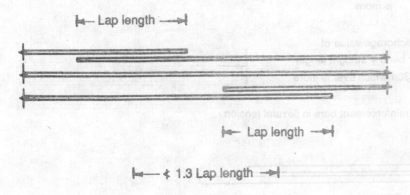

Fig. 6.4 Staggering of lap splices

The lap length of the reinforcement bars in flexural tension including the anchorage value of hook shall be L_d or 30 ϕ whichever is more as shown in Fig. 6.5. The length of the straight portion of the lap shall not be less than 15 ϕ or 200 mm whichever is greater.

The lap length of the reinforcement bars in flexural compression shall be equal to the development length in compression or 24 ϕ whichever is more as shown in Fig. 6.5.

In cases where more than half of the reinforcement bars are spliced at a section or when the splices are made at sections of maximum stress, the following precautions shall be taken.
   (a) Lap length shall be increased.
   (b) Closely spaced stirrups or spirals around the length of the splice shall be provided. The stirrups shall be able to resist tension equal to full tensile force in the lapped bars and shall be provided in the outer one third of the lap length at both ends with atleast three stirrups on either side (Fig. 6.6). In case of reinforcement bars of diameter larger than 28 mm, lap splices shall be completely enclosed by stirrups or spirals at close spacing as shown in Fig. 6.7(a). For reinforcement bars of diameter larger than 36 mm, it may be welded wherever possible or otherwise lap splices shall be completely enclosed by spirals (Fig. 6.7(b)).

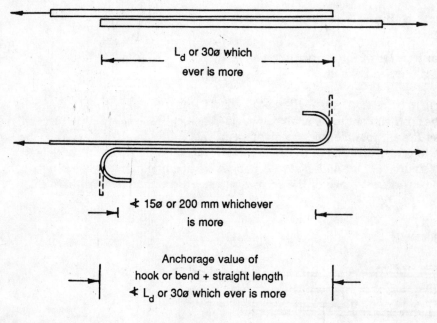

(a) Lap length of reinforcement bars in flexural tension

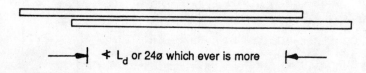

(b) Lap length of reinforcement bars in flexural compression

Fig. 6.5 Lap length of reinforcement bars in flexure

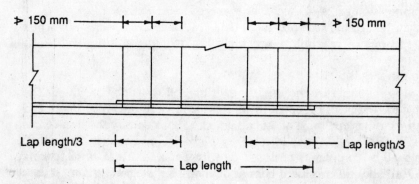

Fig. 6.6 Transverse reinforcement at splice

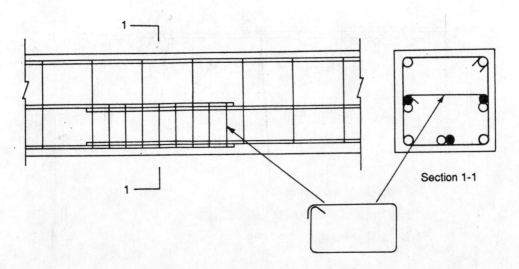

(a) Closely spaced stirrups around lap length

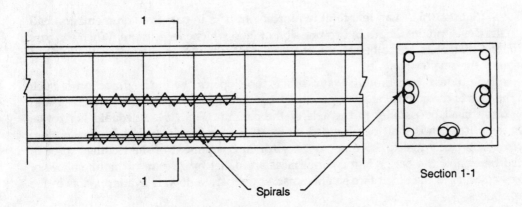

(b) Closely spaced spirals around lap length

Fig. 6.7 Closely spaced stirrups and spirals around lap length

(ii) Splices in tension members: The lap length of the reinforcement bars in direct tension including the anchorage value of hooks shall be $2 L_d$ or 30ϕ whichever is more. The length of the straight portion of the lap shall not be less than 15ϕ or 200 mm whichever is greater. It shall be enclosed in spiral made of bars not less than 6 mm in diameter with pitch not more than 100 mm as shown in Fig. 6.8.

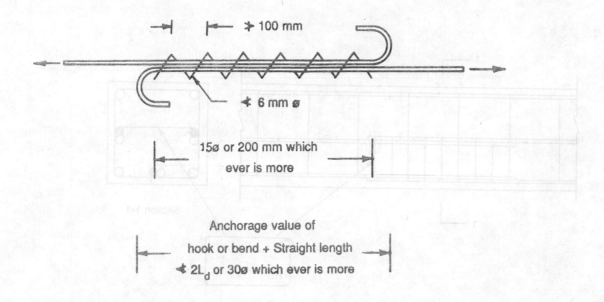

Fig. 6.8 Lap length of reinforcement bars in direct tension

(iii) Splices in column: Lap length of reinforcement bars in compression members shall be equal to the development length in compression or 24 φ whichever is more. When the bars of two different diameters are to be spliced, the lap length shall be calculated on the basis of the diameter of the smaller bar.

When the longitudinal bars are offset at a splice, the slope of the inclined portion of such bars with the axis of the column shall not exceed 1 in 6 and the portion of the bars above and below the offset shall be parallel to the axis of the column (Fig. 6.9). Adequate horizontal support at the offset bends shall be provided by metal ties or spirals. It shall be designed to resist horizontal thrust equal to 1.5 times the horizontal component of the nominal force in the inclined portion of the bar. When column faces are offset by 75 mm or more, splices of vertical bars adjacent to the offset face shall be made by separate dowels overlapped as in the case of lap splices.

6.2.4 Hooks and Bends

Hooks and bends shall be such that it will ensure their adequacy without overstressing the concrete or reinforcement steel. This is achieved by limiting the bearing stress in concrete at bends to be within their ultimate capacity at ultimate limit state. The bearing stress (σ) in concrete at bends (Fig. 6.10) is given by,

$$\sigma = \frac{F_{bt}}{r\phi} \tag{6.2}$$

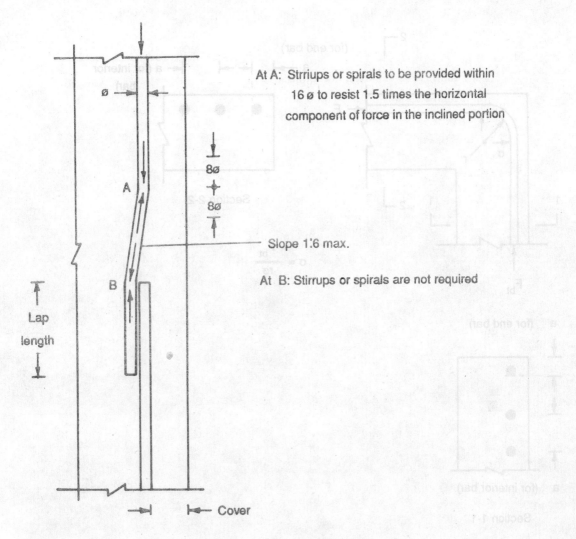

Fig. 6.9 A splice with offset cracked bars in column

where F_{bt} = tensile force in a reinforcement bar or group of bars (in N)

r = internal radius of bend (in mm)

ϕ = diameter of bar or diameter of equivalent area of bars in bundle (in mm)

The ultimate bearing stress (σ_u) of concrete is determined by,

$$\sigma_u = \frac{1.5 f_{ck}}{1 + 2\phi/a} \qquad (6.3)$$

where f_{ck} = characteristic strength of concrete

414 Handbook of Reinforced Concrete Design

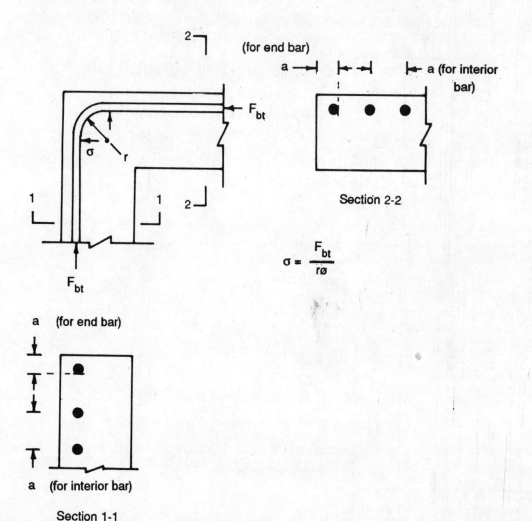

Fig. 6.10 Bearing stress at bend

a = cover plus diameter of bar or diameter of equivalent area of group of bars adjacent to the face of the member or centre to centre distance between bars or groups of bars perpendicular to the plane of bend for interior bars or groups of bars respectively.

Therefore,
$$\frac{F_{bt}}{r \cdot \phi} \leq \frac{1.5 f_{ck}}{1 + 2\phi/a}$$

or
$$r \geq \frac{F_{bt}}{1.5 f_{ck}} \phi \left(1 + \frac{2\phi}{a}\right) \qquad (6.4)$$

The radius of bends or hooks shall be in accordance with Eq. 6.4. However, the bearing stress in concrete in bends or hooks need not be checked if the radius of bends or hooks conform to that of the standard bends or hooks.

6.2.5 Change in Direction of Reinforcement

Change in the direction of tension or compression reinforcement induces a resultant force as shown in Fig. 6.11. The resultant force in closing bends (Fig. 6.11(a)) will be resisted by concrete without causing spalling as the effective thickness of concrete section shall be effective in resisting the resultant force, whereas the resultant force in opening bends (Fig. 6.11(b)) will cause spalling of concrete as the cover to the reinforcement is inadequate to resist the resultant force. Such a force tends to induce kink in the top bar and straightens the bottom bar. This can be prevented by resisting such forces by additional links or stirrups as shown in Fig. 6.12(a). The additional links or stirrups shall be designed for resisting 1.5 times the resul-

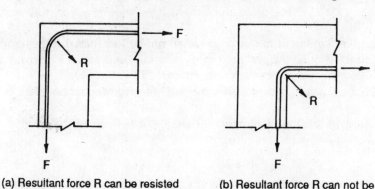

(a) Resultant force R can be resisted by effective depth of concrete section

(b) Resultant force R can not be resisted by concrete cover

Fig. 6.11 Resultant force at bend

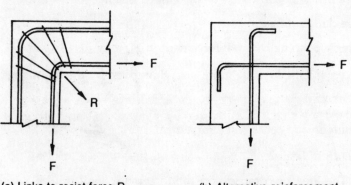

(a) Links to resist force R

(b) Alternative reinforcement detailing to avoid links

Fig. 6.12 Reinforcement detailing at bend

tant force. Alternatively, the detailing of the reinforcement bars shall be made as shown in Fig. 6.12(b) so that no transverse force is induced at the kink.

6.3 REINFORCEMENT DETAILING FOR FOUNDATION

The IS code provisions for area, cover, spacing and diameter of reinforcement bars and detailing for isolated footing and combined footing of two columns, grid, mat and pile foundations are discussed in this section.

6.3.1 Area, Cover, Spacing and Diameter of Reinforcement Bars

The area, cover, spacing and diameter of reinforcement bars shall be provided as follows.

(i) Area of reinforcement bars: Area of reinforcement bars in foundation slab and beam shall be as follows.

In foundation slab, reinforcement bars in either direction shall not be less than 0.15 per cent of the gross sectional area of concrete where plain mild steel bars are used and 0.12 per cent of the gross sectional area of concrete where high strength deformed bars are used.

In foundation beam, area of the tension, compression and side face reinforcement shall be as follows.

(a) *Tension reinforcement:* Area of tension reinforcement shall not be less than that given by the following.

$$A_{st,min} = \frac{0.85\,b\,d}{f_y}$$

where $A_{st,min}$ = minimum area of tension reinforcement

b = width of beam or width of web of flanged beam

d = effective depth of beam

f_y = characteristic strength of reinforcement in N/mm^2

The area of tension reinforcement shall not be more than that given by the following,

$$A_{st,max} = 0.04\,bD$$

where $A_{st,max}$ = maximum area of tension reinforcement

D = total depth of beam

(b) *Compression reinforcement:* The area of compression reinforcement shall not be more than that given by the following

$$A_{sc,max} = 0.04\,bD$$

where $A_{sc,max}$ = maximum area of compression reinforcement.

(c) *Side face reinforcement:* It shall be provided where the depth of the web in a beam exceeds 750 mm. The total area of the side face reinforcement shall not be less than 0.1 per cent of the web area and shall be distributed equally on two faces at a spacing not more than 300 mm or thickness of web whichever is less.

(ii) Cover to reinforcement bars:
The minimum thickness of cover to main reinforcement bars shall not be less than 50 mm for surfaces in contact with the earth and not less than 40 mm for the external exposed face. However, when a levelling course of lean concrete is not provided, a cover of 75 mm is usually provided. In case of raft foundation, whether resting directly on soil or on lean concrete, the cover for reinforcement shall not be less than 75 mm.

(iii) Spacing of reinforcement bars:
Spacing of reinforcement bars in the foundation slab and beam shall be as follows.

In the foundation slab, spacing of main reinforcement bar shall not be more than three times the effective depth of the slab or 450 mm whichever is smaller. The spacing of distribution bars provided for shrinkage and temperature stresses shall not be more than five times the effective depth of slab or 450 mm whichever is smaller.

In the foundation beam, horizontal clear spacing of reinforcement bars shall not be less than the diameter of the bar or maximum nominal size of aggregate plus 5 mm whichever is more. The vertical clear spacing of reinforcement bars shall not be less than 15 mm or diameter of bar or 2/3 maximum or nominal size of aggregate plus 5 mm whichever is more. The maximum spacing between parallel bars and clear distance of corner bars from the corner of beam near the tension face shall not be greater than the values given in Table 2.2 depending on the amount of redistribution carried out in the analysis and the characteristic strength of reinforcement unless the calculation for crack width shows that a greater spacing of bars is acceptable. The spacing given in the table are not applicable to members subjected to particularly aggressive environments unless the ultimate resistance has been calculated by limiting f_y to 300 N/mm².

(iv) Diameter of reinforcement bars:
The diameter of reinforcement bars in foundation slab and beam shall be as follows.

In foundation slab, the diameter of main reinforcement shall not be less than 10 mm. The diameter of the distribution bars shall not be less than 8 mm. The diameter of the reinforcement bars shall not be more than 0.125 times the thickness of the slab.

In foundation beam, the diameter of reinforcement bars shall not be less than 12 mm.

6.3.2 Reinforcement Detailing for Isolated Footing

Reinforcement detailing for isolated footing for walls and columns are made as follows.

(i) Isolated footing for walls:
Isolated footing supporting a wall is reinforced when the projection of the footing beyond the wall exceeds the thickness of the footing as shown in

Fig. 6.13. It consists of main reinforcement in transverse direction and secondary reinforcement in longitudinal direction. Wherever an abrupt change in the magnitude of the load or variation in ground support or local loose pockets occur along the footing, adequate reinforcement in longitudinal direction shall be provided.

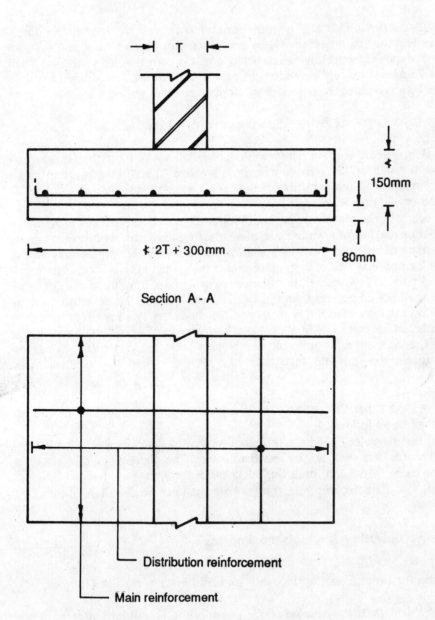

Fig. 6.13 Reinforcement detailing for isolated footing supporting a wall

Reinforcement Detailing 419

(ii) Isolated footing for columns: Isolated footings for a column are generally square but rectangular, circular or any shapes may be provided. In square footing, uniformly distributed reinforcement in each direction shall be provided across the full width of footing as shown in Fig. 6.14(a). In rectangular footing, the reinforcement in the long direction is pro-

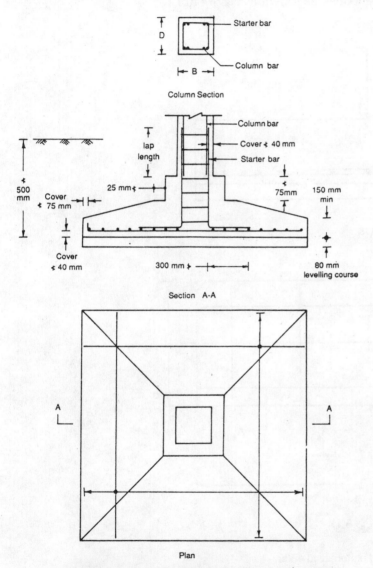

- For footing bars, provide 90° standard bend if required for development length.
- Development length for starter or column bars in footing is considered from top of the pedestal or footing

(a) Isolated square footing for column

Fig. 6.14(a) Reinforcement detailing for isolated square footing for column

vided uniformly distributed across the width of the footing. For reinforcement in the shorter direction, a central band equal to the width of the footing shall be provided with $2/(\beta+1)$ times the total area of reinforcement in the shorter direction where β is the ratio of the long side to the short side of the footing. The remaining reinforcement shall be uniformly distributed in the outer portions of the footing as shown in Fig. 6.14(b).

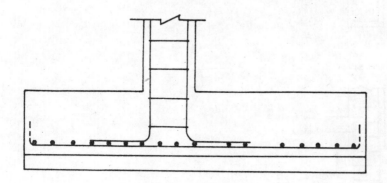

Section A-A

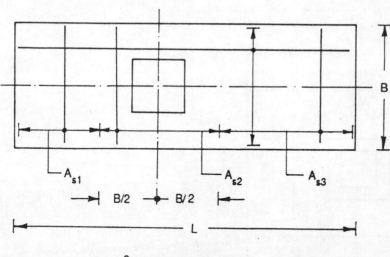

$$A_{s2} = \frac{2}{1+L/B}(A_{s1}+A_{s2}+A_{s3})$$

A_{s1} and A_{s3} distributed uniformly at ends

(b) Transverse reinforcement detailing for isolated rectangular footing for column

Fig. 6.14(b) Reinforcement detailing for isolated rectangular footing for column

The column bars are extended into the footing and then bent horizontally to rest over the bottom layer of footing reinforcement. The vertical length of the column bars inside the footing shall not be less than the development length of bar and its horizontal projection over the footing reinforcement shall not be less than 300 mm. When dowel bars are provided, the column bars rest on the footing or may be tied with the dowel bars. The area of the dowel bars shall not be less than 0.5 per cent of cross-sectional area of the column and there shall not be less than four dowel bars of diameter not exceeding the diameter of column bars by more than 3 mm. The length of dowel bars on both sides of bearing surface shall not be less than the development length of the bar.

Column bars of diameter larger than 36 mm in compression can be dowelled at the footing with bars of smaller diameter of the necessary area. The dowel shall extend into the column, a length equal to the development length of the column bar and into the footing, a length equal to the development length of the dowel. If the depth of the footing or footing and pedestal combined is less than the development length required for the dowel, the diameter of dowels may be suitably reduced and the number of dowels increased to satisfy the required area and development length.

The footings are sloped or stepped towards the edges while satisfying the requirements for bending and punching shear. The thickness at the edges shall not be less than 150 mm for footings on soil, nor less than 300 mm for footings on piles.

6.3.3 Reinforcement Detailing for Combined Footing

Reinforcement detailing of some typical combined footings are shown in Figs 6.15 and 6.16. For trapezoidal footing supporting two columns, main reinforcement is provided in the transverse direction at the bottom of the slab under columns and in longitudinal direction at the bottom and top of slab as shown in Fig. 6.15.

The reinforcement detailing of combined slab-beam footing under two columns has been shown in Fig. 6.16(a). It consists of main reinforcement in transverse direction at the bottom of the slab for its cantilever action under upward soil pressure. The footing beam is reinforced as shown in the figure for its bending action under upward soil pressure from bottom and downward load from columns. The curtailment of reinforcement is made in accordance with bending moment diagram. Figure 6.16(b) shows the reinforcement detailing of strip footing under columns. It consists of main reinforcement in transverse direction at the bottom of slab under columns, and in longitudinal direction for longitudinal bending action of strip footing under upward soil pressure from bottom and downward load from columns.

6.3.4 Reinforcement Detailing for Column on Edge of Footing

When the column is located at the edge of the footing, shear cracks develop along the inclined plane which can be prevented by providing U-type horizontal bars around the vertical starter column bars as shown in Fig. 6.17.

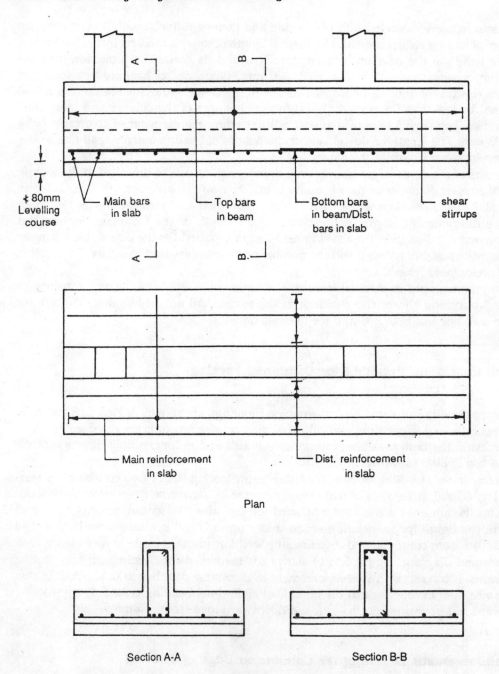

Fig. 6.15(a) Reinforcement detailing for combined slab beam footing under columns

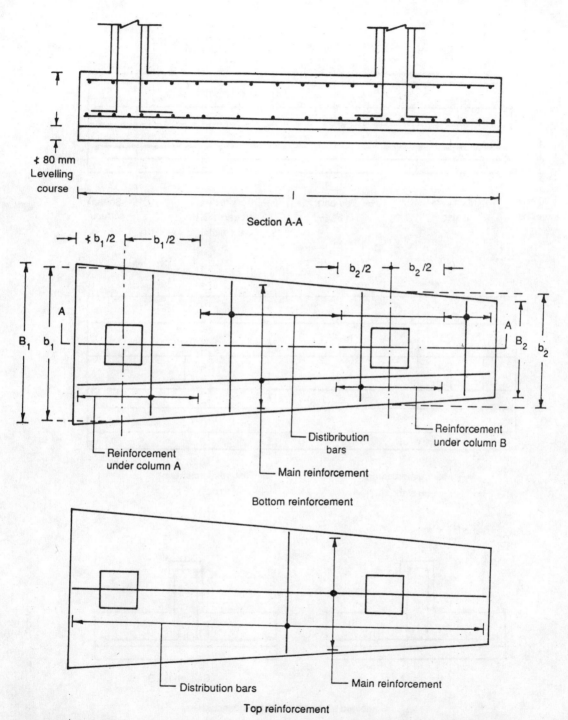

Fig. 6.15(b) Reinforcement detailing for combined trapezoidal footing under columns

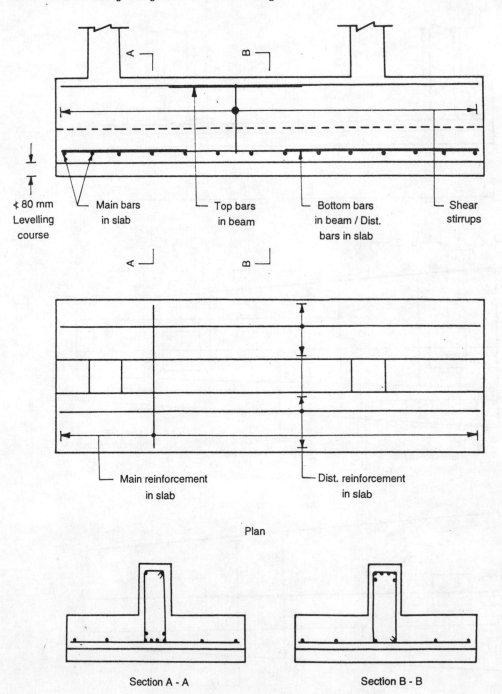

Fig. 6.16(a) Reinforcement detailing for combined slab beam footing under columns

Reinforcement Detailing 425

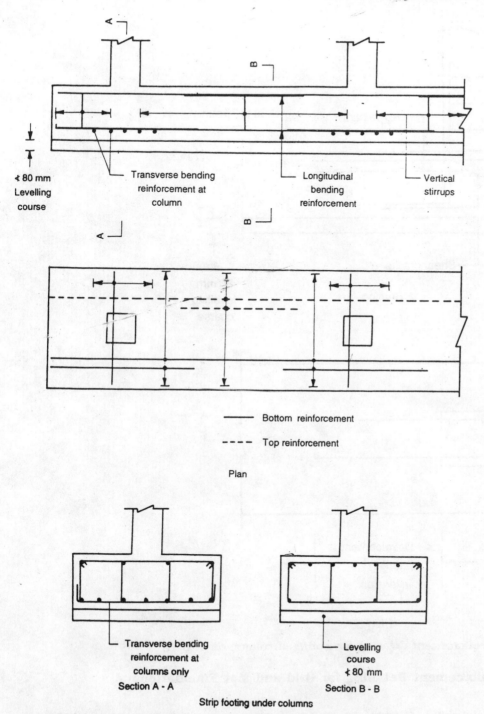

Fig. 6.16(b) Reinforcement detailing of strip footing under columns

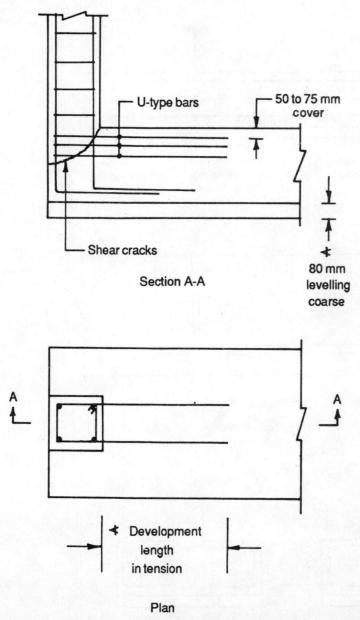

Fig. 6.17 Reinforcement detailing for footing of column on edge

6.3.5 Reinforcement Detailing for Grid and Mat Foundations

Reinforcement detailing for grid and mat foundations are shown in Figs 6.18 and 6.19 respectively. It consists of main reinforcement in the top and bottom layers of foundation slab in

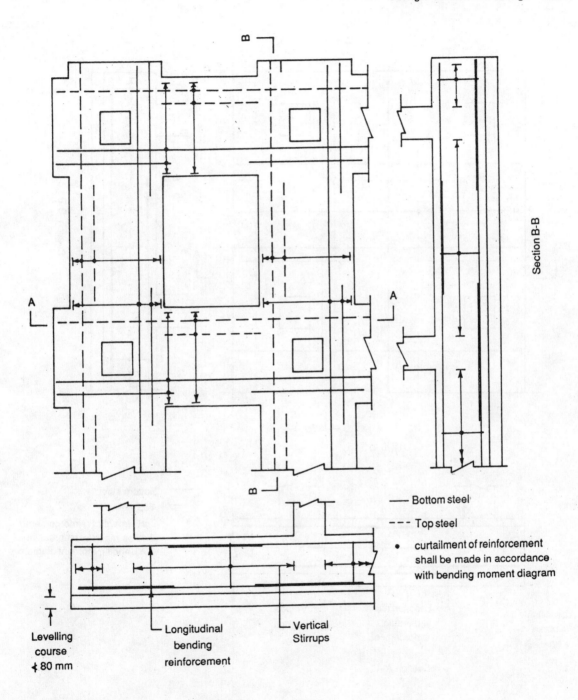

Fig. 6.18 Reinforcement detailing for grid foundation

428 Handbook of Reinforced Concrete Design

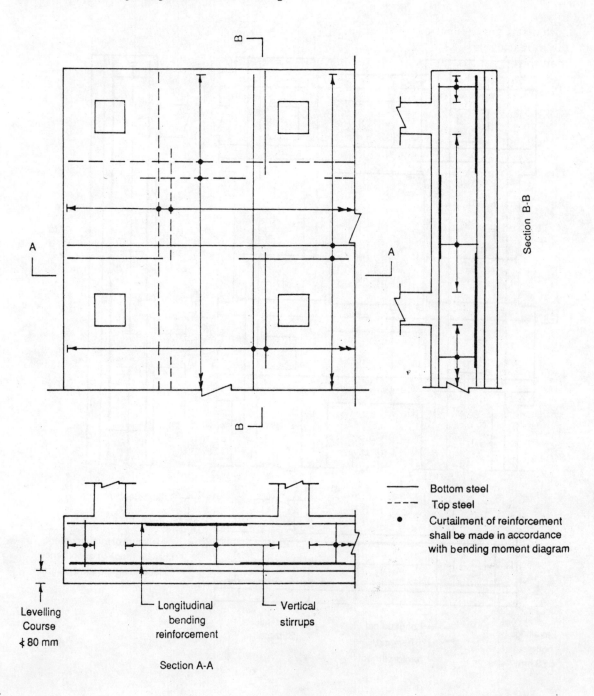

Fig. 6.19 Reinforcement detailing for mat foundation

Reinforcement Detailing 429

both directions of bending. The curtailment of reinforcement is made in accordance with bending moment diagram. If the raft consists of several parts with varying load and depth of raft, than the expansion joint shall be provided between each part of the raft.

6.3.6 Reinforcement Detailing of Under-reamed Piles

Typical reinforcement detailing of single and double under-reamed piles is shown in Fig. 6.20. The area of longitudinal reinforcement in under-reamed piles shall not be less than 0.4 per

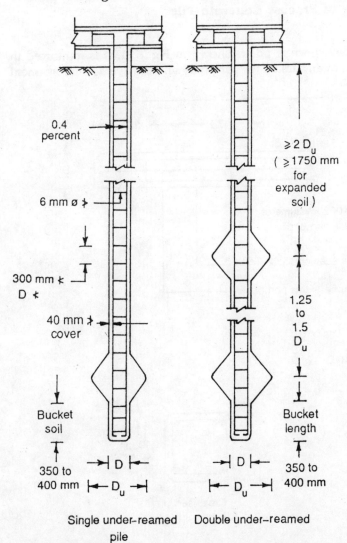

Fig. 6.20 Reinforcement detailing for under-reamed piles

cent of its gross sectional area. Its number shall not be less than 4 but for diameter greater than 400 mm, minimum of 6 bars of diameter not less than 12 mm shall be provided. The transverse reinforcement shall be of diameter not less than 6 mm with the spacing not more than the diameter of stem or 300 mm whichever is more. For piles of length greater than 5 m and diameter greater than 375 mm, transverse reinforcement shall be of diameter not less than 8 mm.

The clear cover to longitudinal reinforcement shall not be less than 40 mm in mild environment and not less than 75 mm in aggressive environment of sulphates etc.

6.3.7 Reinforcement Detailing of Precast Concrete Pile

The reinforcement detailing of precast concrete pile is shown in Fig. 6.21. It is reinforced in the same way as the column with the main bars on the periphery and transverse reinforcement

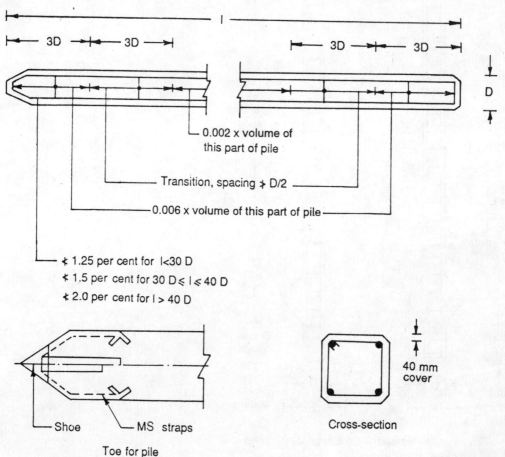

Fig. 6.21 Reinforcement detailing of precast pile

around main bars. The main bars are bent inwards at the lower end and welded to the shoe made of chilled cast iron or steel provided to protect tip of the pile when driven through soil. The area of longitudinal reinforcement shall not be less than the following.

(i) 1.25 per cent of the cross-sectional area for piles of length less than 30 times the least dimension of the pile section.
(ii) 1.5 per cent of the cross-sectional area for piles of length between 30 to 40 times the least dimension of the pile section.
(iii) 2 per cent of the cross-sectional area for piles of length greater than 40 times the least dimension of the pile section.

The transverse reinforcement in the form of hoops and links shall be of diameter not less than 6 mm and its volume shall be as follows.

(i) In each end segments of the pile of length equal to 3 times the least dimension of column section, volume of the transverse reinforcement shall not be less than 0.006 times the gross volume of the end segments of pile.
(ii) In the remaining length of the pile, volume of transverse reinforcement shall not be less than 0.002 times the gross volume of that part of the pile.

The spacing of the transverse reinforcement shall not be greater than the least dimension of the pile section. The transition between the close spacing of transverse reinforcement in the end segments and the maximum spacing in the remaining length of the pile shall be gradual over a length of 3 times the least dimension of the pile section.

The cover of concrete to the longitudinal and transverse reinforcement shall not be less than 40 mm in mild environment and not less than 50 mm when the piles are exposed to sea water or water having other corrosive content.

6.3.8 Pile Group Pattern and Reinforcement Detailing of Pile Cap

Generally two or more than two piles are provided in a foundation because of alignment problem and inadvertent eccentricities. Some typical patterns of pile group for isolated footings and walls are shown in Fig. 6.22(a).

Pile cap is provided over a group of piles to distribute the load from the column to the piles in the group. Column is positioned at the centre of gravity of the pile group so that dowel or column bars are embedded into pile cap in a similar way as provided in the column base.

The loads acting on the pile cap from the column and piles are resisted by the development of bending moment and shear force in the pile cap. The thickness of the pile cap is fixed such that it is adequate to resist shear without shear reinforcement and the bars projecting from the piles and the dowel bars for the column can be provided adequate bond length. The clear overhang of the pile cap beyond the outermost pile in the group shall normally be 100 to 150 mm depending on the size of pile.

The plan arrangement of reinforcement details of some typical pile caps are shown in Fig. 6.22(b). A typical reinforcement detailing of a pile cap supporting a column and resting on three piles is shown in Fig. 6.23. The clear cover to the main reinforcement bars in the bottom layer of the pile cap shall not be less than 60 mm. The reinforcement from the piles

432 Handbook of Reinforced Concrete Design

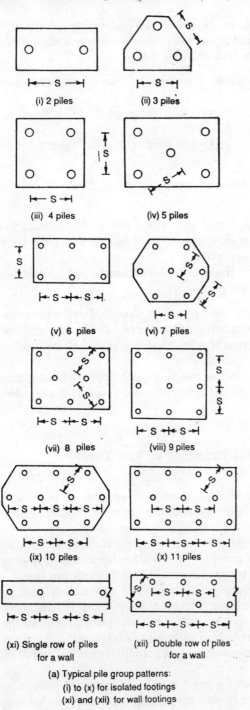

(i) 2 piles
(ii) 3 piles
(iii) 4 piles
(iv) 5 piles
(v) 6 piles
(vi) 7 piles
(vii) 8 piles
(viii) 9 piles
(ix) 10 piles
(x) 11 piles
(xi) Single row of piles for a wall
(xii) Double row of piles for a wall

(a) Typical pile group patterns:
 (i) to (x) for isolated footings
 (xi) and (xii) for wall footings

Fig. 6.22(a) Typical pile group patterns: (i) to (x) for isolated footings and (xi) to (xii) for wall footings

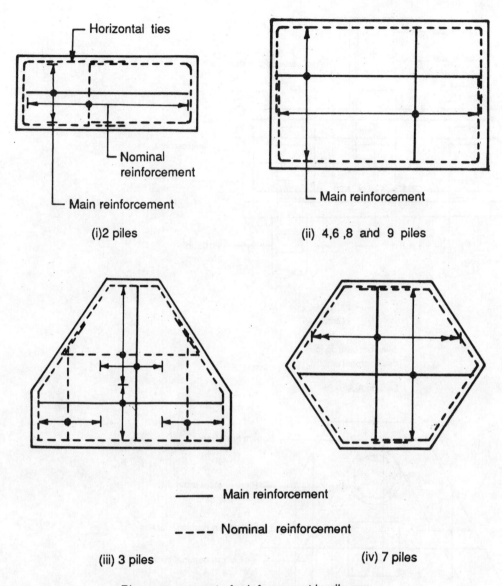

Fig. 6.22(b) Plan arrangement of reinforcement in pile caps

and column shall be properly embedded into the pile cap for a length not less than the development length of bar. It is also provided with reinforcement going round the outer piles in the group to prevent the tendency to fail in bursting due to high principal tension as shown in the figure.

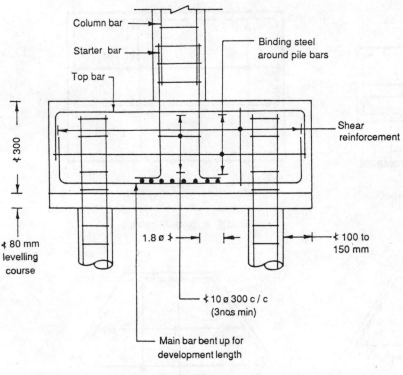
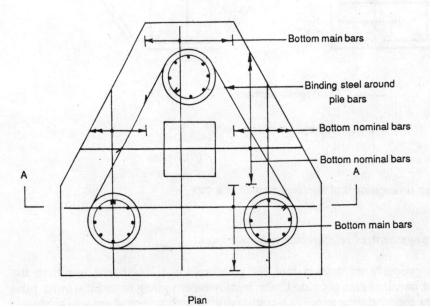

Fig. 6.23 Reinforcement detailing for pile cap of three piles

6.3.9 Reinforcement Detailing of Grade Beams

The grade beam supporting a masonary wall behaves as a deep beam. The total thickness of a grade beam shall not be less than 150 mm. The reinforcement at the bottom is provided continuous and an equal amount of reinforcement is provided at the top, up to a distance of span/4 on both sides of pile or footing as shown in Fig. 6.24. A minimum of three bars of 10 mm φ shall be provided both at the top and bottom and stirrups of diameter not less than 6 mm at a spacing not more than 300 mm is provided.

In expansive soils, grade beam shall be placed 80 mm above the ground level. A ledge projection of 75 mm thickness and extending 80 mm into the ground shall be provided on the outer side of the grade beam over piles in expansive soil. In other soils, grade beam may be provided on the ground over a levelling course not less than 80 mm thick as shown in the figure.

6.4 REINFORCEMENT DETAILING FOR COLUMNS

The reinforcement detailing for longitudinal and transverse reinforcements should be made in accordance with various provisions made in the IS : 456–1978 which are described as follows.

6.4.1 Longitudinal Reinforcement Detailing

The longitudinal reinforcement detailing shall be made in accordance with the following provisions.

(i) Area of reinforcement bars: It shall not be less than 0.8 per cent nor greater than 6 per cent of the gross area of section. If the cross-sectional area is larger than that required to support the load, the minimum percentage of steel shall be based upon the area of concrete required to resist the direct stress and not upon the actual area.

(ii) Diameter of reinforcement bars: It shall not be less than 12 mm.

(iii) Number of reinforcement bars: It shall not be less than one bar at each corner in the non-circular section and not less than 6 bars in the circular section.

(iv) Spacing of reinforcement bars: It shall not be more than 300 mm measured along the periphery of the column.

(v) Cover to reinforcement bars: Cover to the longitudinal reinforcement shall be provided as described below.

    (a) It shall not be less than 40 mm or the diameter of the bar whichever is more. In case of column of minimum dimension of 200 mm or less, whose reinforcement bars do not exceed 12 mm diameter, a cover of 25 mm may be provided.

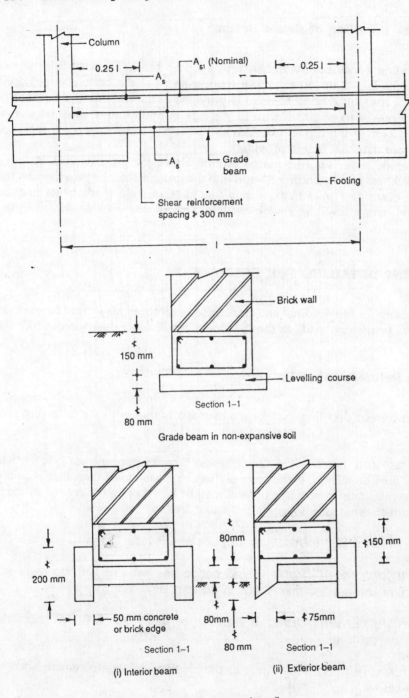

Fig. 6.24 Reinforcement detailing for grade beam

(b) At each end of the reinforcement bar, cover shall not be less than 25 mm or twice the diameter of the bar whichever is more.

(c) For columns subjected to the action of harmful chemicals, increased thickness of 15 mm to 50 mm in addition to the recommended value under normal conditions may be provided. For reinforced concrete members totally immersed in sea water, the additional cover shall be 40 mm, whereas when such members are periodically immersed in sea water or subjected to sea spray, the additional cover shall be 50 mm. When concrete of grade M25 or above are used, the additional thickness of cover as specified above may be reduced to half. In all such cases, the cover shall not exceed 75 mm.

(vi) Splicing of reinforcement bars: It is provided by lapping the bars or by butt welding. The lap length of reinforcement bars in compression members shall be equal to the development length in compression or 24 ϕ, whichever is more. When the bars of two different diameters are to be spliced, the lap length shall be calculated on the basis of the smaller diameter of the bar. When the longitudinal bars are offset at a splice, the slope of the inclined portion of such bars with the axis of the column shall not exceed 1 in 6 and the portions of the bar above and below the offset shall be parallel to the axis of the column (Fig. 6.25).

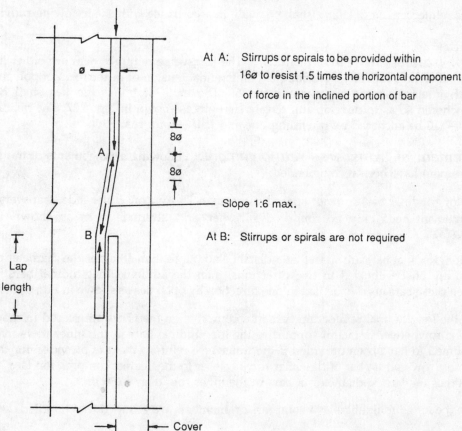

Fig. 6.25 Splice with offset cracked bars in column

Adequate horizontal support at the offset bends shall be provided by ties or spirals. It shall be placed at a distance not more than 8.0 times the diameter of longitudinal bar on both sides from the point of the bend (Fig. 6.25). Usually three closely spaced ties are used, one of which may be part of the regularly spaced ties. Additional ties shall be designed to resist horizontal thrust equal to 1.5 times the horizontal component of the nominal stress in the inclined portion of the bar.

Typical splice details for exterior and interior columns are shown in Fig. 6.26. When the cross-section of a column is reduced in the upper storey such that the offset between the column faces does not exceed 75 mm, the longitudinal bars from the lower storey are offset to be within the upper column as shown in Fig. 6.26(a). When the offset between the column faces exceeds 75 mm, the longitudinal bars in the lower column shall be terminated at the floor slab and the separate dowels shall be provided as shown in Fig. 6.26(b).

6.4.2 Transverse Reinforcement Detailing

The transverse reinforcement detailing shall be made in accordance with the following provisions.

(i) Shape of transverse reinforcement: The transverse reinforcement are either in the form of circular rings capable of resisting circumferential tension or in the form of polygonal links with internal angles not exceeding 135°. The ends of transverse ties shall be adequately anchored so as to develop full tensile strength as shown in Fig. 6.27. The helical reinforcement shall be anchored by providing one and half extra turns.

(ii) Arrangement of transverse reinforcement: Following arrangements of transverse reinforcement have been recommended:

(a) If the longitudinal bars are not spaced more than 75 mm on either side, transverse reinforcement need only go round the corner and alternate bars as shown in Fig. 6.28(a).

(b) If the longitudinal bars are spaced at a distance not more than 48 times the diameter of tie and are effectively tied in two directions, then the aditional longitudinal bars in between these bars need to be tied in one direction by open ties as shown in Fig. 6.28(b).

(c) When the longitudinal reinforcing bars in a compression members are placed in more than one row, effective lateral support to the longitudinal bars in the inner rows may be assumed to have been provided if the transverse reinforcement is provided for the outermost row and no bar of the inner row is closer to the nearest compression face by three times the largest diameter of bars in the inner row (Fig. 6.28(c)).

(d) Where the longitudinal bars in a compression members are grouped (not bundled) and

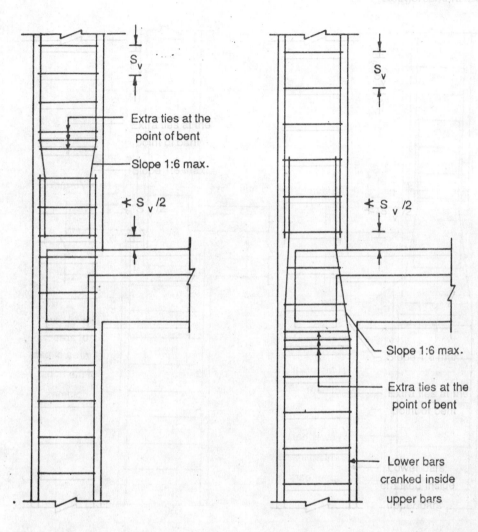

(i) Splice with upper bars cranked into position inside lower bars

(ii) Splice with lower bars cranked into position inside upper bars with stepping of columns on one side

Fig. 6.26(a) Splice details for exterior column

each group is adequately tied with transverse reinforcement, the transverse reinforcement for the compression member as a whole may be provided on the assumption that each group is a single longitudinal bar for the purpose of determining the diameter and spacing of transverse reinforcement. The diameter of such transverse reinforcement need not exceed 20 mm (Fig. 6.28(d)).

Some typical arrangements of transverse reinforcement are shown in Fig. 6.29.

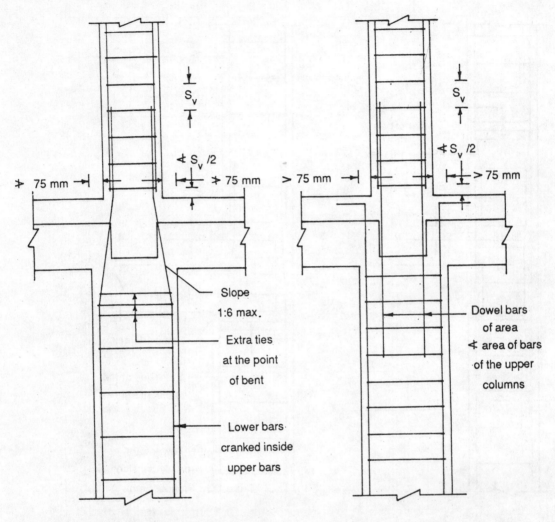

(i) Splicing when the relative displacement of column faces is less than 75 mm

(ii) Splicing when the relative displacement of column faces is more than 75 mm

Fig. 6.26(b) Splice details for interior columns

(iii) Diameter and pitch of transverse reinforcement: The diameter and pitch of the transverse reinforcement shall be as follows:

(a) Diameter of transverse reinforcement: The diameter of transverse reinforcement shall not be less than one-fourth of the diameter of the largest longitudinal bar and in no case less than 5 mm as shown in Fig. 6.30. In case of helical reinforcement, where an increased load on account of helical reinforcement is considered, the diameter of the helical reinforcement shall

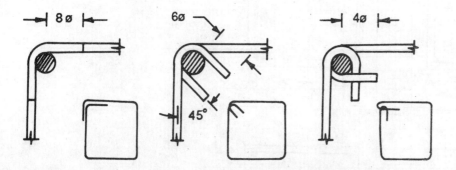

Fig. 6.27 Different ways of anchoring ends of stirrups

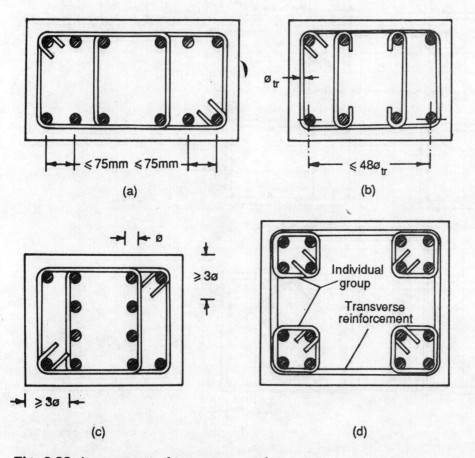

Fig. 6.28 Arrangement of transverse reinforcement

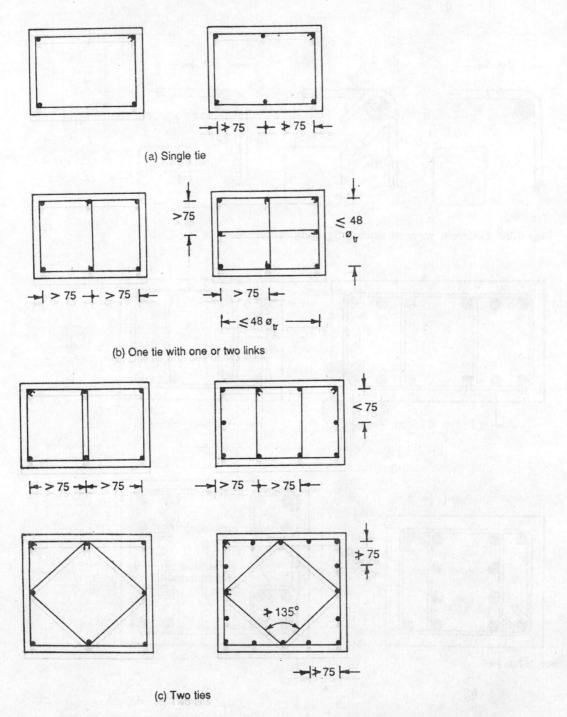

Fig. 6.29(a), (b), & (c) Typical arrangement of transverse reinforcements

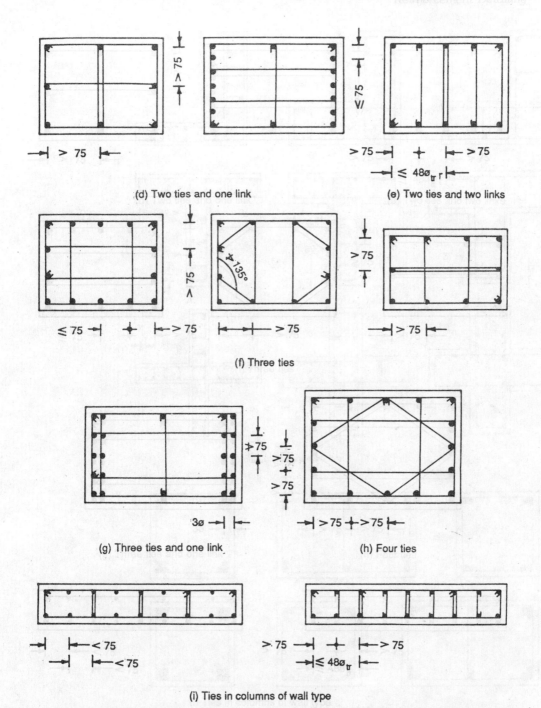

Fig. 6.29(d), (e), (f), (g), (h) & (i) Typical arrangement of transverse reinforcement

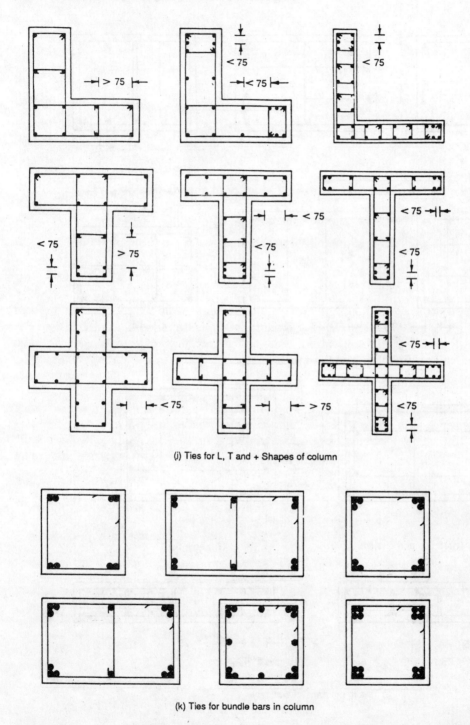

(j) Ties for L, T and + Shapes of column

(k) Ties for bundle bars in column

Fig. 6.29(j), & (k) Typical arrangement of transverse reinforcements

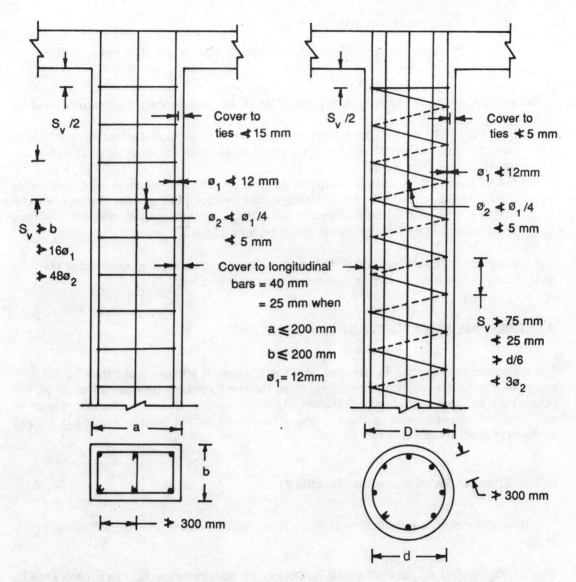

Fig. 6.30 Reinforcement detailing requirements for diameter, spacing and cover of transverse reinforcement

be such that the ratio of the volume of helical reinforcement to the volume of the core shall not be less than $0.36\,(A_g/A_c - 1.0)\,f_{ck}/f_y$ where,

A_g = gross area of the section

A_c = area of the core of the helically reinforced column measured to the outside diameter of the helix

f_{ck} = characteristic strength of concrete

f_y = characteristic strength of helical reinforcement which shall not exceed 415 N/mm^2

(b) Pitch of transverse reinforcement: The pitch of lateral ties and helical reinforcement shall be as follows.

Lateral ties: The pitch of the lateral ties shall not be more than the least lateral dimension of the compression member nor more than 16 times the smallest diameter of the longitudinal reinforcement.

Helical reinforcement: Where an increased load on account of helical reinforcement is considered, the pitch of the helical reinforcement shall not be less than 75 mm, nor more than one sixth of the core diameter of the column, nor less than 25 mm, nor less than 3 times the diameter of the helical reinforcement (Fig. 6.30).

(iv) Cover to transverse reinforcement: Cover to transverse reinforcement shall not be less than its diameter or 15 mm whichever is more (Fig. 6.30).

6.5 REINFORCEMENT DETAILING FOR BEAMS

The reinforcement detailing for flexural, shear and torsional reinforcement shall be made in accordance with the following provisions. Also the reinforcement detailing of some typical beams such as beams of different depth at support, intersecting beams, tie members, corners and junctions of beams and columns, cranked beams, haunched beams, opening in web of beam have been discussed in this section.

6.5.1 Flexural Reinforcement Detailing

The flexural reinforcement detailing shall be made in accordance with the following provisions.

(i) Arrangement of flexural reinforcement: Reinforcement bars in tension shall be arranged conforming to the requirements of cover at the bottom and side faces and horizontal and vertical spacing of the bars. Figure 6.31 shows requirements for minimum cover and clear spacing of longitudinal bars or groups of bars in beams.

The maximum clear distance between parallel bars or groups of bars and clear distance of corner bars from the corner of beam near the tension face (Fig. 6.32) shall not be greater than the values given in Table 6.3 depending on the amount of redistribution carried out in the analysis and the characteristic strength of reinforcement unless the calculation for crack width shows that a greater spacing of bars is acceptable. The spacings given in the table are not applicable to members subjected to particularly aggressive environments unless the ultimate moment of resistance has been calculated by limiting f_y to 300 N/mm^2.

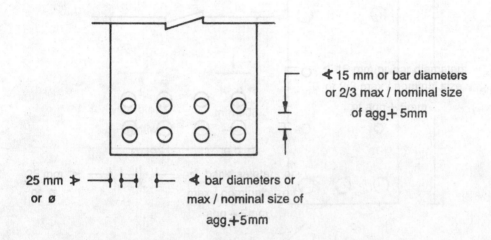

(a) Minimum cover and spacing between individual bars

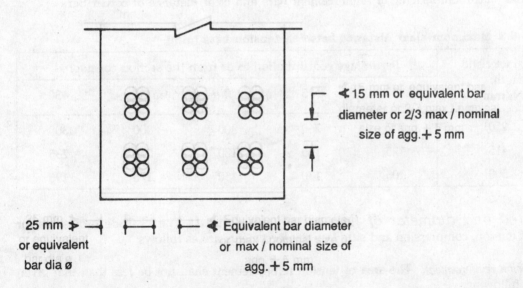

(b) Minimum cover and spacing between groups of bars

Fig. 6.31 Minimum cover and spacing between groups of bars

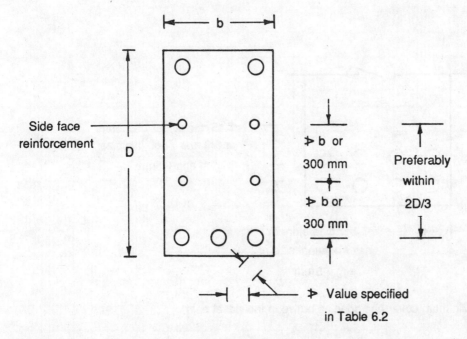

Fig. 6.32 Maximum spacing of reinforcement bars and clear distance of corner bars

Table 6.3 Maximum clear distance between tension bars (mm)

Characteristic strength, f_y (N/mm²)	Percentage redistribution to or from the section considered				
	−30	−15	0	+15	+30
250	215	260	300	300	300
415	125	155	180	210	235
500	105	130	150	175	195

(ii) Area and diameter of flexural reinforcement: The IS Code provisions for area of tension, compression and side face reinforcements are as follows:

(a) Tension reinforcement: The area of tension reinforcement shall not be less than that given by the following,

$$A_{st,\,min} = \frac{0.85\,b\,d}{f_y} \tag{6.5}$$

where $A_{st,\,min}$ = minimum area of tension reinforcement

b = width of beam or width of web of flanged beam

d = effective depth of beam

f_y = characteristic strength of reinforcement in N/mm^2

The area of tension reinforcement shall not be more than that given by the following.

$$A_{st,\,max} = 0.04\,bD \qquad (6.6)$$

where $A_{st,\,max}$ = maximum area of tension reinforcement

D = total depth of beam

(b) Compression reinforcement: The area of compression reinforcement shall not be more than that given by the following.

$$A_{sc,\,max} = 0.04\,bD \qquad (6.7)$$

where $A_{sc,\,max}$ = maximum area of compression reinforcement

(c) Side face reinforcement: It shall be provided where the depth of the web in a beam exceed 750 mm. The total area of the side face reinforcement shall not be less than 0.1 per cent of the web area and shall be distributed equally on two faces at a spacing not more than 300 mm or thickness of web whichever is less (Fig. 6.33). The diameter of reinforcement bar shall not be less than 12 mm.

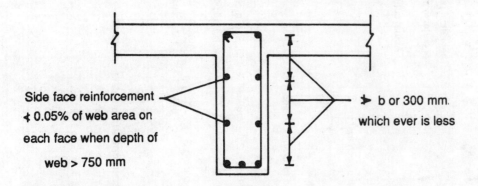

Fig. 6.33 Side face reinforcement in beams

(iii) Curtailment of flexural reinforcement: IS : 456 –1978 has recommended the following conditions for the curtailment of reinforcement bars:

(a) Extension beyond the theoretical cut-off point: A reinforcement bar to be curtailed is extended to a length equal to the effective depth of a member or 12 φ whichever is more beyond the point at which it is not required theoretically except at simple support or end of cantilever (Fig. 6.34).

450 Handbook of Reinforced Concrete Design

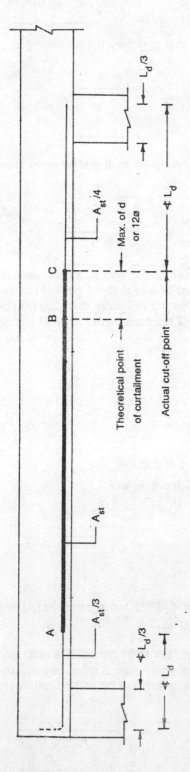

The bar diameter (∅) shall be limited such that :
- At simply supported end, $L_d \not> 1.33 \, M_1/V + l_d$
- At the point of inflection, $L_d \not> M_1/V + l_d$

At the cut-off points, any one of the following conditions be specified :

(i) $\tau_c + \tau_b \geq 1.5 \tau_v$

(ii) For ∅ ≯ 36mm, (a) A_{st} continuing ≮ 2 A_{st} required and (b) $\tau_c + \tau_b \geq 1.33 \, \tau_v$

(iii) Excess area of shear stirrups in a distance 0.75 d from the cut-off point along the terminated bar,
$$A_{sv} \geq \frac{0.4 b \, S_v}{f_y} \text{ where spacing } S_v \not> \frac{d}{8 \beta_b}, \beta_b = \frac{\text{Area of bars cut-off}}{\text{Total area of bars}}$$

Fig. 6.34 Curtailment of reinforcement for positive moment

(b) Provisions for maintaining the shear capacity at cut-off region: Flexural reinforcement shall not be curtailed in tension zone unless any one of the following conditions is satisfied.

1. Shear capacity shall be atleast 1.5 times the applied shear at the point of curtailment,

   i.e., $\tau_c + \tau_b > 1.5 \tau_v$

   where τ_c = shear capacity of concrete which is determined by considering continuing reinforcement only

   τ_b = shear capacity of shear reinforcement

   τ_v = nominal shear stress

2. For diameter of bars equal to 36 mm and smaller, the following requirements shall be satisfied.
   i) Area of continuing reinforcement shall be atleast twice that required to resist the applied bending moment.
   ii) Shear capacity shall be atleast 1.33 times the applied shear at the point of curtailment, i.e.,

   $$\tau_c + \tau_b \geq 1.33 \tau_v$$

3. Excess shear stirrups than that required for shear and torsion is provided over a distance equal to 0.75 times the effective depth from the point of curtailment. Excess area of shear stirrups, A_{sv} is given by:

   $$A_{sv} \geq \frac{0.4 \, b S_v}{f_y}$$

   where b = width of beam

   f_y = characteristic strength of reinforcement in N/mm^2

   S_v = spacing of stirrups

   $\not> 0.125 \, d / \beta_b$

   d = effective depth of beam

   $\beta_b = \dfrac{\text{area of bars curtailed at the section}}{\text{total area of bars at the section}}$

(c) Provision for curtailment of reinforcement for positive moment: Following conditions shall be satisfied for the curtailment of reinforcement for positive moment.

1. Atleast one third of the reinforcement for the maximum positive moment in the simply supported member and one fourth of the reinforcement for the maximum positive moment in the continuous member shall extend to a length equal to $L_d / 3$ along the same face of the member into the support, where L_d is the development length based on full stress in the reinforcement bars (Fig. 6.34).
2. At simple supports where the ends of the reinforcement are confined by compressive

reaction and at points of inflection, the diameter of the tension reinforcement for positive moment shall be limited to such a value that the following conditions shall be satisfied (Fig. 6.34).

At simply supported end, $\quad L_d < 1.33 M_1/V + l_d$

At point of inflection, $\quad L_d < M_1/V + l_d$

where
$\quad M_1$ = Moment of resistance of the section

$\quad V$ = shear force at the section

$\quad L_d$ = At simple support, sum of anchorage beyond the centre of the support and the equivalent anchorage value of any hook or mechanical anchorage

$\quad l_d$ = At point of inflection, 12 times the diameter of the bar or effective depth of beam whichever is more

3. When flexural member is a part of a primary lateral load resisting system, the reinforcement for positive moment required to be extended into the support shall be anchored to develop full design stress in tension at the face of the support (Fig. 6.35).

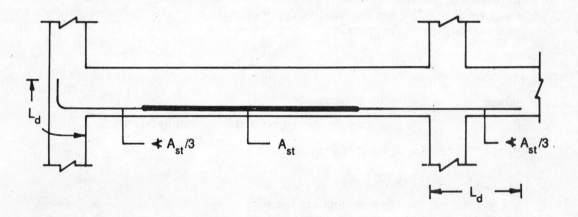

Fig. 6.35 Anchorage of reinforcement for positive moment when beams are part of lateral load resisting system

(d) Provision for curtailment of reinforcement for negative moment: For negative moment reinforcement, atleast one third of the total reinforcement provided at the support shall extend beyond the point of inflection for a distance not less than the effective depth of the member or 12 times the diameter of the bar or one-sixteenth of the clear span whichever is greater (Figs 6.36 and 6.37).

IS : 456 –1978 has also recommended the simplified curtailment rules for beams designed for predominantly uniformly distributed loads. The simplified curtailment rules for simply

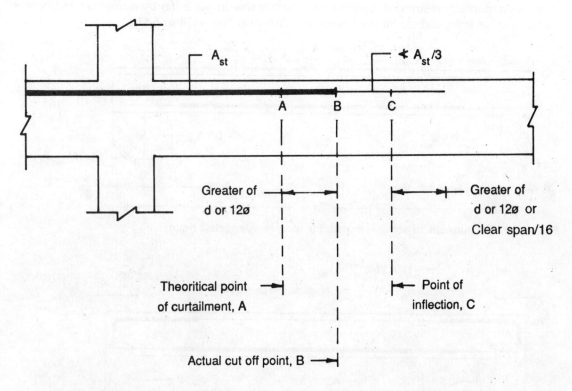

Fig. 6.36 Curtailment of reinforcement for negative moment at continuous edge

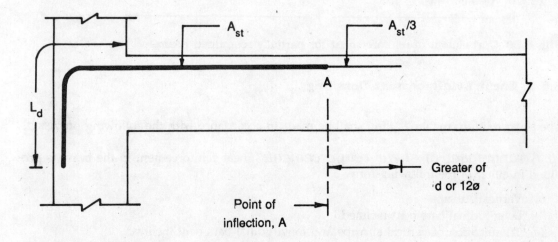

Fig. 6.37 Curtailment of reinforcement for negative moment at discontinuous ends

supported, partially restrained, continuous whose spans do not differ by more than 15 per cent of the longest span and cantilever beams are shown in Figs. 6.38 to 6.41.

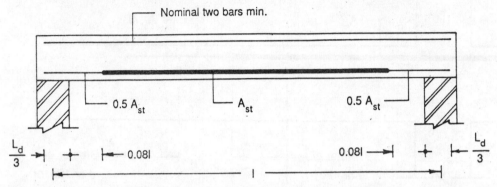

Fig. 6.38 Curtailment of reinforcement for simply supported beam

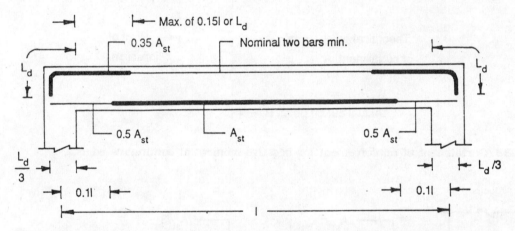

Fig. 6.39 Curtailment of reinforcement for partially restrained beam

6.5.2 Shear Reinforcement Detailing

The shear reinforcement detailing shall be made in accordance with the following provisions.

(i) Arrangement of shear reinforcement: Shear reinforcement in the beam is provided in either of the following form:

   (a) Vertical stirrups
   (b) Longitudinal bars bent inclined
   (c) Combination of vertical stirrups and longitudinal bars bent inclined.

A vertical stirrups shall be anchored adequately at both its ends as shown in Fig. 6.42. In flanged or I beams, it shall go round the longitudinal bars located close to the outer face of the flange. The different possible ways of providing shear reinforcement are shown in Fig. 6.43.

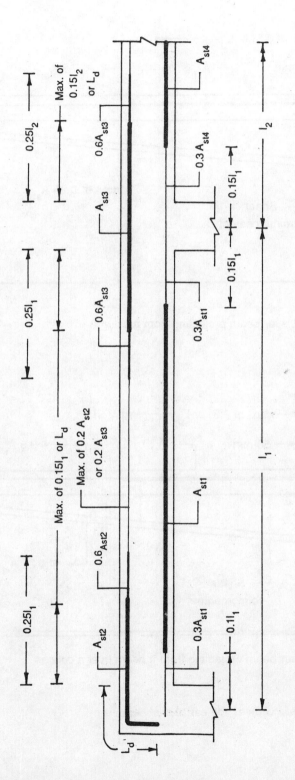

Fig. 6.40 Curtailment of reinforcement for continuous beams

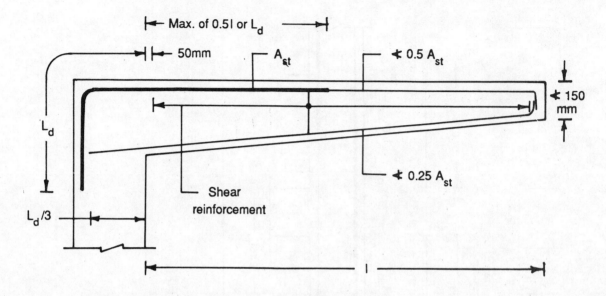

(a) Cantilever beam projecting from a column

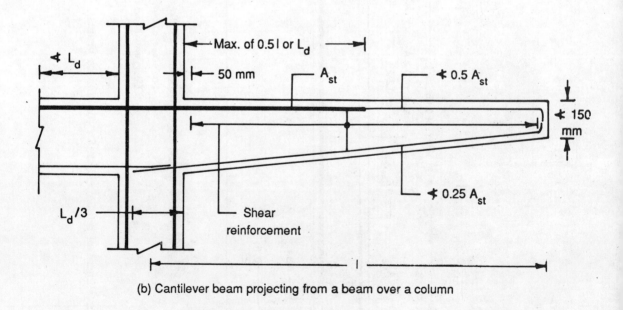

(b) Cantilever beam projecting from a beam over a column

Fig. 6.41 Curtailment of reinforcement for cantilever beam

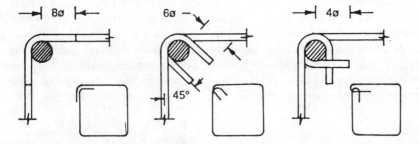

Fig. 6.42 Different ways of anchoring ends of stirrups

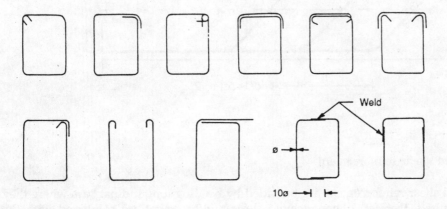

(a) Double legged stirrups

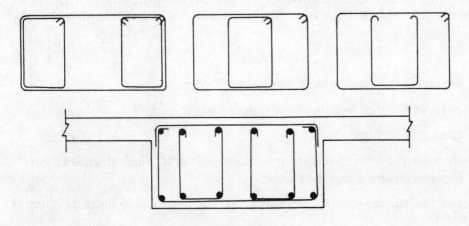

(b) Multilegged stirrups

Fig. 6.43 Different shapes of vertical stirrups

Their ends shall be properly embedded to satisfy the anchorage requirement. An appropriate form of shear reinforcement shall be chosen depending on constructional requirements keeping in view the requirements for end anchorage.

The shear reinforcement provided by bending the longitudinal bars inclined through the depth of beam is shown in Fig. 6.44. Generally two bars are bent at a section where they are no more required for resisting moment. The spacing and anchorage requirements of such inclined shear reinforcement are also shown in the figure.

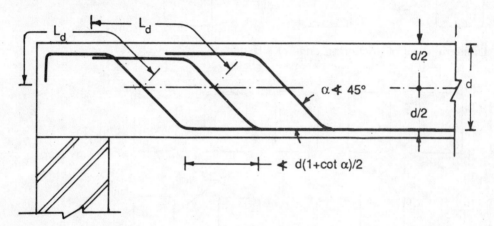

Fig. 6.44 Inclined shear reinforcement

As the inclined shear reinforcement are provided by bending longitudinal bars where they are no more required, therefore these inclined shear reinforcement are combined with the vertical stirrups to resist shear force.

(ii) Area, spacing, cover and diameter of shear reinforcement: The minimum area of shear reinforcement in the form of stirrups shall not be less than that given by the following,

$$A_{sv} < 0.4\, b S_v / f_y$$

where A_{sv} = total cross-sectional area of stirrup legs effective in shear

b = width of beam or breadth of the web of flanged beam

S_v = spacing of stirrups

f_y = characteristic strength of stirrup reinforcement in N/mm² which shall not be taken greater than 415 N/mm²

The spacing of vertical stirrups shall not be greater than minimum of 0.75 times the effective depth of beam and 450 mm.

The clear cover to vertical stirrups shall not be less than 15 mm.

The diameter of shear reinforcement shall not be less than 6 mm or 0.25 times the maximum diameter of longitudinal bars whichever is more.

(iii) Stirrups in edge beams: The closed form of stirrups shall be provided in the edge or spendrel beams and atleast one longitudinal bar shall be placed at each corner of the beam as shown in Fig. 6.45(a). The diameter of the longitudinal bars at the corners shall not be less than the diameter of stirrups or 12 mm whichever is more. However, for easier placing of longitudinal bars, 90° stirrup hook may be provided along with atleast one longitudinal bar per stirrup of diameter not less than the diameter of stirrup and bent as shown in Fig. 6.45(b) for proper anchorage. Alternatively, two piece closed stirrups may be provided as shown in Fig. 6.45(c).

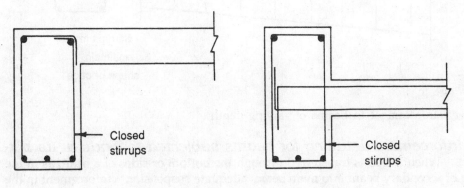

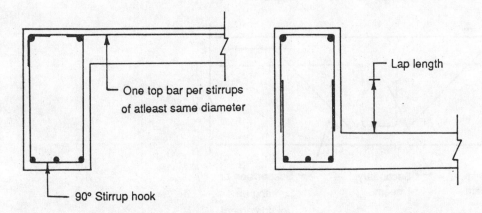

Fig. 6.45 Shear reinforcement in edge and spendrel beam

(iv) Shear stirrups in beams of varying depth: As the detailing of individual shear stirrups of varying depth required for beams of varying depth is quite combursome, therefore shear reinforcement in the form of two piece closed stirrups, known as concertina stirrups, may be provided as shown in Fig. 6.46. However, usual shear stirrups at intervals may be provided to maintain the shape of the beam as shown in the figure.

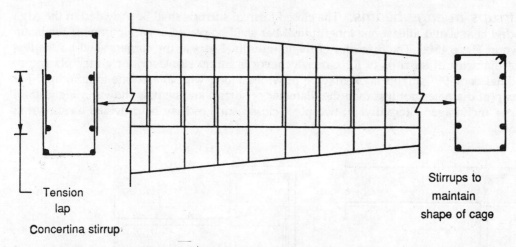

Fig. 6.46 Sheer reinforcement in beams of varying depth

(v) Shear reinforcement detailing for beams subjected to loads at its bottom or sides: When a load is transferred through the bottom or sides of a beam as in the case of framing of secondary beam into main beam, adequate suspension reinforcement in the form of stirrups is provided in the main beam at its junction with the secondary beam to transfer the load to the top of the main beam. If the load is large, bent up bars may also be used in addition to the stirrups as shown in Fig. 6.47.

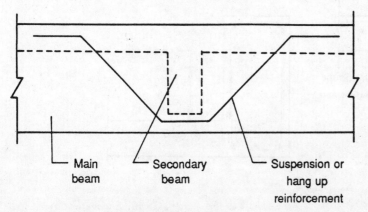

Fig. 6.47 Shear reinforcement detailing for beams subjected to loads at its bottom or side

6.5.3 Torsion Reinforcement Detailing

A member subjected to torsion shall be designed for equivalent bending moment and shear force. The longitudinal reinforcement shall be placed at the corners of the cross-section and atleast one longitudinal bar shall be placed at each corner as shown in Fig. 6.48 for rectangular and flanged sections.

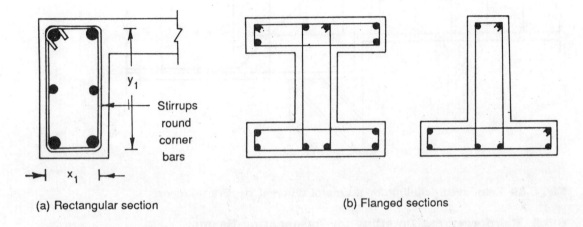

- Corner bars properly anchored at support
- Side face reinforcement of 0.05 percent of web area on each face at spacing ≯ 300 mm when depth of beam > 450 mm
- Stirrups round corner bars at spacing ≯ x_1
  ≯ $(x_1+y_1)/4$
  ≯ 300 mm

Fig. 6.48 Reinforcement detailing for torsion in beam

The transverse reinforcement shall be closed rectangular stirrups placed perpendicular to the axis of the member. In case of flanged sections, rectangular stirrups shall be provided for each rectangular part of the section as shown in Fig. 6.48(b). The spacing of stirrups shall not exceed the least of x_1, $(x_1 + y_1)/4$ and 300 mm where x_1 and y_1 are short and long dimensions of the stirrups respectively (Fig. 6.48(a)). Where mutilegged stirrups are provided, only the stirrups going round the outer face shall be considered for resisting torsional moment.

When the depth of the section exceeds 450 mm, additional longitudinal reinforcement of 0.05 per cent of web area shall be provided on each face of the section at a spacing not exceeding 300 mm or web thickness whichever is lower (Fig. 6.48(a)).

6.5.4 Reinforcement Detailing for Beams of Different Depth at Support

Reinforcement detailing for beams of different depth over the support shall be made as shown in Fig. 6.49.

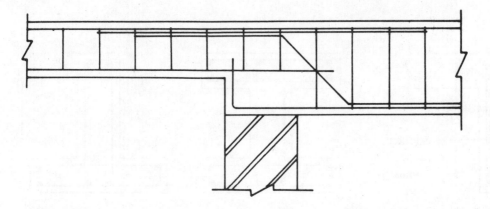

Fig. 6.49 Reinforced detailing for beams of different depth at support

6.5.5 Reinforcement Detailing for Intersecting Beams

Reinforcement detailing for intersecting beams shall be made such that the layers of reinforcement in two beams are at different levels. Generally the top bars of the secondary beam are placed above the top bars of the main beam because the reinforcement of the secondary beam are of smaller diameter and requires less cover (Fig. 6.50). However, when the main beam is heavily reinforced, it may be economical to place the top reinforcement of main beam over the top reinforcement of the secondary beam.

The bottom bars may be placed conveniently in the two layers by providing different depth of the secondary and main beams. This can be achieved even by reducing the depth of the secondary beam by 50 mm than that of the main beam. Where the sofits of both beams are at the same level, the bottom reinforcement of secondary beam can be draped over the bottom reinforcement of main beam as shown in Fig. 6.50.

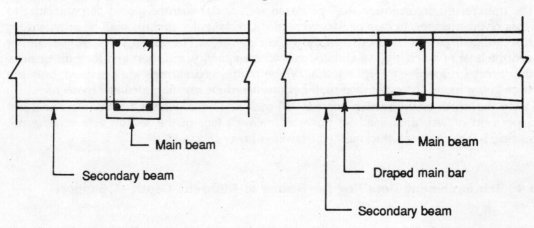

Fig. 6.50 Reinforcement detailing for interacting beams

6.5.6 Reinforcement Detailing for Tie Members

The longitudinal reinforcement of the tie members subjected to pure tension shall be adequately anchored at its ends as shown in Fig. 6.51. Generally a small splay at the ends of the tie member subjected to small load is made to spread the axial force and to allow for any moment that may induce at the ends of the tie member (Fig. 6.51(a)). When the tie member is subjected to large load, a large splay shall be provided at the ends as shown in Fig. 6.51(b).

Transverse reinforcement in tie member is not required except nominal links to form the longitudinal bars into a cage.

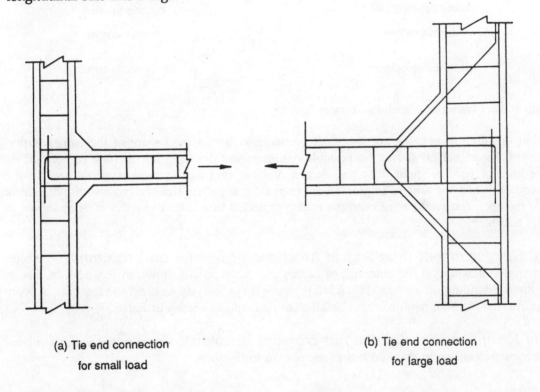

(a) Tie end connection for small load

(b) Tie end connection for large load

Fig. 6.51 Reinforcement detailing for tie end connections

6.5.7 Reinforcement Detailing at Corners, Junctions of Beams and Columns and in Cranked Beams

Reinforcement detailing at corners, junctions of beams and columns and in cranked beams are discussed as follows.

(i) Reinforcement detailing at corners: Resultant of forces in reinforcement bars in closing corners does not cause failure as shown in Fig. 6.52(a) whereas the resultant of forces in reinforcement bars in the opening corners causes failure (Fig. 6.52(b)). Therefore, reinforce-

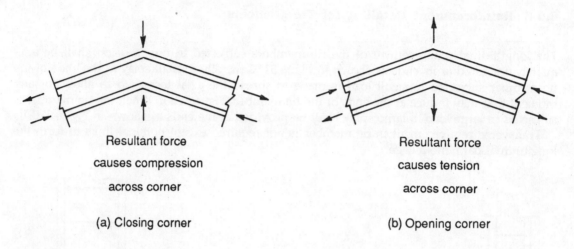

Fig. 6.52 Closing and opening corners

ment detailing in opening corners shall be made such that the failure of the corner is prevented. Figure 6.53(a) shows the possible reinforcement detailing for opening corners where the area of reinforcement does not exceed one per cent of the cross-sectional area of the member. When the area of reinforcement exceeds one per cent of the cross-sectional area of the member, transverse reinforcement is also provided in addition to splay steel as shown in Fig. 6.53(b).

(ii) Reinforcement detailing at junctions of beams and columns: The reinforcement detailing at the junctions of beams and columns are shown in Fig. 6.54. In case of column extending up a beam (Fig. 6.54(c)) where it is necessary to bend the top bars of beam up into the column, reinforcement detail shall be made as shown in Fig. 6.55.

(iii) Reinforcement detailing in cranked beams: Different possible ways of reinforcement detailing in cranked beams are shown in Fig. 6.56.

6.5.8 Reinforcement Detailing for Haunched Beam

Heavily loaded beams subjected to large shear force and bending moment at supports are provided with haunches as shown in Fig. 6.57. The detailing of the longitudinal bars in haunches are shown in Fig. 6.58(a). In addition to main bars carried through the haunches and curtailed wherever it is no more required, additional bars are provided parallel to the haunches. These bars are placed inside the outer bars of the beam as shown in Fig. 6.58(a). The stirrups in the haunches can either be positioned normal to the haunches or placed vertically as shown in Fig. 6.58(b).

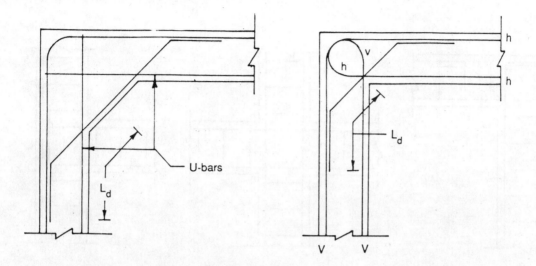

(a) Reinforcement detailing for 90° opening corners with reinforcement ≤1 percent of cross sectional area

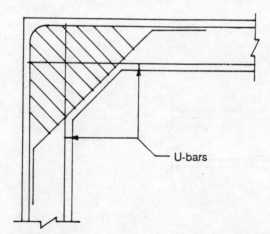

(b) Reinforcement detailing for 90° opening corners with reinforcement >1 percent of cross sectional area

Fig. 6.53 Reinforcement detailing for 90° opening corners

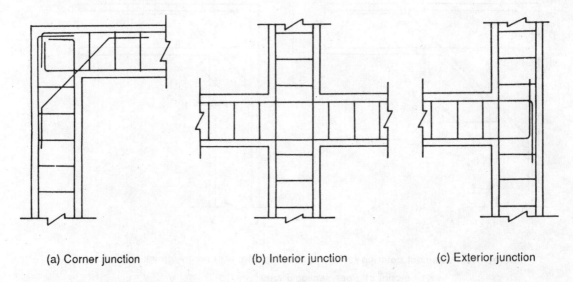

(a) Corner junction (b) Interior junction (c) Exterior junction

Fig. 6.54 Reinforcement detailing at beam-column junctions

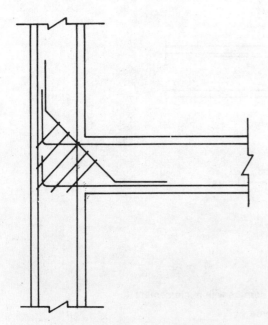

Fig. 6.55 Reinforcement detailing at beam-column junction when top bars of the beam are bent upward

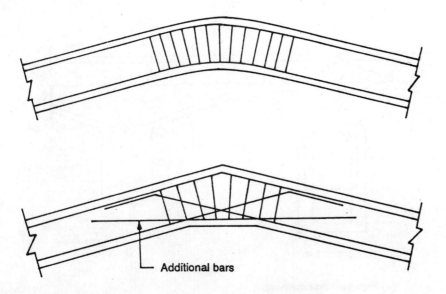

Fig. 6.56 Reinforcement detailing for cranked beams

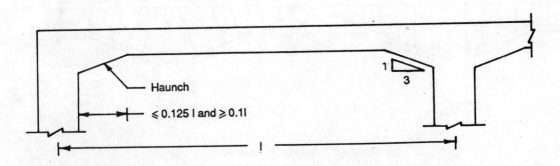

Fig. 6.57 Haunched beams

6.5.9 Reinforcement Detailing at Opening in Web of Beam

Reinforcement detailing at opening shall depend on whether it is small or large opening. The area of the small openings shall not be greater than 1000 mm^2 for members having effective depth not exceeding 500 mm or 0.004 d^2 when the effective depth, d is greater than 500 mm. It should be placed in the mid depth of the web while maintaining the requirements of cover to longitudinal and transverse reinforcement. Such openings are considered not to interfere with the development of strength of the member. The edge of the openings should not be within 0.33 d from the compression face of the beam so that they do not encroach into the compression zone of the member. When two or more small openings are placed transversely

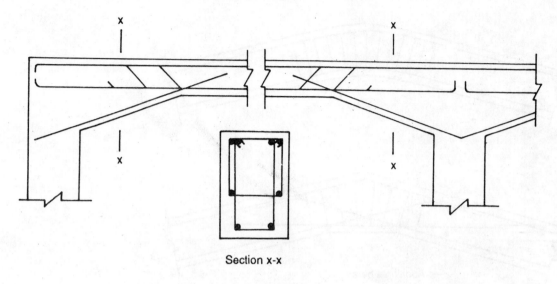

(a) Flexural reinforcement

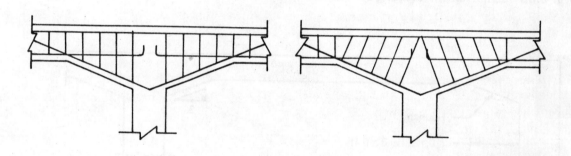

(b) Shear reinforcement in haunches

Fig. 6.58 Reinforcement detailing for haunched beam

in the web, the distance between the outer edges of the small opening shall be considered as the height of the equivalent one large opening. The adjacent openings should be arranged such that no potential failure plane pass throuh several openings. Generally the clear distance between such openings, measured along the length of the member, shall not be less than 150 mm. A typical reinforcement detailing around such an opening is shown in Fig. 6.59.

Large openings have their largest dimension greater than one quarter of the effective depth of beam. The edge of the opening should not be within $0.33\,d$ from the compression face of the beam and the height of the opening should not exceed 0.4 times the effective depth of the beam. Such openings should not be placed at locations where they effect the flexural or shear strength of the member nor in the zone of potential plastic hinge nor at locations where the shear stress is greater than $0.33\,\sqrt{f_{ck}}$ where f_{ck} is the characteristic strength of concrete.

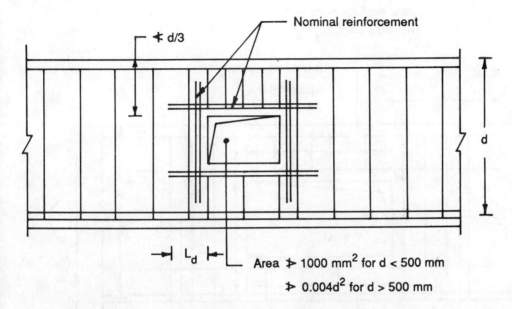

Fig. 6.59 Reinforcement detailing at small opening in the web of beam

The design of the compression side of the web is made to resist 1.5 times the shear accross the opening as shown in Fig. 6.60. Transfer of shear by the tension side of the web is neglected. Shear reinforcement to resist at least twice the design shear force in the opening shall be provided on both sides of the opening over a length not exceeding 0.5 times the effective depth of beam, as shown in Fig. 6.60.

6.6 REINFORCEMENT DETAILING FOR SLABS

Reinforcement detailing for solid, flat and waffle slabs are described as follows.

6.6.1 Reinforcement Detailing for Solid Slabs

The requirements for area, spacing, cover and diameter of reinforcement bars and reinforcement detailing for simply supported, restrained, continuous and cantilever slabs and for re-entraint corners and opening in slabs are described as follows.

(i) Area, spacing, cover and diameter of reinforcement bars: The IS code provisions for area, spacing, cover and diameter of reinforcement bars are as follows.

(a) Area of reinforcement bars: Area of reinforcement bars in either directions shall not be less than 0.15 per cent of the gross sectional area of concrete where plain mid steel bars are used

470 *Handbook of Reinforced Concrete Design*

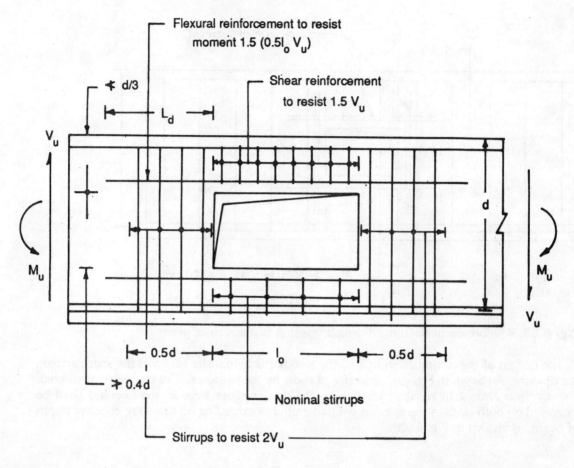

Fig. 6.60 Reinforcement detailing requirements at large opening in the web of beam

and 0.12 per cent of gross sectional area of concrete where high strength deformed bars are used.

(b) Spacing of reinforcement bars: The spacing of main reinforcement bars shall not be more than the three times the effective depth of slab or 450 mm, whichever is smaller. The spacing of distribution bars provided for shrinkage and temperature stresses shall not be more than five times the effective depth of slab or 450 mm whichever is smaller.

(c) Cover to reinforcement bars: The cover to reinforcement bars shall not be less than 15 mm or the diameter of bar whichever is more. The cover at each end of the reinforcement bars shall not be less than 25 mm or twice the diameter of bar whichever is more.

(d) Diameter of reinforcement bars: The diameter of the main bars shall not be less than 8 mm for high strength deformed bars and 10 mm for plain mild steel bars. The diameter of the

distribution bars shall not be less than 6 mm. The diameter of the reinforcement bars shall not be more than 0.125 times the thickness of the slab.

(ii) Reinforcement detailing for simply supported slabs: Simply supported slabs may be classified as one way or two way slabs. Slabs supported on two parallel sides or on four sides if the effective length of the slab is atleast two times larger than the effective width of the slab behave as one way slab. The IS Code provisions of reinforcement detailing of one way and two way simply supported slabs are described below.

(a) One way slabs: The reinforcement detailing of one way simply supported slab is shown in Fig. 6.61. The main reinforcement in the shorter direction is placed in the bottom layer. At least 50 per cent of the main reinforcement provided at mid-span shall extend into the support to a length equal to $L_d/3$ along the same face of the slab, where L_d is the development length based on full design stress in the reinforcement. The remaining 50 per cent of the reinforcement shall be cranked up and extended to within $0.1\, l_x$ of the support where l_x is the effective shorter span.

The distribution reinforcement shall be placed above the main reinforcement. It shall be extended up to the end if the slab is supported on two parallel sides only as shown in Fig. 6.61(a). If the slab is supported on four sides, then 50 per cent of the distribution reinforcement shall extend along the same face of the slab into the support and the remaining 50 per cent of the distribution reinforcement may be cranked up and extended to within $0.1\, l_x$ of the support where l_x is the effective shorter span (Fig. 6.61(b)).

(b) Two-way slabs: The reinforcement detailing of a two-way simply supported slab is shown in Fig. 6.62. The reinforcement in the shorter direction is placed in the bottom layer. At least 50 per cent of the reinforcement provided at mid-span in each direction shall extend into the supports up to a length of $L_d/3$ where L_d is the development length based on full design stress in the reinforcement. The remaining 50 per cent of the reinforcements shall be cranked up and extended to within $0.1\, l_x$ or $0.1\, l_y$ of the supports, as appropriate, where l_x and l_y are the effective spans in the shorter and longer directions respectively.

(iii) Reinforcement detailing for restrained slabs: Reinforcement detailing of restrained slabs, whose corners are prevented from lifting and are designed for predominantly uniformly distributed load, shall be made in accordance with the following IS code provisions.
  (a) Slabs are considered as divided into middle and edge strips in each direction as shown in Fig. 6.63. The width of the middle strip shall be equal to 3/4 of the total width and the width of each edge strip shall be equal to 1/8 of the total width.
  (b) The tension reinforcement provided at mid-span in the middle strip shall extend along the same face of the slab to within 0.25 times the effective span of a continuous edge or 0.15 times the effective span of a discontinuous edge, and at least 50 per cent of the reinforcement shall extend into the support, the remaining 50 per cent of the reinforcement shall be terminated or cranked.
  (c) The tension reinforcement over the continuous edge of a middle strip shall extend in the upper part of the slab a distance of 0.15 times the effective span from the support, and at least 50 per cent shall extend a distance of 0.3 times the effective span.
  (d) At discontinuous edge, negative moment may arise depending upon the degree of fixity

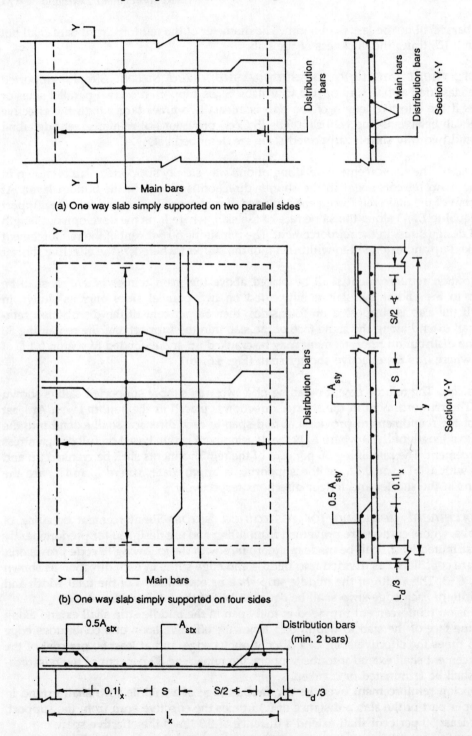

Fig. 6.61 Reinforcement detailing for one-way slab

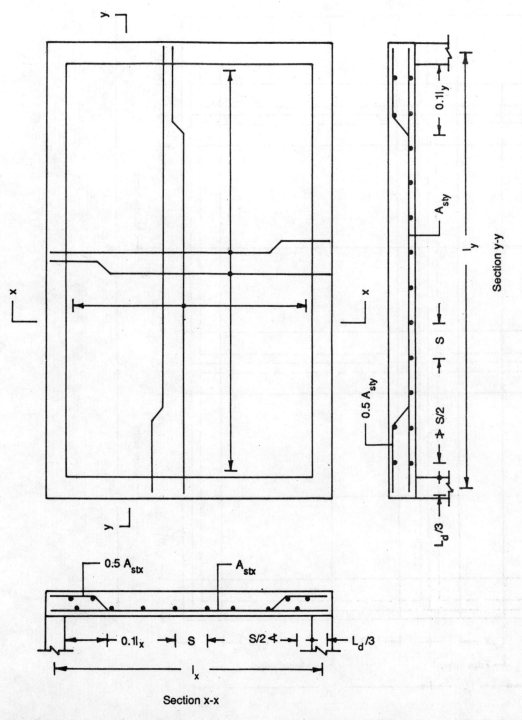

Fig. 6.62 Reinforcement detailing for two-way simply supported slab

474 *Handbook of Reinforced Concrete Design*

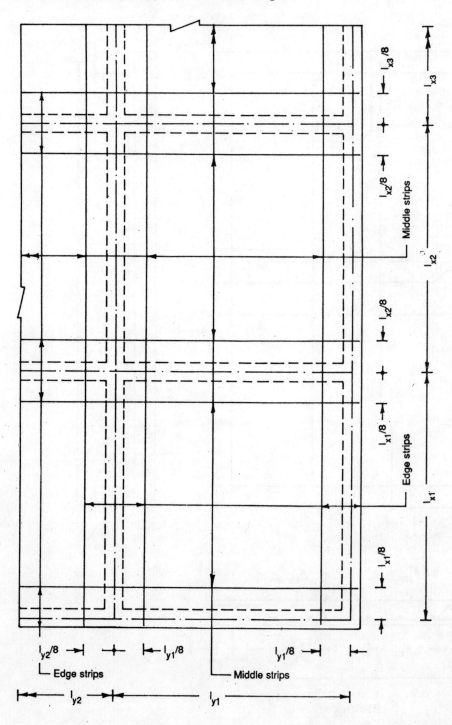

Fig. 6.63 Edge and middle strips

at the support. In general, tension reinforcement equal to 50 per cent of that for span moment shall be provided in the upper part of the slab extending a distance of 0.1 times the effective span from the support.

(e) Reinforcement in the edge strip parallel to the edge shall comply with the requirement of minimum reinforcement of 0.15 per cent of plain mild steel or 0.12 per cent of high strength deformed bars of the total cross-sectional area.

(f) Torsional reinforcement shall be provided at corners where the slab has either one or both edges discontinuous. It shall consist of top and bottom reinforcement, each with layers of bars placed parallel to the sides of the slab and extending from the edges a minimum distance of one-fifth of the shorter span. At corners where the slab is discontinuous on both the edges, the area of reinforcement in each of these four layers shall be three-quarters of the area required for the maximum span moment per unit width in the slab. At corners where slab is discontinuous on one edge only, the area of torsional reinforcement in each of the four layers shall be three - eighth of the area required for the maximum span moment per unit width in the slab.

Based on the above provisions, reinforcement detailing of slab is shown in Fig. 6.64. Detailing for torsional reinforcement and reinforcement in edge strip are shown in Fig. 6.64(a). Detailing for main reinforcement in the middle strips with bent up and straight bars are shown in Figs 6.64(b) and 6.64(c) respectively.

(iv) Reinforcement detailing for cantilever slabs: The main reinforcement in the upper part of cantilever slab designed predominantly for uniformly distributed load shall extend to a distance of 0.5 times the span of slab or development length of bar whichever is more from the face of the support and at least 50 per cent of reinforcement shall extend up to the end of the slab. All reinforcement bars shall be anchored into the support to a length not less than the development length of bar. The secondary reinforcement shall be provided below the main reinforcement.

Based on the above provisions, reinforcement detailing of cantilever slabs continuous over a support and cantilevering from a beam are shown in Figs 6.65 and 6.66 respectively. For slabs cantilevering from a beam, the main reinforcement shall extend to a sufficient length into the supporting beam and then back into the slab. When the slab is cantilevering from the bottom or mid-depth of a beam, additional inclined bars not less than half of the main reinforcement in cantilever slab shall be provided as shown in Fig. 6.66(b).

(v) Reinforcement detailing for re-entrant corners: Re-entrant corners are strengthened by providing diagonal reinforcement as shown in Fig. 6.67 so that the possibility of development of diagonal cracks at re-entrant corners are prevented.

(vi) Reinforcement detailing for openings in slabs: Openings in the floor slabs shall be strengthened by special beams or additional reinforcement around the openings. When the openings are small and the slab is not subjected to any additional load around the opening or vibration condition, the detailing of reinforcement around the openings shall be made as follows:

(a) At least one half of the reinforcement intersected by the opening shall be provided on each side of the opening parallel to the direction of reinforcement (Fig. 6.68). These

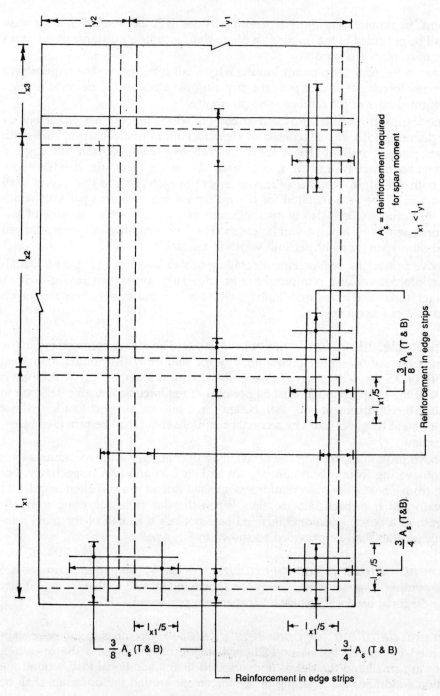

(a) Reinforcement details for edge strips and corners

Fig. 6.64(a) Reinforcement detailing for edge strips and at corners

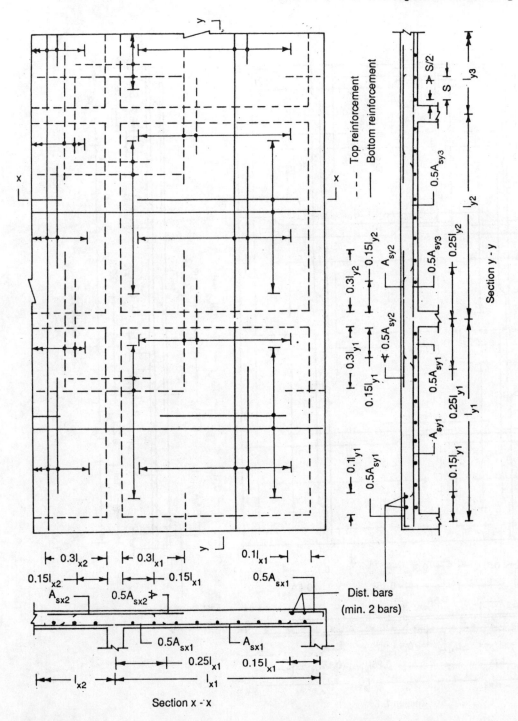

Fig. 6.64(b) Reinforcement detailing for middle strips with bent up bars

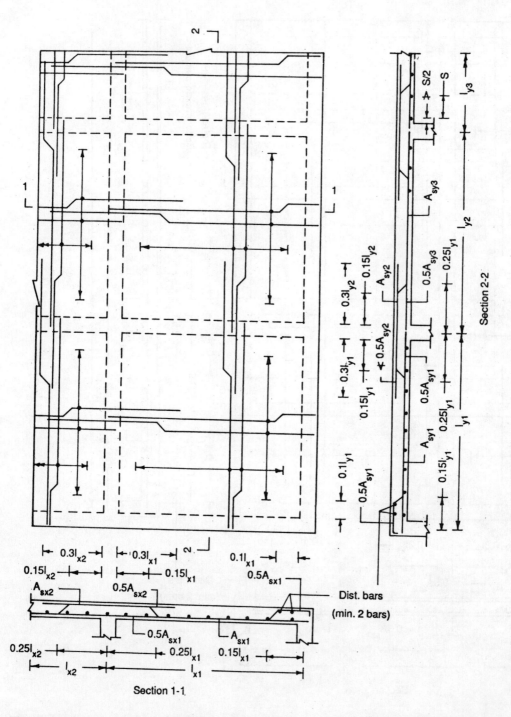

Fig. 6.64(c) Reinforcement detailing for middle strips with straight bars

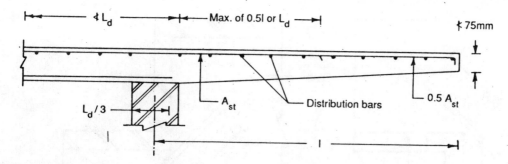

Fig. 6.65 Reinforcement detailing for cantilever slab continuous over a support

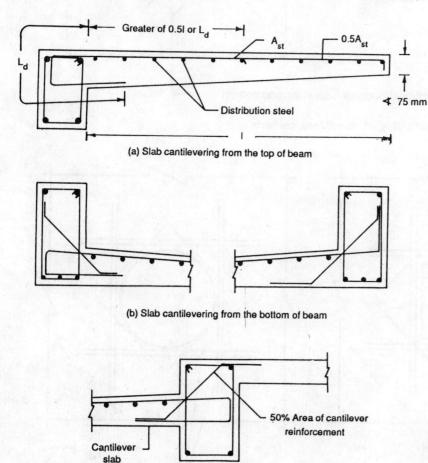

(a) Slab cantilevering from the top of beam

(b) Slab cantilevering from the bottom of beam

(c) Slab cantilevering from the mid depth of beam

Fig. 6.66 Reinforcement detailing for slabs cantilevering from top, bottom and mid-depth of a beam

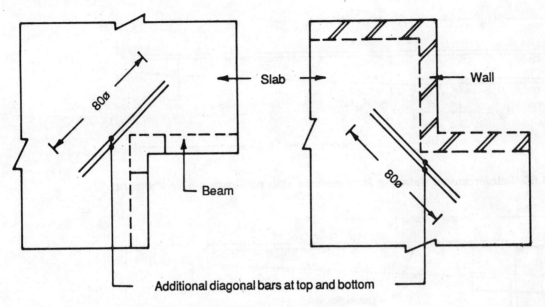

Fig. 6.67 Reinforcement detailing at re-entrant corners

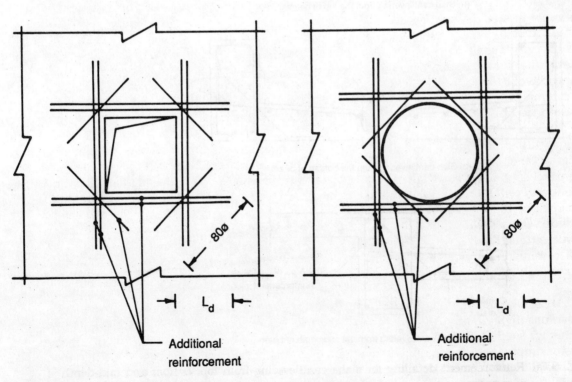

Fig. 6.68 Reinforcement detailing around openings in slabs

reinforcement bars shall be extended to a distance of development length beyond the edges of the opening (Fig. 6.68).

(b) Diagonal stitching bars shall be placed across the corners of rectangular opening. In case of circular opening, it shall be placed to frame the circular opening as shown in Fig. 6.68(b). The diameter of such bars shall not be less than that of the slab reinforcement and their length shall not be less than 80 times its diameter. It shall be placed at the top and bottom if the thickness of slab exceeds 150 mm.

6.6.2 Reinforcement Detailing for Flat Slabs

Slabs supported directly on the columns without beams are known as flat slabs. Different types of flat slabs shown in Fig. 6.69 are:

(i) Flat slab with drop and column head.
(ii) Flat slab with column head.
(iii) Flat slab with drop.
(iv) Flat slab without drop and column head.

The dimensions of different components of flat slab are proportioned as described below.

(i) Slab: Generally the thickness of slab is governed by the serviceability requirement of deflection and it shall not be less than 125 mm.

(ii) Drop: It shall be rectangular in plan and have length in each direction not less than one-third and not greater than half of the panel length in that direction. For exterior panel, the width of drop at right angle to the discontinuous edge and measured from the centre line of column shall be equal to half of width of drop for the adjacent interior panel. The thickness of drop shall be 1.25 to 1.5 times the thickness of slab elsewhere.

(iii) Column head: The shape of the column head may be rectangular or circular. The length of the rectangular column head in each direction shall not be more than one fourth of the panel length in that direction. In case of circular column head, the diameter shall not exceed one fourth of the average of panel lengths in each direction. The portion of the column head lying within the largest right angle cone or pyramid that has a vertex angle of 90° and can be included entirely within the outline of the column and column head is considered for design purposes (Fig. 6.69).

The reinforcement detailing of the flat slab shall be made in accordance with the following provisions.

(i) Area, spacing, cover and diameter of reinforcement bars: The IS Code provisions for area, spacing, cover and diameter of reinforcement bars for flat slabs are same as that for solid slabs discussed in Section 6.6.1(i) except that in flat slab with drop, the thickness of drop panel for determination of area of reinforcement shall be lesser of the thickness of drop and thickness of slab plus one-quarter the distance between edge of drop and edge of capital.

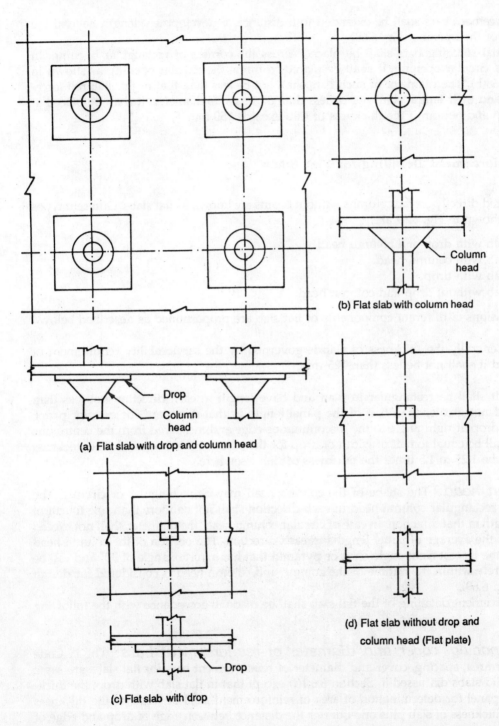

Fig. 6.69 Types of flat slabs

(ii) Flexural reinforcement detailing: For flexural reinforcement detailing, slabs are considered as divided into middle and column strips in each direction as shown in Fig. 6.70. The reinforcement for maximum moment is provided along the span of slab and necessary curtailment is done wherever it is not required. Curtailment of reinforcement can be made either by cutting the bottom reinforcement for positive moment resulting in straight bars or by bending alternate bars up which are extended into the adjacent span up to the required length. When the design is based on the direct design method as specified in IS : 456–1978 for predominantly uniformly distributed load, simplified detailing rules for straight and bent up bars of flat slabs without drop and with drop are shown in Figs 6.71 and 6.72 respectively.

(iii) Shear reinforcement detailing: The possible ways of providing shear reinforcement for slab at column head are shown in Fig. 6.73. The shear reinforcement detailing by vertical stirrups in one direction and bars in the other direction laid under and on top of reinforcement bars forming the cage with the shear stirrups as shown in Fig. 6.73(a) is very convenient and most preferred.

(iv) Reinforcement detailing for openings in flat slabs: Opening of larger size in the flat slab shall be strengthened by special beams or additional reinforcement around the opening based on the results of analysis by considering opening in the flat slab. However, for opening of smaller size conforming to the following, analysis by considering opening in the flat slab is not required.

   (a) Opening of any size placed within the middle half of the span in each direction.
   (b) In the area common to two column strips, opening of size not more than one-eighth of the width of column strip in either span.
   (c) In the area common to one column strip and one middle strip, opening of size that does not interrupt more than one quarter of the reinforcement in either strip.

Reinforcement detailing around such openings shall be made as discussed in Section 6.6.1 and shown in Fig. 6.68.

6.6.3 Reinforcement Detailing for Waffle Slabs

A slab supported directly on columns without beams and having recesses on the soffit so that the soffit comprises of a series of ribs in the two directions is known waffle slab (Fig. 6.74(a)). Generally the width of web in the waffle slab shall not be more than 100 mm, the clear spacing shall not exceed 1000 mm and depth shall not be more than 3.5 times the minimum width of web of rib. It is provided with solid head around the column.

For flexural reinforcement detailing, waffle slab is considered as divided into the column and middle strips as in the case of flat slab. The curtailment of the reinforcement in the column and middle strips shall be made similar to that of the flat slab except that at least 50 per cent of the total tension reinforcement is extended at the bottom into the support and anchored as shown in Fig. 6.74(b). Often the reinforcement detailing of ribs is made as beam by extending at least one bar at each corner of the ribs throughout for holding transverse reinforcement in the form of vertical stirrups.

484 Handbook of Reinforced Concrete Design

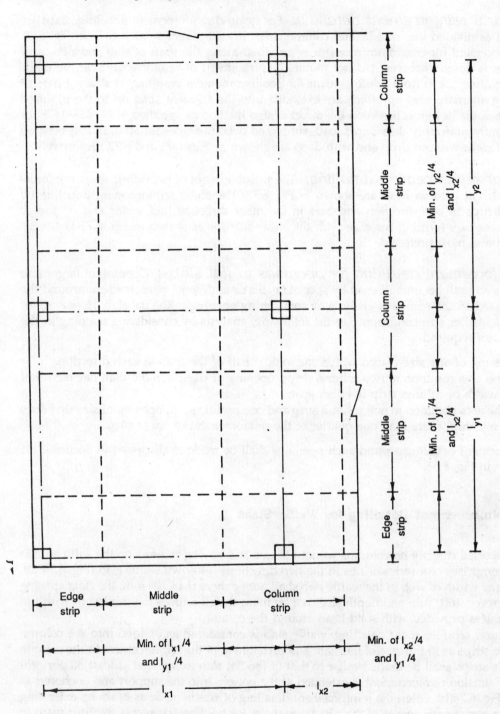

Fig. 6.70 Column and middle strips

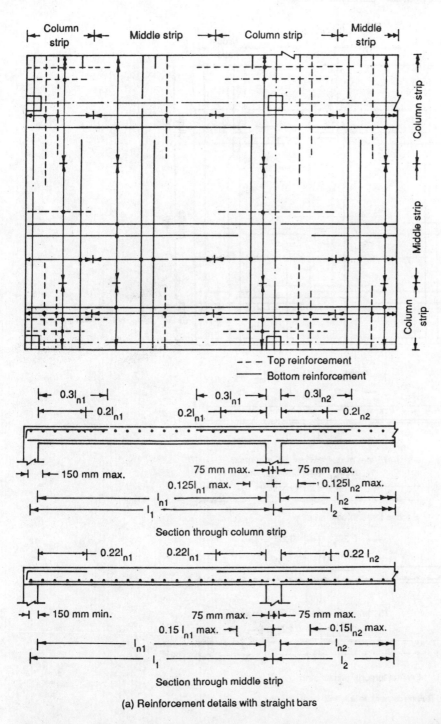

Fig. 6.71(a) Reinforcement detailing for flat slab without drop with straight bars

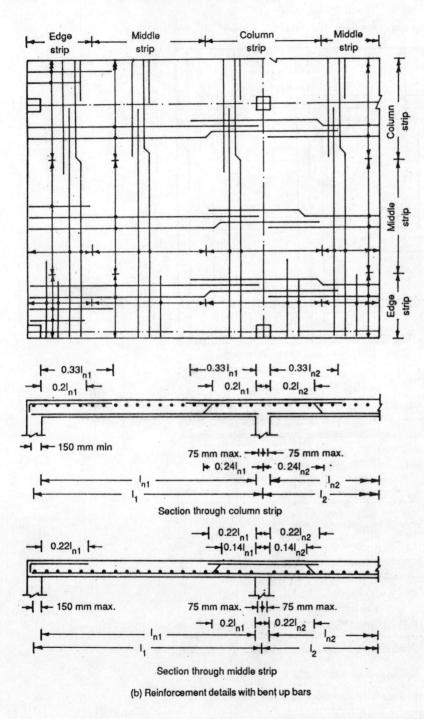

Fig. 6.71(b) Reinforcement detailing for flat slab without drop with straight bars

Reinforcement Detailing

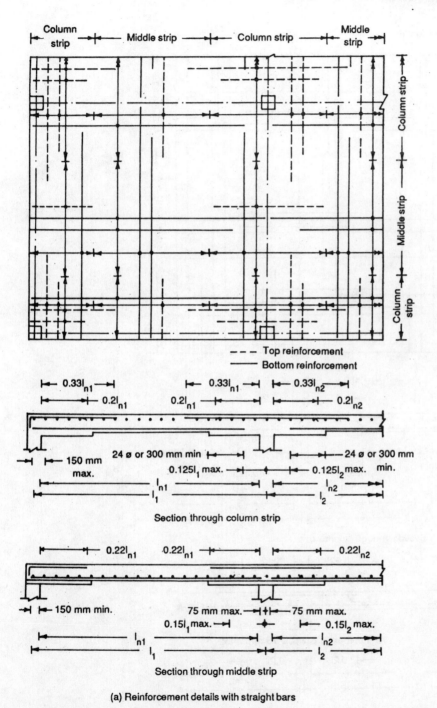

(a) Reinforcement details with straight bars

Fig. 6.72(a) Reinforcement detailing for flat slab with drop with straight bars

488 Handbook of Reinforced Concrete Design

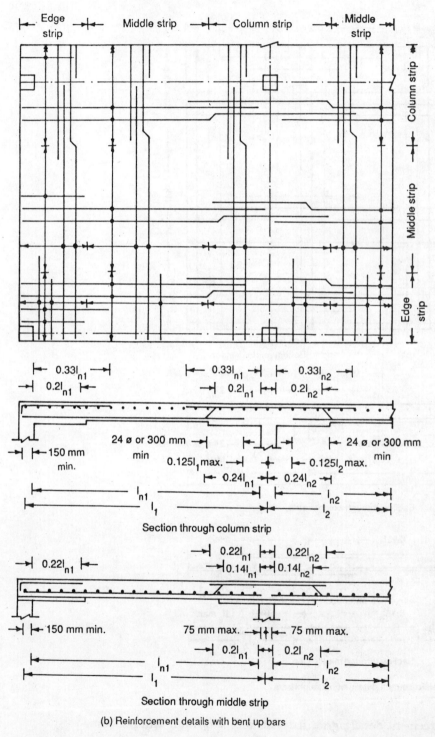

(b) Reinforcement details with bent up bars

Fig. 6.72(b) Reinforcement detailing for flat slab with drop with bent up bars

Reinforcement Detailing **489**

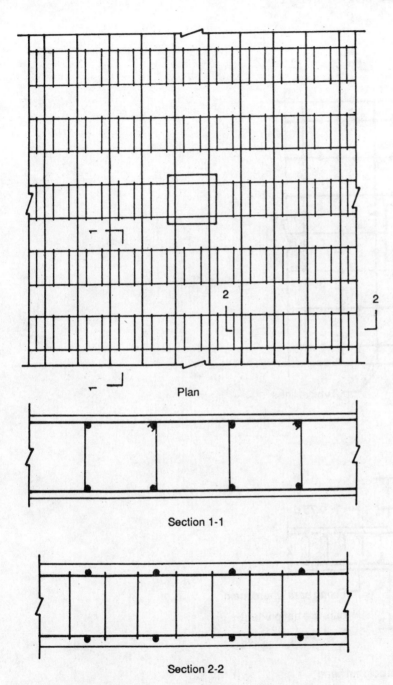

(a) Two legged stirrups for slab at column head

Fig. 6.73(a) Two legged shear stirrups for flat slab at column head

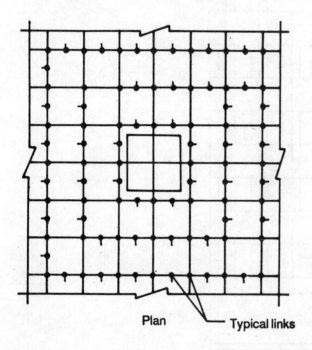

Plan — Typical links

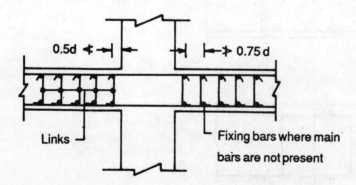

(b) Vertical links for slab at column head

Fig. 6.73(b) Vertical links for flat slab at column head

Reinforcement Detailing 491

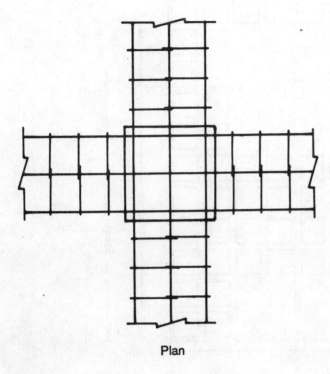

Plan

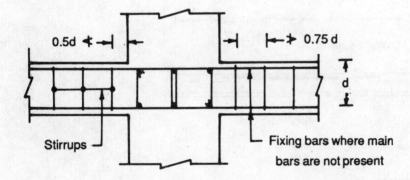

(c) Two legged stirrups forming beam cage
for slab and column head

Fig. 6.73(c) Two-legged stirrups forming beam cage for flat slab at column head

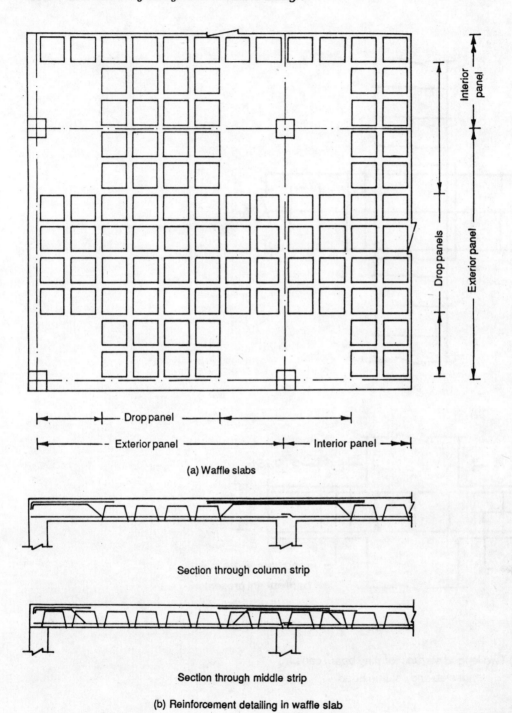

Fig. 6.74 Wiffle slab

6.7 REINFORCEMENT DETAILING FOR STAIRCASE

The requirements for area, spacing, cover and diameter of reinforcement bars and reinforcement detailing of different types of staircase are described as follows.

6.7.1 Area, Spacing, Cover and Diameter of Reinforcement Bars

The IS code provisions for area, spacing, cover and diameter of reinforcement bars are as follows.

(i) Area of reinforcement bars: The area of reinforcement bars in either directions shall not be less than 0.15 per cent of gross-sectional area of concrete where plain mild steel bars are used and 0.12 per cent of gross sectional area of concrete where high strength deformed bars are used.

(ii) Spacing of reinforcement bars: The spacing of main reinforcement bars shall not be more than three times the effective depth of the slab or 450 mm whichever is smaller. The spacing of distribution bars provided for shrinkage and temperature stresses shall not be more than five times the effective depth of slab or 450 mm whichever is smaller.

(iii) Cover to reinforcement bars: The cover to the reinforcement bars shall not be less than 15 mm or the diameter of bar whichever is more. The cover at each end of the reinforcement bars shall not be less than 25 mm or twice the diameter of bar whichever is more.

(iv) Diameter of reinforcement bars: The diameter of the main bars shall not be less than 8 mm for high strength deformed bars and 10 mm for mild steel bars. The diameter of the distribution bars shall not be less than 6 mm. The diameter of the reinforcement bars shall not be more than 0.125 times the thickness of the slab.

6.7.2 Reinforcement Detailing for Staircase Supported on Side Beam

The reinforcement detailing of staircase supported on side beams is shown in Fig. 6.75. The main reinforcement for span moment is placed in the bottom layer. Atleast half the area of main reinforcement is provided in the upper layer up to a distance of 0.1 l to 0.15 l from the face of the support. Two alternate arrangements for the placing of reinforcement is shown in the figure.

6.7.3 Reinforcement Detailing for Cantilever Steps from Reinforced Concrete Wall

The reinforcement detailing of cantilever steps from R.C. wall is shown in Fig. 6.76. The main reinforcement is provided in the upper layer which is extended into the support up to a

494 Handbook of Reinforced Concrete Design

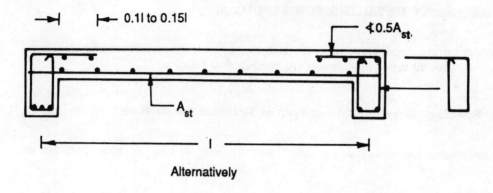

Alternatively

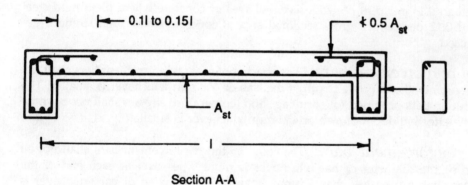

Section A-A

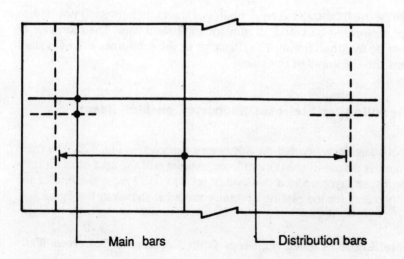

Plan

Fig. 6.75 Reinforcement detailing for staircase supported on side beams

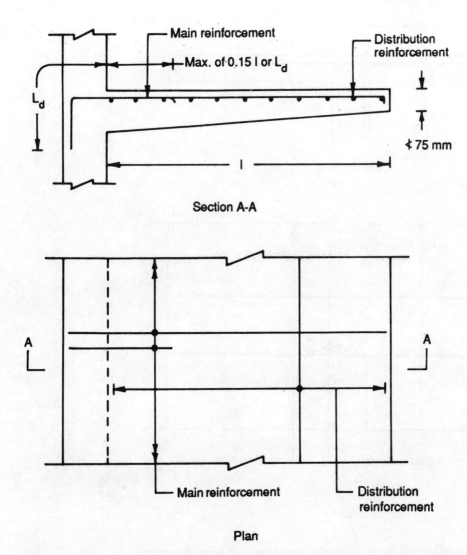

Fig. 6.76 Reinforcement detailing for cantilever steps from reinforced concrete wall

distance of development length of the bar. Alternate reinforcement bars may be curtailed at a distance not less than 0.5 l or L_d whichever is more, from the face of support.

6.7.4 Reinforcement Detailing for Flight Supported on Central Beam

Figure 6.77 shows reinforcement detailing for flight supported on central beam. The reinforcement bars shall be bent at the ends for adequate development length as shown in the figure.

496 *Handbook of Reinforced Concrete Design*

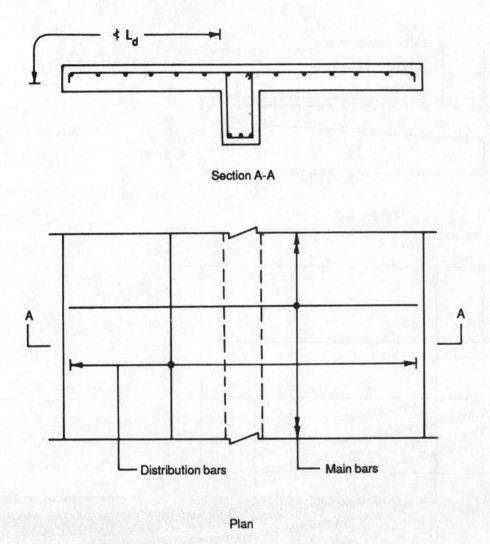

Fig. 6.77 Reinforcement detailing for flight supported on central beam

6.7.5 Reinforcement Detailing for Staircase Simply Supported at ends of Landing Slabs

Reinforcement detailing of staircase simply supported at ends of landing is shown in Fig. 6.78. Main reinforcement for span moment is provided at the bottom of waist slab which is extended either at the bottom or at the top of the landing slab as appropriate and then embedded into the support. In places where main reinforcement of waist slab is extended at the bottom of the landing slab, half the area of reinforcement for span moment is provided at the top of the landing slab to resist negative moment induced due to partial fixity of the support. These bars

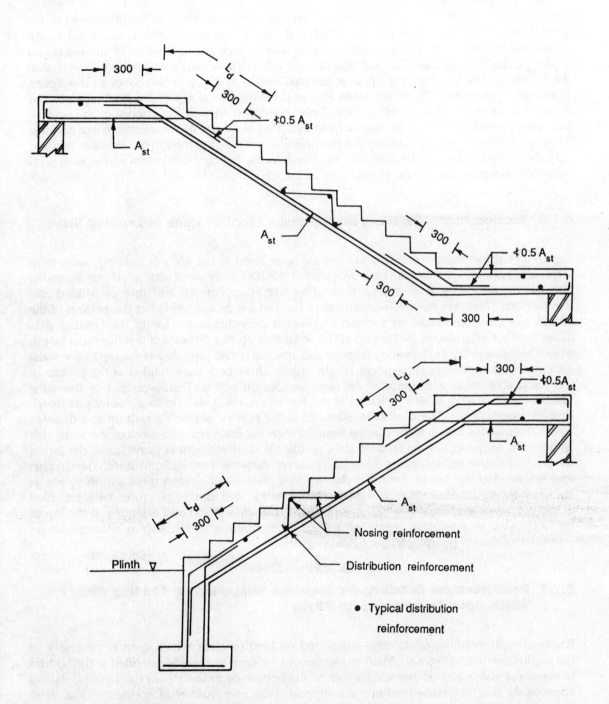

Fig. 6.78 Reinforcement detailing for staircase simply supported at ends of landing slabs

are extended at the bottom of waist slab to a nominal length of 300 mm. Reinforcement of the same diameter and spacing is also provided at the top of the waist slab for a nominal length of 300 mm to resist any negative moment that may induce at the junction of the waist and landing slabs because these are inclined at each other. These reinforcement are extended at the bottom of the landing slab up to a nominal length of 300 mm as shown in the figure. Wherever reinforcement from the waist slab is extended at the top of the landing slab, half the area of reinforcement as that for span moment is provided at the bottom of the landing slab which is embedded into the support and extended at the top of the waist slab to a nominal length of 300 mm. Same area of reinforcement is also provided at the top of the waist slab and extended at the top of the landing slab as shown in the figure. Distribution and nosing reinforcement are provided as shown typically in the figure.

6.7.6 Reinforcement Detailing for Staircase Fixed at Ends of Landing Slabs

Figure 6.79 shows reinforcement details of staircase fixed at the ends of landing. Main reinforcement bars for span moment are provided at the bottom of waist slab which are extended either at the bottom or at the top of the landing slab as appropriate and then embedded into the support. Wherever main reinforcement of the waist slab is extended at the bottom of the landing slab, reinforcement for support moment is provided at the top of the landing slab. These bars are extended at the bottom of the waist slab up to a distance of development length of bar. Reinforcement of the same diameter and spacing is also provided at the top of the waist slab up to a distance of development length of bar. These bars are extended at the bottom of the waist slab up to a distance of development length of bar. Reinforcement of the same diameter and spacing is also provided at the top of the waist slab up to a distance of development length of bar. These bars are extended at the bottom of landing slab up to a distance of development of bar as shown in the figure. Wherever main reinforcement of the waist slab is extended at the top of the landing slab, additional reinforcement is provided at the top of the landing slab to resist support moment. These reinforcement are embedded into the support and extended at the top of the waist slab up to a distance of development length of bar as shown in the figure. Half the area of the reinforcement as that for the span moment is provided at the bottom of landing slab which is embedded into the support and extended at the top of the waist-slab to a nominal length of 300 mm. Distribution and nosing reinforcement are provided as shown typically in the figure.

6.7.7 Reinforcement Detailing for Staircase Supported on Landing Slabs which Span Transversely to Flight

Reinforcement detailing of staircase supported on landing slabs which span transversely to the filight is shown in Fig. 6.80. Main reinforcement for span moment is provided at the bottom of the waist slab which is extended either at the bottom or at the top of the landing slab as appropriate and then embedded into the support. Wherever main reinforcement of the waist slab is extended at the bottom of the landing slab, half the area of reinforcement as that for span moment is provided at the top of the landing slab. These bars are extended at the bottom

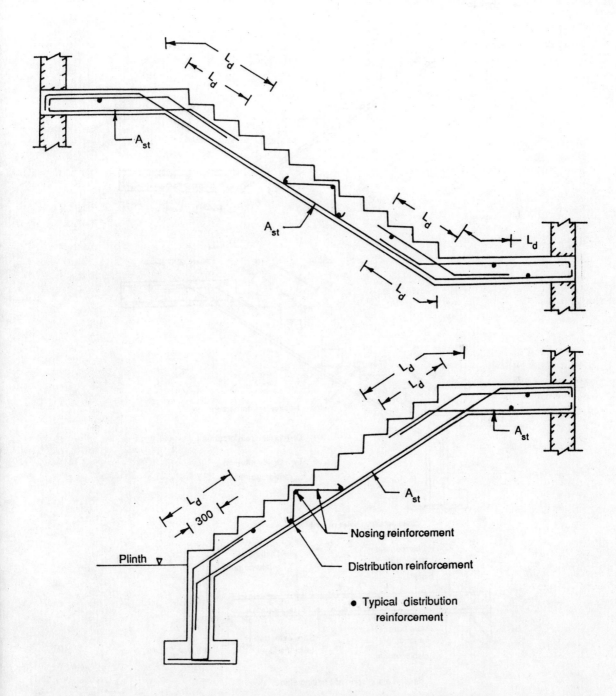

Fig. 6.79 Reinforcement detailing for staircase fixed at ends of landing slabs

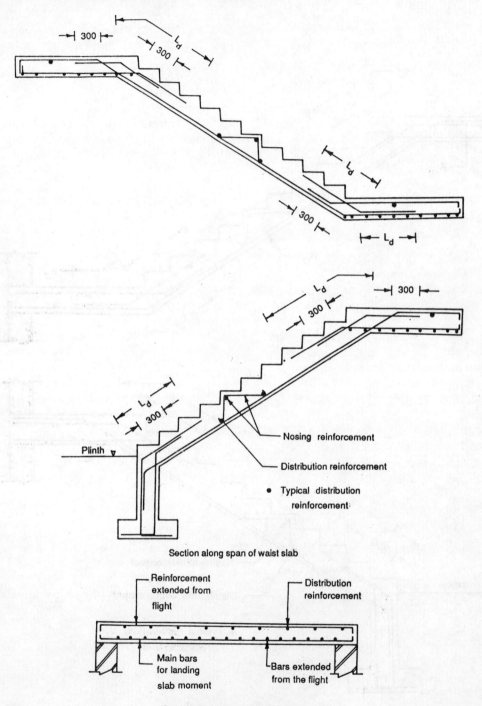

Fig. 6.80 Reinforcement detailing for staircase supported on landing slabs which span transversely to the flight

of the waist slab to a nominal length of 300 mm. Reinforcement of the same diameter and spacing is also provided at the top of the waist slab up to a distance of development length of bar. These bars are extended at the bottom of the landing slab up to a distance of development length of bar as shown in the figure. Wherever main reinforcement of the waist slab is extended at the top of the landing slab, half the area of reinforcement as that for span moment is provided at the bottom of the landing slab which is embedded into the support and extended at the top of the waist slab up to a distance of development length of bar. Same area of reinforcement is also provided at the top of the waist slab and extended at the top of the landing slab as shown in the figure. Distribution and nosing reinforcement are provided as shown typically in the figure.

6.7.8 Reinforcement Detailing for Double Cantilever Staircase with Support at Junction of Landing and Waist Slabs

Figure 6.81 shows reinforcement detailing of double cantilever staircase with support at the junction of landing and waist slabs. Main reinforcement for span moment is provided at the bottom of waist slab which is extended at the top of the landing slabs either up to its end or up to a nominal length of 300 mm as shown in the figure. Wherever bottom reinforcement in the waist slab is extended up to the end of the landing slab, additional reinforcement is provided at the top of the landing slab to resist cantilever moment which is extended at the bottom of waist slab up to a distance of development length of the bar. Reinforcement is also provided at the top of the waist slab up to a distance of $0.3\ l$ which is extended at the bottom of landing slab up to a distance of development length of bar. Wherever bottom reinforcement of waist slab is extended at the top of the landing slab up to a distance of 300 mm, reinforcement at the top of the landing slab is provided up to its end to resist cantilever moment which is extended at the top of the waist slab up to a distance of $0.3\ l$ as shown in the figure. Distribution and nosing reinforcement are provided as shown typically in the figure.

6.7.9 Reinforcement detailing for Tread-Riser Staircase

Two alternate arrangements of reinforcement detailing of tread-riser staircase are shown in Fig. 6.82. The reinforcement in tread and riser can be in the form of closed loop with standard hook at ends or bars bent up with overlap so that bars are stressed to their full strength. In alternate arrangement of reinforcement detailing, stepped tension bars are tied with pins and stirrups that resist resultant tension tending to cause cracks.

502 *Handbook of Reinforced Concrete Design*

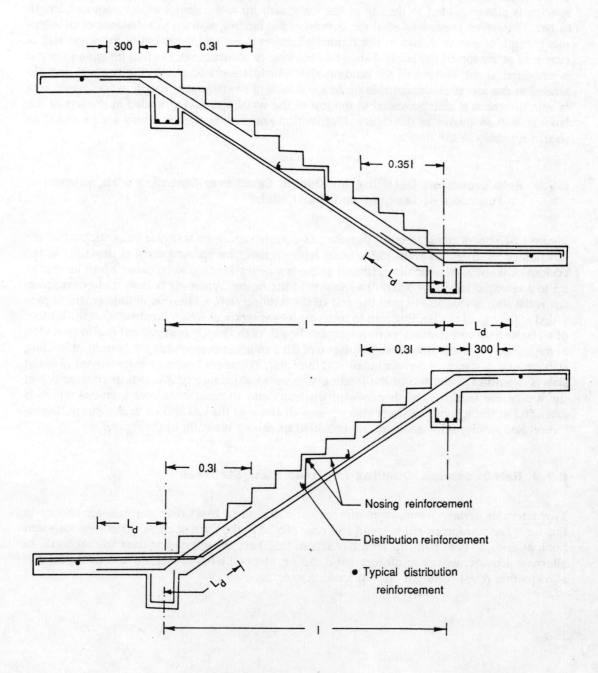

Fig. 6.81 Reinforcement detailing for double cantilever staircase with support at the junction of landing and waist slabs

Reinforcement Detailing 503

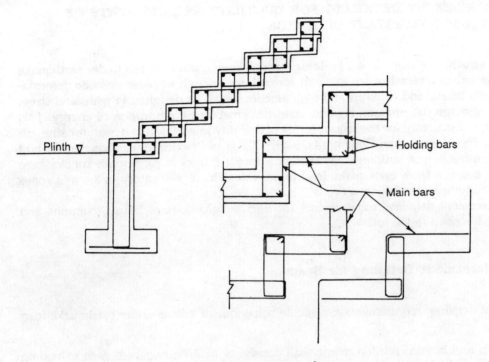

(a) Closed links with pins

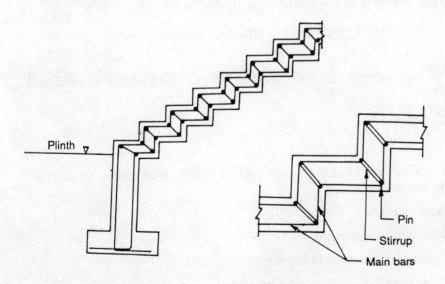

(b) Stepped bars with pins

Fig. 6.82 Reinforcement detailing for slabless tread riser staircase

6.8 REINFORCEMENT DETAILING FOR DUCTILITY REQUIREMENTS OF EARTHQUAKE-RESISTANT BUILDINGS

The primary structural elements of a building such as beams and columns under earthquake load are subjected to reversal of stresses and forces large enough to cause inelastic deformations. Therefore, beams and columns and their junctions shall be designed to withstand stress reversal and undergo deformation beyond yield deformation by absorption of energy. This requires design of structure for the requirements of ductility in addition to design for strength requirements. Therefore, Indian Code IS : 4326–1976 'Code of Practice for earthquake resistant design and construction of buildings' recommends ductile frames in earthquake zones where risk of major damage from earthquake load is possible. This is obligatory in seismic zones where seismic coefficient is 0.05 or more.

The reinforcement detailing requirements for ductile behaviour of beams, columns and junctions are described in the following.

6.8.1 Reinforcement Detailing for Beams

Reinforcement detailing requirements for ductile behaviour of beams under earthquake load are as follows:

1. The top and bottom reinforcement shall consist of at least two bars each extending throughout the length of member. The percentage area of such reinforcement at each face of the member shall not be less than that given by the following.

    p_{min} = 0.35 for concrete of grade M 15 and plain mild steel bars

    = $\dfrac{6 f_{ck}}{f_y}$ for concrete of grades other than M15 and steel other than mild steel

    where $p_{min} = \dfrac{100 A_{s,min}}{bd}$

    $A_{s,min}$ = Minimum area of steel on each face of beam which extend throughout the length of beam

    b = width of beam

    d = effective depth of beam

    f_{ck} = characteristic strength of concrete

    f_y = characteristic strength of reinforcement

2. The area of tension reinforcement on any face at any section shall not exceed the following:

$p_{t,\text{max}} = p_c + 1.1$ for concrete of grade M15 and mild steel reinforcement

$$= p_c + \frac{19 f_{ck}}{f_y} \text{ for concrete of grade other than M15 and mild steel reinforcement}$$

$$= p_c + \frac{15 f_{ck}}{f_y} \text{ for cold worked deformed bars}$$

where $p_{t,\text{max}} = \dfrac{100 A_{st,\text{max}}}{bd}$

$A_{st,\text{max}}$ = maximum area of tension steel
b = width of beam
d = effective depth of beam

$$p_c = \frac{100 A_{sc}}{bd}$$

A_{sc} = area of compression steel
f_{ck} = characteristic strength of concrete
f_y = characteristic strength of steel

3. The top and bottom reinforcements of beam shall be adequately anchored in column as shown in Fig. 6.83, so as to develop their full stress at the face of the column. For beams framing on one side of column, anchorage length of bar shall not be less than the development length of bar. For beams framing on both sides of column, reinforcement bars on both faces of the beam shall be taken continuously through the column.
4. The tensile reinforcement cannot be spliced at sections of maximum moment and the splice shall be contained within at least two closed stirrups as shown in Fig. 6.83
5. The transverse reinforcement in the form of vertical stirrups shall be provided to ensure that shear failure shall precede the flexural failure of beam. It shall be designed to resist the shear due to ultimate load acting on the beam plus shear due to plastic moment capacities at the ends of beam. Vertical stirrups shall be preferred to resist shear as they are effective for both upward and downward shears which may develop due to reversal of shear under earthquake load. Therefore, where diagonal bars are also used to resist shear, their maximum shear capacity shall be restricted to 50 per cent of the design shear so that vertical stirrups provided for resisting remaining shear can resist reversal of shear under earthquake load. The spacing of vertical stirrups shall not exceed 0.25 times the effective depth of beam in a length equal to 2.0 times the effective depth of beam near each end of beam and at spacing of 0.5 times the effective depth of beam in the remaining length of beam (Fig. 6.83).

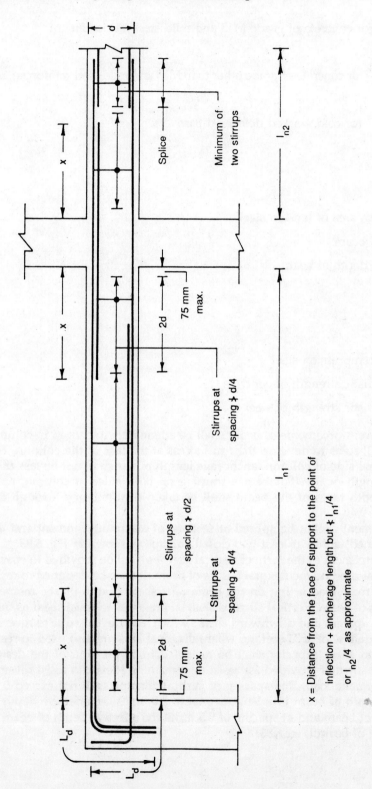

Fig. 6.83 Reinforcement detailing for ductile behaviour of beam

6.8.2 Reinforcement Detailing for Columns

Reinforcement detailing requirements for ductile behaviour of columns under earthquake load are as follows:

1. If the uniform axial stress on the column under earthquake load condition is less than 0.1 times the characteristic strength of concrete, then the column shall be designed according to requirements for beam as discussed in Section 6.8.2.
2. If the uniform axial stress on column under earthquake load condition is greater than 0.1 times the characteristic strength of concrete, then the transverse reinforcement shall be provided as given below:
   (a) Transverse reinforcement in the column shall be provided to resist ultimate shear force resulting from the lateral and vertical loads at ultimate stage. The spacing of transverse reinforcement shall not exceed 0.5 times the effective depth of column section measured from the compression fibre to tension reinforcement.
   (b) The special confining reinforcement shall be provided above and below the beam junction in a length of column at each end which shall not be less than maximum of the clear height of column/6, larger lateral dimension of column section and 450 mm as shown in Fig. 6.84. The spacing of confining reinforcement shall not exceed 100 mm (Fig. 6.84).
3. For circular column, the cross-sectional area of the reinforcement bars forming circular hoops of spiral used for confining concrete is obtained by,

$$A_{sh} = 0.08 \, S_v \, D_k \left(\frac{A_g}{A_k} - 1\right)\frac{f_{ck}}{f_y}$$

where A_{sh} = cross-sectional area of circular hoops or spiral reinforcement

S_v = spacing of hoops or pitch of spiral

D_k = diameter of core measured to the outer of the hoops or spiral

A_g = gross area of column section

A_k = area of core

f_{ck}, f_y = chracteristic strength of concrete and steel respectively

4. For rectangular column, the cross-sectional area of reinforcement bars forming rectangular hoops used for confining concrete is given by,

$$A_{sh} = 0.16 \, S_v \, h \left(\frac{A_g}{A_k} - 1\right)\frac{f_{ck}}{f_y}$$

where h = longer dimension of hoop as shown in Fig. 6.85

A_k = area of concrete core in the rectangular stirrup measured to its outside dimensions

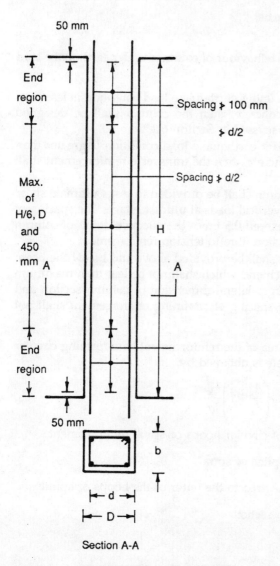

Fig. 6.84 Shear reinforcement detailing in column

6.8.3 Reinforcement Detailing at Beam-column Joints

Beam-column joints at exterior or corner columns shall be confined by the transverse reinforcement through the joint as shown in Fig. 6.86 for their ductile behaviour under earthquake load. The area of transverse reinforcement and its spacing shall be same as that of confining transverse reinforcement at the ends of column. In case of beam-column joints at interior columns, joint is confined by beams on all sides. Therefore, transverse reinforcement is not required at such joints.

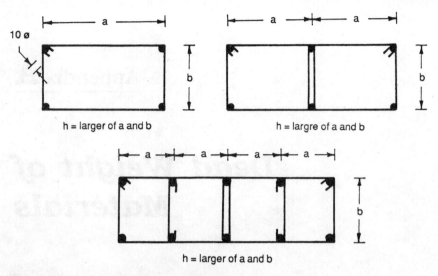

Fig. 6.85 Dimension of beam rectangular tie

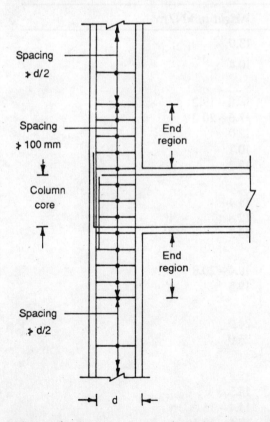

Fig. 6.86 Reinforcement detailing of beam column joint at external column

Appendix A

Dead Weight of Materials

Materials	Weight in kN/m³
Asphalt	13.0
Bitumen	10.4
Bricks	
clay bricks	16.0 – 19.2
refractory bricks	17.6 – 20.0
brick ballast	12.0
brick dust (surkhi)	10.1
brick masonry	19.2
Cement	
normal	14.4
loose	12.8
in bags	13.2
Clay	
clay fills	10.4 – 20.8
clay undisturbed	19.5
Concrete	
plain concrete	24.0
reinforced concrete	25.0
Earth	
dry and loose earth	12.8
dry and compact earth	15.5
moist and loose earth	14.4 – 16.0
moist and compact earth	17.6 – 18.4

(contd.)

Materials	Weight in kN/m³
Gravel	
loose gravel	16.0
rammed gravel	19.2 – 20.8
Granite	26.4 – 28.0
Lime	
limestone in lumps uncalcined	14.4
freshly burnt lime pieces	8.8
slacked lime	11.2
Marble	27.2
Mortar	
Cement sand mortar	20.8
lime mortar	17.6
Sandstone	22.4 – 24.0
Sand	
dry and clean sand	15.4 – 16.0
river sand	18.4
wet sand	17.6 – 20.0
Steel	78.50
Stone masonry	20.8 – 27.0
Timber	
chir	0.575
cypress	0.515
deodar	0.56
fir	0.465
sisso	0.77
sal	0.8
teak	0.625
white cedar	0.735
Water	
fresh water	10.0
salt water	10.25

Appendix B

Live Loads on Floors and Roofs

Table B.1 Live loads on floors

Loading Class Number (kN/m^2)	Description	Minimum live load (kN/m^2)	Alternate minimum live load
2.0	Floors in dwelling houses, tenements, hospital wards, bedrooms and private sitting rooms in hostels and dormitories	2.0	Subject to a minimum total load of 2.5 and 6.0 times the values given in Col. 3 for any given slab panel and beam respectively. This load shall be assumed uniformly distributed on the entire area of the slab panel or the entire length of beam.
2.5	Office floors other than entrance halls, floors of light workrooms: (a) With separate storage (b) Without storage	 2.5 4.0	
3.0	Floors of banking halls, Office entrance halls and reading rooms	3.0	
4.0	Shop floors used for the display and sale of merchandise, work rooms, classrooms in schools, places of assembly with fixed seating, restaurants, circulation space in machinery halls, power stations etc.	4.0	
5.0	Floors of warehouses, workshops, factories, office floors for storage and filling purposes, place of assembly without fixed sitting, public rooms in hotels, dance halls, waiting halls etc.	5.0	

(contd.)

Table B.1 *(contd.)*

Loading Class Number (kN/m²)	Description	Minimum live load (kN/m²)	Alternate minimum live load
7.5	Floors of warehouses, workshops, factories and other buildings or parts of buildings of similar category for medium weight loads	7.5	
10.0	Floors of warehouses, workshops, factories and other buildings or parts of buildings of similar category for heavy weight loads, floors of book stores and libraries, roofs and pavement lights over basements projecting under the public footpath	10.0	
Garage light	Floors used for garages for vehicles not exceeding 25 kN gross weight: (a) Slabs (b) Beams	4.0 2.5	The worst combination of actual wheel loads, whichever is greater
Garage heavy	Floors used for garages for vehicles not exceeding 40 kN gross weight	7.5	Subject to a minimum one-and-half times maximum wheel load but not less than 0.9 kN considered to be distributed over 75 cm square
Stairs	Stairs, landing and corridors for class 2.0 loading: (a) Not liable to overcrowding (b) Liable to overcrowding and for all other purposes	3.0 5.0	Subject to minimum of 1.3 kN concentrated load at the unsupported end of each step for stairs constructed out of structurally independent cantilever steps
Balcony	Balconies for: (a) Class 2.0 loading and not liable to overcrowding (b) All other classes and liable to overcrowding	3.0 5.0	

(contd.)

Table B.2 Live loads on roofs

Description	Minimum live load (kN/m²)	Alternate minimum live load (kN/m²)
Flat, sloping or curved roof with slopes up to and including 10 degrees		
(a) Access provided	1.5 kN/m²	0.375 kN uniformly distributed over any span of one meter width of the roof slab and 0.9 kN uniformly distributed over the span in the case of all beams
(b) Access not provided, except for maintenance	0.75 kN/m²	0.19 kN uniformly distributed over any span of 1 m width of the roof slab and 0.45 kN uniformly distributed over the span in the case of all beams
Sloping roof with slope greater than 10 degrees		
(a) Roof membrane, sheets or purlines	0.75 kN/m² less 0.02 kN/m² for every degree increase in slope over 10 degrees	
(b) Trusses, beams, girders etc. that support the roof membrane and roof purlines	2/3 of load as above	
Curved roofs with slope at spring greater than 10 degrees	$0.75 - 3.45\,(h/l)^2$ kN/m²	0.4 kN/m²

$h =$ height of the highest point of the structure measured from the springing

$l =$ chord width of the roof if singly curved and shorter of the two sides if doubly curved

Appendix C

Maximum Bending Moments and Reactions in Beams

Table C.1 Maximum bending moments and reactions in single span beams

Bending moment = coefficient × $W \times l$
Reaction = coefficient × W
W = total load on beam
l = effective span

Type of load and support conditions	Moment coefficients			Reaction coefficients	
	A	B	C	A	C
(single point load W at al from A)	0	$a'(1-a)$	0	$1-a$	a
For n equally spaced n=2	0	1/6	0	1/2	1/2
Equal loads n=3	0	1/6	0	1/2	1/2
Total load $W = nw$ n=4	0	3/20	0	1/2	1/2
n=5	0	3/20	0	1/2	1/2

(contd.)

Table C.1 (contd.)

Coefficients for maximum bending moment and reactions					
Type of load and support conditions	Moment coefficients			Reaction coefficients	
	A	B	C	A	C
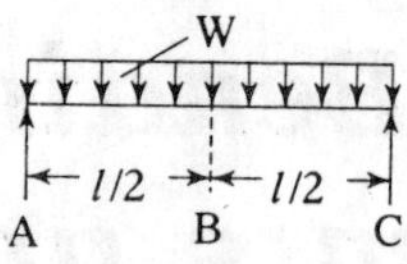 W, l/2 + l/2, A B C (uniform load)	0	1/8	0	1/2	1/2
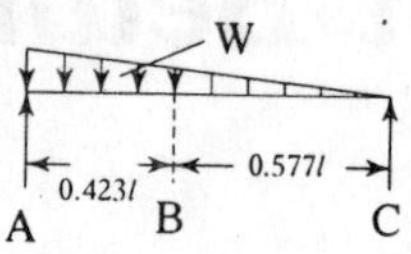 W, 0.423l + 0.577l, A B C (triangular)	0	0.128	0	1/2	1/3
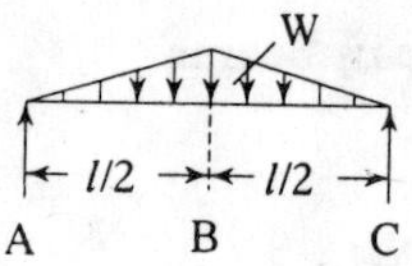 W, l/2 + l/2, A B C (triangular peak)	0	1/6	0	1/2	1/2
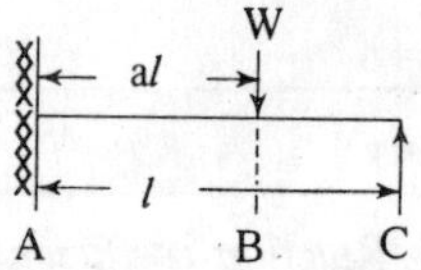 W, al, l, A B C (fixed-A, point load)	$\dfrac{a}{2}(1-a)\times(2-a)$	$\dfrac{a^2}{2}(1-a)\times(3-a)$	0	$(1-a)\times\left(1+a-\dfrac{a^2}{2}\right)$	$\dfrac{a^2}{2}(3-a)$
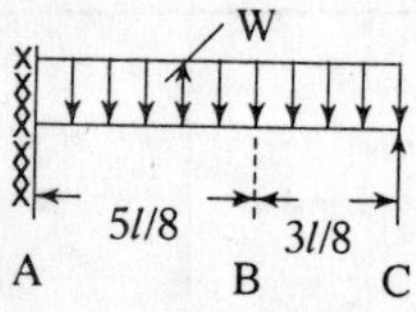 W, 5l/8 + 3l/8, A B C (fixed-A, UDL)	$-1/8$	$9/128$	0	$5/8$	$3/8$
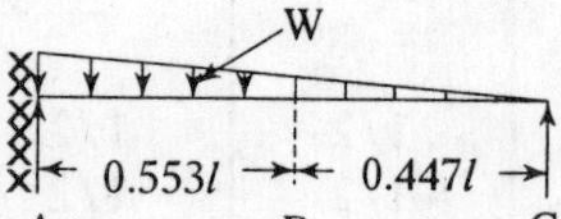 W, 0.553l + 0.447l, A B C (fixed-A, triangular)	$-2/15$	0.059	0	$4/5$	$1/5$

(contd.)

Table C.1 (contd.)

Type of load and support conditions	Moment coefficients			Reaction coefficients	
	A	B	C	A	C
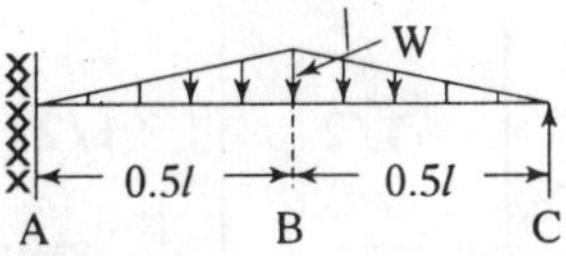 Triangular load, peak W at B, 0.5l each side, fixed A and C	−5/32	2/29	0	21/32	11/32
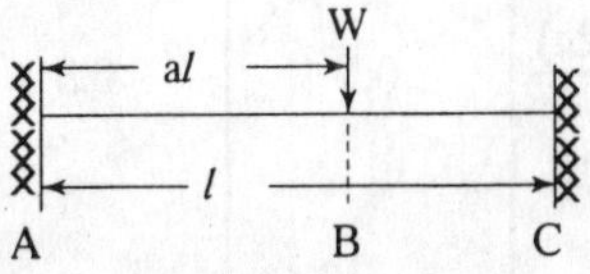 Point load W at B, distance al from A, fixed A and C	$-a(1-a)^2$	$2a^2 \times (1-a)^2$	$-a^2(1-a)$	$(1-a^2) \times (1+2a)$	$a^2 \times (3-2a)$
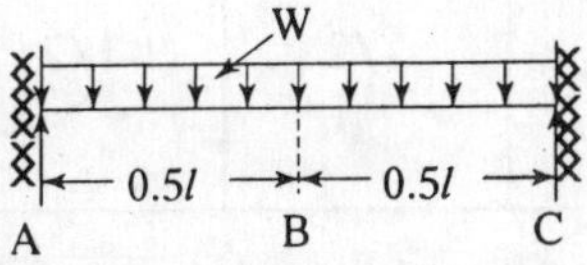 UDL W over full span, fixed A and C	−1/2	1/24	−1/12	1/2	1/2
For n equally spaced Equal loads, Total load W = nw; n = 1	−0.125	0.125	−0.125	0.5	0.5
n = 2	−0.111	0.055	−0.111	0.5	0.5
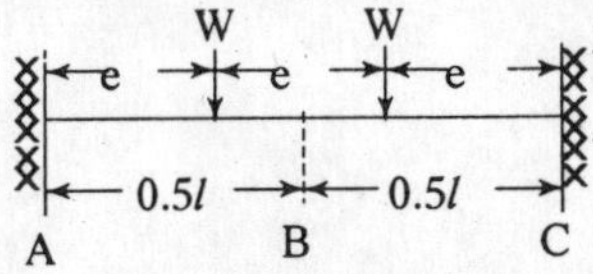 n = 3	0.104	0.062	−0.104	0.5	0.5
n = 4	0.100	0.050	−0.100	0.5	0.5
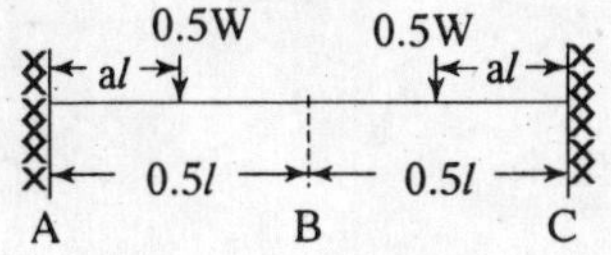 Two point loads 0.5W each at al from supports, fixed A and C	$-\dfrac{a}{2}(1-a)$	$\dfrac{a^2}{2}$	$-\dfrac{a}{2}(1-a)$	1/2	1/2

(contd.)

Table C.1 (contd.)

Type of load and support conditions	Moment coefficients			Reaction coefficients	
	A	B	C	A	C
Triangular load W, A-B-C spans 0.5l + 0.5l, fixed ends	$-5/48$	$1/16$	$-5/48$	$1/2$	$1/2$
Trapezoidal load W/2 with al ramps, spans 0.5l + 0.5l, fixed ends	$-\dfrac{a(3-2a)}{12}$	$\dfrac{a^2}{6}$	$-\dfrac{a(3-2a)}{12}$	0.5	0.5
Triangular loads with peak $2W/l$, spans 0.5l + 0.5l, fixed ends	$-1/16$	$1/48$	$-1/16$	$1/2$	$1/2$

Table C.2 Maximum positive and negative moments and reactions in continuous beams of equal spans under uniformly distributed loads

Moment   = Coefficient × W × l
Reaction = Coefficient × W
where W = total uniformly distributed load on one span only
      l = effective span

Coefficients for moments and reactions

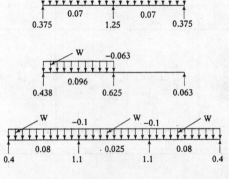

(Contd.)

Table C.2 *(contd.)*

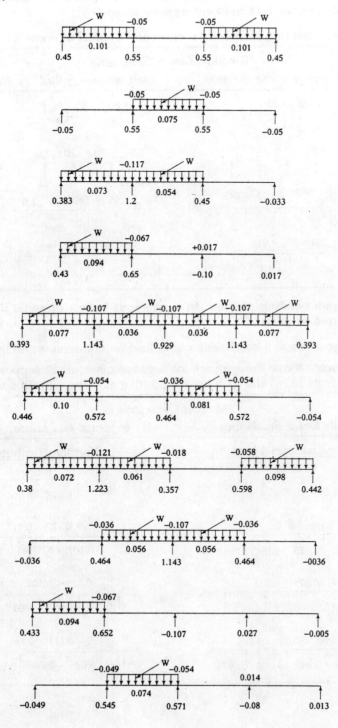

Table C.3 Bending moment in continuous beams of equal spans freely supported at ends due to unit moment applied at ends

Bending moment = coefficient × applied moment

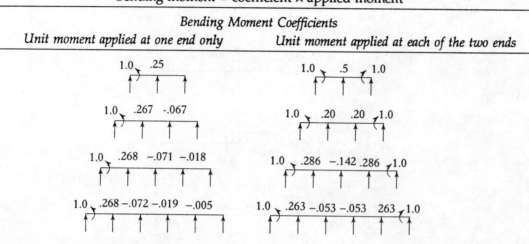

Table C.4 Maximum bending moment in continuous beams of equal span freely supported at ends

Bending moment = coefficient × total load on one span × effective span

Coefficients above the span are for negative moments at supports
Coefficients below the span are for positive moments at mid spans

Moment Coefficients

All spans equally loaded simultaneously — Uniformly distributed load

```
        0.125
    ↑0.070  ↑0.070  ↑
```

```
     0.10      0.10
  ↑0.080 ↑0.025 ↑0.080 ↑
```

```
     0.107   0.071   0.107
  ↑0.077 ↑0.036 ↑0.036 ↑0.077 ↑
```

```
     0.105  0.08   0.08  0.105
  ↑0.078 ↑0.033 ↑0.046 ↑0.033 ↑0.078 ↑
```

For incidental load causing the worst effect — Uniformly distributed load

```
        0.125
    ↑0.096  ↑0.096  ↑
```

```
     0.117     0.117
  ↑0.101 ↑0.075 ↑0.101 ↑
```

```
     0.121   0.107   0.121
  ↑0.099 ↑0.081 ↑0.081 ↑0.099 ↑
```

```
     0.120  0.111  0.111  0.120
  ↑0.100 ↑0.080 ↑0.086 ↑0.080 ↑0.100 ↑
```

(contd.)

Appendix C

Table C.4 (contd.)

Centre point load	Centre point load

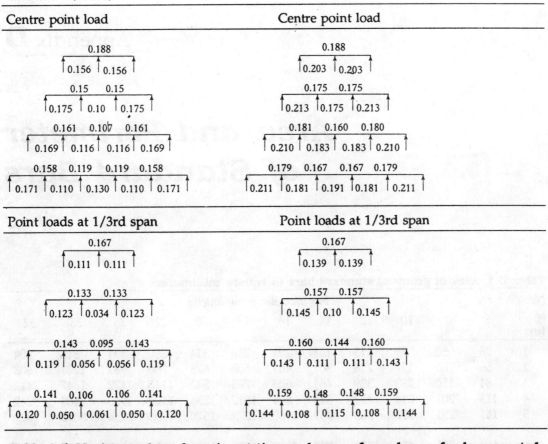

Table C.5 Maximum shear force in continuous beams of equal span freely supported at ends

Shear force = coefficient × total load on one span

Shear force coefficients

All spans equally loaded simultaneously	For incidental load causing the worst effect
Uniformly distributed load	Uniformly distributed load

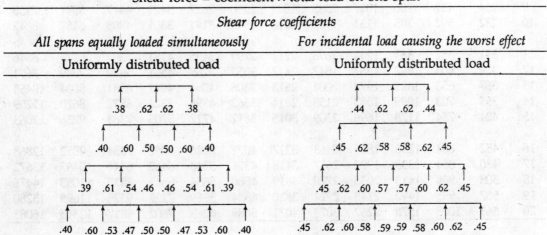

Appendix D

Area and Perimeter of Standard Bars

Table D.1 Area of group of standard bars in square millimeters

No. of bars	Bar diameter in millimetre											
	6	8	10	12	14	16	18	20	22	25	28	32
1	28	50	79	113	154	201	254	314	380	491	616	804
2	56	100	157	226	307	402	508	628	760	981	1231	1608
3	84	150	235	339	461	603	763	942	1140	1472	1847	2412
4	113	201	314	452	615	804	1017	1256	1520	1963	2463	3217
5	141	251	392	565	769	1005	1272	1570	1900	2454	3078	4021
6	169	301	471	678	923	1206	1526	1885	2280	2945	3694	4825
7	197	351	549	791	1077	1407	1781	2199	2660	3436	4310	5629
8	226	402	628	904	1231	1608	2035	2513	3041	3927	4926	6434
9	254	452	706	1017	1385	1809	2290	2827	3421	4417	5541	7238
10	282	502	785	1131	1539	2010	2544	3141	3801	4908	6157	8042
11	311	552	863	1244	1693	2211	2799	3455	4181	5399	6773	8846
12	339	603	942	1357	1847	2412	3053	3769	4561	5890	7389	9651
13	367	653	1021	1470	2001	2613	3308	4084	4941	6381	8004	10455
14	395	703	1099	1583	2155	2814	3562	4398	5321	6872	8620	11259
15	424	754	1178	1696	2309	3015	3817	4712	5703	7363	9236	12063
16	452	804	1256	1809	2463	3217	4071	5026	6082	7854	9852	12868
17	480	854	1335	1922	2617	3418	4326	5340	6462	8344	10467	13672
18	508	904	1413	2035	2770	3619	4580	5654	6842	8835	11083	14476
19	537	955	1492	2148	2924	3820	4834	5969	7222	9326	11699	15280
20	565	1005	1570	2262	3078	4021	5089	6283	7602	9817	12315	16085

Table D.2 Area of bars at given spacing in square millimeters/meter width

Spacing mm	\multicolumn{12}{c}{Bar diameter in millimeter}											
	6	8	10	12	14	16	18	20	22	25	28	30
50	565	1005	1571	2262	3079	4021	5089	6283	7603	9817	12315	16085
60	471	838	1309	1885	2566	3351	4241	5236	6336	8181	10268	13404
70	404	718	1122	1616	2199	2872	3635	4488	5430	7012	8796	11489
80	353	628	982	1414	1924	2513	3181	3927	4752	6136	7697	10053
90	314	558	873	1257	1710	2234	2827	3491	4224	5454	6842	8936
100	283	503	785	1131	1539	2011	2545	3142	3801	4909	6157	8042
110	257	457	714	1028	1399	1828	2313	2856	3456	4462	5548	7311
120	236	419	654	942	1283	1675	2121	2618	3168	4091	5131	6702
130	217	387	604	870	1184	1547	1957	2417	2924	3776	4737	6186
140	202	359	561	808	1100	1436	1818	2244	2715	3506	4398	5745
150	188	335	524	754	1026	1340	1696	2094	2534	3272	4105	5362
160	177	314	491	707	962	1257	1590	1963	2376	3968	3848	5927
170	166	296	462	665	905	1183	1497	1848	2236	2887	3622	4731
180	157	279	436	628	855	1117	1444	1745	2112	2727	2421	4468
190	149	265	413	595	810	1058	1330	1653	2001	2584	3241	4233
200	141	251	393	565	770	1005	1272	1571	1901	2454	3079	4021
210	135	239	374	539	733	957	1212	1496	1810	2337	2932	3830
220	128	228	357	514	700	914	1157	1428	1728	2231	2799	3656
230	123	218	341	492	669	874	1106	1366	1653	2134	2677	3497
240	118	209	327	471	641	838	1060	1309	1584	2054	2566	3351
250	113	201	314	452	616	804	1018	1257	1520	1963	2463	3217
260	109	193	302	435	592	773	979	1208	1462	1888	2368	3093
270	105	186	291	419	570	745	942	1164	1408	1818	2281	2979
280	101	179	280	404	550	718	909	1122	1358	1753	2199	2876
290	97	173	271	390	531	693	877	1083	1311	1693	2123	2773
300	94	168	262	377	513	670	848	1047	1267	1636	2052	2681
320	88	157	245	353	481	628	795	982	1188	1534	1924	2513
340	83	148	231	333	453	591	748	924	1118	1444	1811	2365
360	78	140	218	314	428	558	707	873	1056	1363	1710	2234
380	74	132	207	298	405	529	670	827	1000	1292	1620	2116
400	71	126	196	283	385	503	636	785	950	1227	1539	2011

Table D.3 Perimeter of group of standard bars in millimeters

No. of bars	Bar diameter in millimeter											
	6	8	10	12	14	16	18	20	22	25	28	32
2	37	50	62	75	87	100	113	125	138	157	175	201
3	56	75	94	113	131	150	169	188	207	235	263	301
4	75	100	125	150	175	201	226	251	276	314	351	402
5	94	125	157	188	219	251	282	314	345	392	439	502
6	113	150	188	226	263	301	339	376	414	471	527	603
7	131	175	219	263	307	351	395	439	483	549	615	703
8	150	201	251	301	351	402	452	502	552	628	703	804
9	169	226	282	339	395	452	508	565	622	706	791	904
10	188	251	314	376	439	502	565	628	691	785	879	1005
11	207	276	345	414	483	552	622	691	760	863	967	1105
12	226	301	376	452	527	603	678	753	829	942	1055	1206
13	245	326	408	490	571	653	735	816	898	1021	1143	1306
14	263	351	439	527	615	703	791	879	967	1099	1231	1407
15	282	376	471	565	659	753	848	942	1036	1178	1319	1507
16	301	402	502	603	703	804	904	1005	1105	1256	1407	1608
17	320	427	534	640	747	854	961	1068	1174	1335	1495	1709
18	339	452	565	678	791	904	1017	1130	1244	1413	1583	1809
19	358	477	596	716	835	955	1074	1193	1313	1492	1671	1910
20	376	502	628	753	879	1005	1130	1256	1382	1570	1759	2010

Appendix E

SI Conversion Factors

Table E.1 SI conversion factors

Geometry			
spans	1 ft	=	0.3048 m
displacement	1 in	=	25.4 mm
surface area	1 ft^2	=	0.0929 m^2
volume	1 ft^3	=	0.0283 m^3
	1 yd^3	=	0.765 m^3
Structural properties			
cross-sectional dimensions	1 in	=	25.4 mm
area	1 in^2	=	645.2 mm^2
section modulus	1 in^3	=	16.39 × 10^3 mm^3
moment of inertia	1 in^4	=	0.4162 × 10^6 mm^4
Material properties			
density	1 lb/ft^3	=	16.03 kg/m^3
modulus and stress	1 lb/in^2	=	0.006895 MPa
	1 kip/in^2	=	6.895 MPa
Loadings			
concentrated load	1 lb	=	4.448 N
	1 kip	=	4.448 kN
density	1 lb/ft^3	=	0.1571 kN/m^3
linear loads	1 lb/ft	=	14.59 kN/m
	1 kip/ft	=	14.59 kN/m
surface load	1 lb/ft^2	=	0.0479 kN/m^2
	1 kip/ft^2	=	47.9 kN/m^2
Stress and moments			
stress	1 lb/in^2	=	0.006895 MPa
	1 kip/in^2	=	6.895 MPa
moment or torque	1 lb-ft	=	1.356 N.m
	1 kip-ft	=	1.356 kN.m

Appendix F

Selected References

F.1 Indian Standard Code of Practice

F.1.1 *Cement*

IS : 4845 - 1968   Definitions and terminology relating to hydraulic cement
IS : 6452 - 1972   High alumina cement for structural use
IS : 8112 - 1976   High strength ordinary Portland cement
IS : 3535 - 1966   Hydraulic cements, method of sampling
IS : 4031 - 1968   Hydraulic cement, physical tests
IS : 4032 - 1968   Method of chemical analysis of hydraulic cement
IS : 269 - 1976   Ordinary and low heat Portland cement
IS : 1489 - 1976   Portland-pozzolana cement
IS : 8041 - 1978   Rapid-hardening Portland cement
IS : 6909 - 1973   Supersulphated cement

F.1.2 *Aggregates*

IS : 383 - 1963   Aggregates, coarse and fine, from natural sources for concrete
IS : 6461 - 1972   Glossary of terms relating to concrete:
    Part I   Concrete aggregates
IS : 5640 - 1970   Method of test for determining aggregate impact value of soft coarse aggregates
IS : 2386 - 1963   Methods of test for aggregates for concrete:
    Part I   Practice size and shape
    Part II   Estimation of deleterious materials and organic impurities
    Part III   Specific gravity, density, voids, absorption and bulking
    Part IV   Mathematical properties
    Part V   Soundness
    Part VI   Measuring mortar making properties of fine aggregates
    Part VII   Alkali aggregate reactivity
    Part VIII   Petrographic examination
IS : 650 - 1966   Standard sand for testing of cement